Journey of a Single Cell to a Plant

Editors

S.J. MURCH

Institute for Ethnobotany
National Tropical Botanical Garden
Kalaheo, Hawaii
USA

P.K. SAXENA

Department of Plant Agriculture
University of Guelph
Guelph, Ontario
Canada

Science Publishers, Inc.

Enfield (NH), USA **Plymouth, UK**

SCIENCE PUBLISHERS, INC.
Post Office Box 699
Enfield, New Hampshire 03748
United States of America

Internet site: *http://www.scipub.net*

sales@scipub.net (marketing department)
editor@scipub.net (editorial department)
info@scipub.net (for all other enquiries)

Library of Congress Cataloging-in-Publication Data

Journey of a single cell to a plant/editors, S.J. Murch, P.K. Saxena.
p. cm.
Includes bibliographical references and index.
ISBN 1-57808-352-4
1. Plant cells and tissues, 2. Plant protoplasts, I. Murch, S. J. (Susan J.) II. Saxena, P. K. (Praveen K.)

QK725.J656 2004
571.8'2—dc22

2004049154

ISBN 1-57808-352-4

Published by Science Publishers, Inc., Enfield, NH, USA
Printed in India.

Recollections

Toshiyuki Nagata
Department of Biological Sciences, Graduate School of Science
University of Tokyo, Hongo, Bunkyo-ku
Tokyo 113-0033, Japan

It was on November 5, 1969 when I first observed the cell division of tobacco mesophyll protoplasts. This date is never forgettable for me. Before this date I had devoted my possible efforts to try to culture the protoplasts, which turned out to be almost in vain. As it was written explicitly in the famous textbook by Philip White[1] that mesophyll cells are not a suitable material for culture, this affected my psychological situation to some extent. Nonetheless, I dared to continue this study. On the other hand, there have been reported a few positive results for the culture of mesophyll cells from peanut and *Macleaya cordata*, which encouraged my trial.

This difficulty was, however, completely overcome when I could see the cell division of the protoplasts. In fact, the mesophyll protoplasts were proven to be very good sources for culture. This success was mainly due to the choice of the right materials, which is still true and has not been correctly understood. Although this is not included in the original paper[2, 3], and has been cited in the essay that appeared in the Citation Classics of the Current Contents[4], I noted that my findings gave a significant implication on the totipotency of plant cells. Although totipotency was discovered with carrot as somatic emybryogenesis and with various plant species as organ formations by the balance of auxin and cytokinin concentration, it was not clear whether most of cells retained this characteristics or only some of fractions could show this characteristics. In fact, cell division in the carrot root tissues was reported to be only at the frequency of 5% or so in the earlier publication[5]. Then one critical question is whether 95% of other cells do not retain such characteristics. In this context, what we showed in the 1971 paper[3] is that almost all mesophyll protoplasts could be cultured and organ formation from them was possible. As this is cited frequently, I will not go into further details, but somatic cell genetics using plant protoplasts have launched from our findings.

As a kind of recollections, I would like to add a few sentences on this initial work. As the fate of development of protoplasts can be traced in the database of Institute of Scientific Information, we can now follow the fate of citation of these papers. The papers that appeared in 1970[2] and 1971[3] have been cited 383 and 365, respectively, as of March 2003 and are still being cited in recent papers. It is also mentioned in the article that appeared in the Current Contents[6] that our initial two papers were shown to be core articles in the development of somatic cell genetics of plants. Currently we are studying the effects of plant hormones of auxin and cytokinin upon the induction of cell division of tobacco mesophyll protoplasts. Although both plant hormones are required to induce the cell division, the roles that are conducted by either of these two have not been clarified yet. In particular, the role of cytokinin was found to be more important than that of auxin, which has not been notified by other peoples. Simultaneously cell biological studies using highly synchronizable tobacco BY-2 cell system are being conducted[7].

References

P. White: The cultivation of plant and animal cells. 2nd Ed., Ronald Press (1964)
T. Nagata and I. Takebe: Planta 92, 301 (1970)
T. Nagata and I. Takebe: Planta 99, 12 (1971)
Current Contents, AB & ES. 16 (#9) 1985
Israel, H.W. and Stewards, F.C.: Ann. Bot. 30, 63 (1966)
Current Contents, LS. 25 (#14) (1982)
Nagata, T. et al. (eds.): Tobacco BY-2 cell line, Vol. 53, Biotechnology in Agriculture and Forestry, Springer-Verlag, Heidelberg, Berlin, in press.

Preface

The concept of a totipotent cell, a single cell with the potential to form an individual, is among the most intriguing ideas in modern science. In mammalian systems, the potential of stem cells is both promising and controversial for new treatments for diseases. In plants, the ability to regenerate identical individuals from single cells is the basis for modern agriculture. These scientific advancements have given us virus-free stocks, novel germplasms, clonal propagation systems, and the commercial introduction of difficult to propagate plant species. The timescale for breeding programs was dramatically reduced as plant tissue culture technologies were developed to shorten the time between generations and reduce the number of generations required for a line to be developed. Without the capacity to regenerate plants, it would not have been possible for plant biotechnology and genetic engineering to have advanced this far. Most of the research in this area has been possible because of the significant advances made within the last 50 years.

This book contains detailed reviews by the leading scientists who made these discoveries. Scientists were asked to recount their experiences and ideas that led to a particular breakthrough or change in thinking that allowed new directions to be formed. The focus is on the important advancements in the ability to grow plant cells, the capacity to redirect plant growth, and the competence of individual cells to undergo plant morphogenesis. Morphogenesis, literally a change in the shape of a plant or cell, describes the change in form between individual isolated cells and identical plant embryos. Understanding of this process provides many lessons that have led to advancements in other fields as well as a greater understanding of the principles of how plants grow.

As this book was coming together, we began to realize that it reflects a very special aspect, namely an examination of the process of science from the perspectives of leading scientists. Science is commonly defined as the hypothesis-driven investigation of the natural world but often scientific investigations are initiated or advanced by serendipitous findings. We would all like to pretend that we predicted the outcome of a set of experiments or an important discovery but often it is the unpredicted results that are most interesting. For this volume, we asked scientists who made significant advances to recount their experiences, to revisit the thought processes that led to the major discoveries and to reexamine their own ideas and feelings about both the development of the current thinking in plant morphogenesis and the future for this field. We were fortunate in receiving contributions from many pioneers in the field, all of whom have been both candid and honest in describing their personal experiences and perceptions. These insights provide wonderful examples for the next generation of scientists; younger scientists sometimes feel they are the only ones who have faced many of the challenges that are perhaps universal. Therefore, it is our hope that this book will give new insights into both the process of regeneration of a whole plant from a single cell and the process of scientific discovery.

August, 2003

Susan J. Murch
Praveen K. Saxena

Contents

List of Contributors

Editors

P.K. Saxena

Department of Plant Agriculture, University of Guelph, Guelph, Ontario, Canada, N1G 2W1, e-mail: psaxena@uoguelph.ca

S.J. Murch

Institute for Ethnobotany, National Tropical Botanical Garden, 3530 Papalina Road, Kalaheo, Hawaii, 97641, USA, e-mail: smurch@ntbg.org

Corresponding Authors

A.C. Cassells

Department of Plant Science, National University of Ireland, Cork, Ireland, e-mail: a.cassells@ucc.ie

M.R. Davey

Plant Sciences Division, School of Biosciences, University of Nottingham Sutton Bonington Campus, Loughborough, LEI12 5RD, UK, e-mail: Mike.Davey@nottingham.ac.uk

D. Dudits

Laboratory of Functional Cell Biology, Institute of Plant Biology, Biological Research Center, H.A.S. Temesvari krt 62, H-6726, Szeged, Hungary, e-mail: dudits@nucleus.szbk.u-szeged.hu

A. Fehér

Laboratory of Functional Cell Biology, Institute of Plant Biology, Biological Research Centre, H.A.S. Temesvari krt. 62, H-6726, Szeged, Hungary, e-mail: fehera@nucleus.szbte.u_szeged.hu

S. M. Jain

International Atomic Energy Agency, P.O. Box 100, Wagramer Strasse 5 A-1400 Vienna, Austria, e-mail: smjain@iaea.org

K.J. Kasha
Department of Plant Agriculture, University of Guelph, Guelph, Ontario, Canada, N1G 2W1 e-mail: kkasha@vogvelph.ca

A. Komamine
Research Institute of Evolutionary Biology, 2-4-28, Kamiyoga, Setagaya-ku, Tokyo, 158-0098, Japan, e-mail: khf10654@nifty.ne.jp

A.D. Krikorian
Department of Biochemistry and Cell Biology, State University of New York at Stony Brook, Stony Brook, New York, 11704-5215, USA, e-mail: adkrikorian@earthlink.net

S. C. Maheshawari
International Centre For Genetic Engineering & Biotechnology, Aruna Asaf Ali Marg, 110 067, New Delhi, India, e-mail: maheshwarisc@hotmail.com

K.P. Pauls
Department of Plant Agriculture, Qntario Agricultural College, University of Guelph, Guelph, Ontario, Canada, N1G 2W1
e-mail:kppauls@vogvelph.ca

V. Raghavan
Department of Plant Cellular and Molecular Biology. The Ohio State University. 318 West 12th Avenue, Columbus, OH 43210 USA, e-mail: raghavan.1@osu.edu

L. Fowke
Department of Biology, University of Saskatchewan, 112 Science Place; Saskatchewan SK S7N 5E2, Canada, e-mail: Larry.Fowke@usask.ca

1

Totipotency and Proof of the Concept: An Historical Perspective

A.D. Krikorian
Department of Biochemistry and Cell Biology
State University of New York at Stony Brook
Stony Brook, New York 11794-5215 USA
e-mail: adkrikorian@optonline.net

> Scientists are the bricklayers of knowledge. Each brick in the wall rests on the ones laid underneath it and gains support from the ones laid on each side of it. Hence, all scientific investigators who contribute so much as a brick or the mortar between bricks share credit for the entire wall.
> —Greer Williams (1960)

TOTIPOTENCY: AN ANCIENT CONCEPT, A RELATIVELY NEW WORD

Part of the credo espoused by modern plant biologists is that all nucleated cells are potentially **totipotent**. The Oxford Dictionary of Biology 4th ed. 2000, defines totipotent as: "1. Describing differentiated plant cells that, when isolated, have the ability to develop into an entire new plant if provided with the suitable growing medium. 2. Describing embryonic cells at a stage before their fate is irreversibly determined, when they have the ability to develop into any differentiated cell given the appropriate stimulation." To go one step further, biologists know that under appropriate conditions, both natural and artificial, not only can plant cells give rise to a new plant but that an isolated somatic cell *in vitro* can even yield an embryo! Such embryos are usually referred to as **somatic embryos** although some scientists prefer to call them **embryoids**,

adventitious embryos, adventive embryos, asexual embryos, pseudo-embryos, embryo-like structures, or even the oxymoron, **vegetative embryos.** Somatic embryos are morphologically similar, if not identical, to zygotic embryos and have all the potential for growth inherent in an embryo derived from a sexual union.

For many years it was maintained that a major difference between plants and animals was that all plants could regenerate a new individual from any of their generally totipotent cells whereas higher animals could not. Even today after the much publicized sheep "Dolly" (a Finn Dorset ewe)— the first mammal cloned from an adult animal— the generalization still holds because although derived from a "reconstructed egg", i.e., an enucleated egg into which a nucleus from a cultured udder cell had been transplanted (Wilmut et al. 1997), she did not develop directly from a somatic cell. Thus a ram could not have been cloned "exactly" since mitochondrial DNA is transmitted maternally. This method of cloning by nuclear transfer was first achieved back in 1952 by Robert Briggs and Thomas King and has been widely used to study early development (McKinnel and DiBerardino, 1999).

The concept of totipotency is inherent in the cell theory even as first put forward by Matthias Jacob Schleiden (1804-1881) (see Plate II, 1) in his *Contribution to Phytogenesis* (1838). In fact, he defended for many years the erroneous view that plants reproduce only asexually from the pollen tube tip and that the ovule simply nourished an embryo's growth. Theodor Schwann (1810-1882) in his *Microscopical Researches on the Resemblance of Structure and Growth of Animals and Plants* (1839), unequivocally emphasized the "elementary organism" nature of plant cells. Schleiden wrote that every plant and animal is "an aggregate of fully individualized, independent, separate beings", i.e. cells. Similarly, according to Schwann, "some of these elementary parts [cells] which do not differ from others, are capable of being separated from the organism and continuing to grow independently. Hence we can conclude that each of the other elementary parts, each cell, must possess the capacity to gather new molecules to itself and to grow, and that therefore each cell possesses a particular force, an independent life, as a result of which it too would be capable of developing independently if only there be provided the external conditions under which it exists in the organism" (Schwann 1847–see Schel 1989 for exact Schleiden and Schwann citations). Obviously, the means to test these concepts of the independent nature of elementary organisms effectively were not available until considerably later. Even so, there is a fair amount of early literature dealing with the regeneration of plants from cuttings of varying size, and some investigators even made attempts to assess the limits of divisibility that could be imposed before regeneration ceased (Krikorian 1982 for examples).

The German botanist Herman Vöchting (1847-1917) was one of those who early attempted an analysis of the factors controlling organ formation. His works *Organ Formation in the Plant Kingdom* (1878, 1884) and *Transplantation of Plant Parts* (1892) are replete with careful observations. One conclusion drawn by Vöchting which has special significance even today is: "In every fragment, be it ever so small, of the organs of the plant body, rests the elements from which, by isolating the fragment, under proper external conditions, the whole body can be built up" (Vöchting 1878—see Krikorian 1982 for exact references). Statements like these embody the essentials of what came to be termed **totipotency**. Vöchting also expressed the belief that what we now call the "fate" of a cell is a function of its position; this view is now well established in developmental biology (Plate II,3–5). Authors such as the German experimental plant morphologist Karl Goebel [later von Goebel] (1855-1932) and the German-American physiologist Jacques Loeb (1859-1924) frequently wrote about the phenomenon of regeneration in plants and their writings make it clear that they believed that at least certain cells were totipotent (Loeb, 1924) (Plate II, 6–9). The monumental treatise on *Regeneration and Transplantation* by Eugen Korschelt (1858-died sometime post-1939), Director of the Institute of Zoology at the University of Marburg in Germany, contains an insightful chapter entitled "Loss and Replacement of Parts of Plants" that is still worth reading (Korschelt, 1927).

Thus, botanists appreciated long before aseptic laboratory techniques became routinely available that the development of the higher plant body involves suppression of many, and the development of relatively few, either actual or potential primordia. What factors precisely controlled the expression of this potential remained, and to all intents and purposes has remained to this day, a challenge for the developmental botanist and the plant propagator alike (see Fink 1999; Hartmann et al., 2002).

Totipotent cells capable of producing new growing regions, tissues, organs, and even new plants seemed to be held in check by a host of correlative influences and unidentified inhibitions. Sometimes these could be released easily. For instance, Hans Winkler (1877-1945) showed in 1903 that the adventitious buds commonly formed upon leaves of the wishbone plant, *Torenia asiatica*, may originate from either a single epidermal cell or from a very small group of cells. In 1975 this was confirmed in another species of *Torenia* (Trân Than Vân and Bui Vân Lê 2000).

A similar situation exists in those *Begonia* species which yield plantlets when the larger veins of their detached leaves are cut. It is also true of African violets, genus *Saintpaulia*, which freely form new plants from single epidermal cells at the basal portions of petioles and leaf blades

(Plate II, 8). *Bryophyllum* species which produce foliar embryos (Plate II, 2) and adventitious bud formations from such species as flax (*Linum usitatissimum*) are additional cases (Krikorian 1982; van Harten 1998 for a discussion of the single cell origin of buds). Many examples from the literature of abnormal development in plants (teratology) and plant pathology also serve to corroborate the expression of totipotency by plant cells (Fink 1999). Mention must also be made of the many natural instances of **apomixis** wherein nonzygotic embryos develop in a seed from egg cells or other embryo sac cells or the surrounding nucellus (see Plate II, 13; Nogler 1984; Asker and Jerling 1992; Savidan et al. 2001 and references cited).

TOTIPOTENCY, THE WORD

Thomas Hunt Morgan (1866-1945), embryologist and geneticist, appears to have been the first to use the word **totipotence** in English (Morgan 1901, p. 243). The German word **totipotenz** originated in the last quarter of the 19th century. However, Totipotence was used only occasionally in botanical literature subsequently. Erwin F. Smith (1854-1927), one of the founders of modern plant pathology, working at the U.S. Department of Agriculture, presented a lucid account of totipotency in the context of experimental teratosis in his *An Introduction to Bacterial Diseases of Plants* (Smith 1920). Smith investigated adventive embryo formation on leaves and stems of *Begonia phyllomaniaca*— so named because of the profusion of tiny plantlets they bear. He used the term totipotent unambiguously-- "My experiments show that the surface has, rather uniformly distributed in it, thousands of germinal or totipotent cells, most of which ordinarily remain dormant but which can be shocked into development, if the shock is applied early enough, that is, while the tissues are still very young. These shoots are not the development of preformed buds. They are not branches, but independent organisms" (Smith 1920, p. 583). See Plate II, 11 and 12.

Charles E. Allen (1872-1954) of the University of Wisconsin repeatedly used the words totipotency, totipotent, and equipotent in a very interesting paper initially presented at the Botanical Society of America entitled "The potentialities of a cell" (Allen 1923). As early as 1912 the word had firmly entered the terminology of developmental biology to the extent it was listed in the authoritative dictionary entitled *Terminologie der Entwicklungsmechanik der Tiere und Pflanzen* [Terminology of Animal and Plant Developmental Mechanisms, Roux et al. 1912, pp. 409-410] under the word **Totipotenz** and cross-referenced with **Omnipotenz**, omnipotent).

Somewhat later, Henri Pratt (1902-1981) gave a very cogent account of totipotency in a critical review of histophysiological gradients and plant organogenesis (Pratt 1948, 1951). He saw all plant cell qualities as a reflection of some sort of gradient or other: "Some cells retain the capability of regenerating an entire plant ...The young cell, **totipotent** when in the meristem, gradually grows into a senile cell whose possibilities are irremediably restricted. It becomes '**partipotent**' and finally '**impotent**'. It will require long researches to determine the intimate process of this transformation and the factors that influence it" (Pratt 1948, p. 699 ff.).

Similarly, the morphologist Edmund W. Sinnott (1888-1968) in his *Cell and Psyche,* while reflecting on regeneration in plants and animals which yields entirely new individuals from single cells, stated that "the general conclusion, with all its far-reaching implications, seems justified that every cell, fundamentally and under proper conditions is totipotent, or capable of developing by regeneration into a whole organism" (Sinnott 1960, p. 30).

These several examples from the writings of major scientists of the day suffice to show that the concept of totipotency, if not the widespread use of the word, was firmly established in the outlook of those versed in plant morphology and physiology. For workers today, insertion of the adjective *potentially* or *theoretically* in front of the word totipotent renders the concept more precise.

TOTIPOTENCY: *IN VITRO* STUDIES

Gottlieb Haberlandt and the Early Investigators

For us today the most significant expression of faith in the possibility of cultivating plant cells *in vitro* and in their totipotency was published in 1902 by Gottlieb Haberlandt (1854-1945). Working at the University of Graz in Austria, he concluded, albeit with no experimental evidence nor even using using the word totipotent, that he was not making too bold a prediction by pointing to the possibility that one day "one could successfully cultivate artificial embryos from vegetative cells" (see Krikorian and Berquam 1969 or Laimer and Rücker 2002 for a complete translation into English of Haberlandt's classic paper).

Haberlandt, a distinguished plant physiologist, was not able to achieve successful cultures by today's standards and indeed, no cell division was ever observed (Plate III, 1 and 2). Nevertheless, the work had some heuristic value and stimulated others to pursue similar studies. It is of no small significance that the pioneer of animal tissue culture, Ross G. Harrison (1870-1959) credited Haberlandt as "the first to cultivate isolated cells of higher organisms" and further stated that his "paper sets forth very fully...the purposes and possibilities of modern tissue culture" (Harrison 1928, p. 6)]. As it turned out, and for a number of reasons, successes in plant tissue culture lagged considerably behind progress made by those working with animal tissues. Only in the 1930s did plant tissue culture establish itself with any degree of reliability (see White 1936, 1943; Gautheret 1982, 1983, 1985; Plate III, 3–5). But Haberlandt was always appreciated by the early tissue culture workers for his foresight. Indeed, successful cultures of excised root tips were carried out in the very early 1920s in Haberlandt's laboratory in Berlin by his assistant Walther Kotte (1893-1970). William J. Robbins (1890-1978) at the University of Missouri published a paper on root tip culture a little before Kotte. Neither Robbins nor Kotte knew of each other's work (see Robbins 1957 for a history of the root culture work). Today, the two are credited with having carried out the first culture work on an excised plant organ.

Only after the 'real' successes of plant callus culture (i.e., entailing sustainable subculture) became firmly entrenched in the literature does one increasingly encounter the word totipotency in various publications. For example, Philip R White (1901-1968), one of the pioneers of modern plant tissue culture, used totipotent, totipotency, etc. repeatedly in his lucid discussion of its foundations and potentials, *A Handbook of Plant Tissue Culture* (White, 1943). Curiously, White's definition of a tissue culture was "any preparation of one or more isolated, somatic plant cells which grows and functions normal *in vitro* **without** giving rise to an entire plant" (White 1936; emphasis mine). Harrison explained as early as 1928 that "the expression tissue culture ...is in some degree a misnomer, for it fails to convey in full the idea it represents. Moreover, the use of the phrase might be taken to imply that this method of research – in reality a technique – is an object in itself, whereas it is but a means of investigating the properties of cells under conditions not previously available. What was originally the designation of a method has in fact come to connote the field of knowledge it has brought to light" (Harrison 1928, pp. 5-6).

I.W. Bailey (1884-1967) at Harvard University tried to steer workers to refine their terminology in the early 1940s. He complained that the term "tissue culture" was inappropriate insofar as plants are concerned (Bailey 1943). "Callus-like proliferations derived from excised parts of plants when grown under controlled conditions *in vitro*, obviously provide extremely significant material for the experimental investigation of various morphological problems. It is misleading, however, to refer to abnormal proliferations from heterogeneous bits of stem, roots or other organs as cultures of cambium, procambium, cork cambium, etc." [Readers will know that Bailey's legitimate argument apparently went unheeded and the term "tissue culture" has been universally adopted and has even taken on a broader definition. The term "tissue culture" is now used generically to include protoplast, cell, tissue, organ, and even whole plant culture, the only apparent common factor being that they are in an aseptic environment!]

By 1946 Ernest A. Ball (1909-1997) was using the words totipotent and totipotency liberally in a now classic paper on the aseptic culture of excised stem tips of very small size (Arditti and Krikorian 1996).

F.C. Steward, Marion O. Mapes, and Kathryn A. Mears — Carrot Cultures

By the 1950s some rather dramatic events had begun to take place. In late fall 1954 to very early 1955 (it is not possible to say exactly when) the formation of carrot suspensions was observed in the laboratory of F. C. Steward at Cornell University. This was followed by the production of embryoids in late 1956 – early 1957. Both discoveries, like so many breakthroughs in science, were serendipitous and, in retrospect, the findings clearly represent a watershed not only in plant cell culture work, but in the understanding of plant development.

The nominal facts of the production of somatic embryos from single cells of carrot has today been woven into the now vast literature not only of plant tissue culture, but of developmental biology in general. Detailing the events ranges from fairly accurate descriptions of what was achieved and how it was done, to misunderstood and hence confused and distorted accounts published in textbooks, handbooks, and encyclopedias of all places! Be that as it may, the full story remains to be reconstructed but some highlights need to be noted here.

In the period following World War II a few plant physiologists began to use tissue culture as a tool to probe problems rather than as an end unto itself. F.C. Steward (1904-1993), first at the University of Rochester, then at Cornell, was one pioneer who had at the outset very little interest in the technical details of tissue culture. He wanted to use it as a technique. The trail of events that eventually led to the production of embryoids began at the University of Rochester, with the addition of the liquid endosperm of coconut (coconut water [CW], then widely but imprecisely referred to as coconut "milk") as an additive to semisolid culture media adopted to support growth of explants of carrot root secondary phloem. It proved to be a powerful stimulant to cell division in carrot explants. The quest to isolate and identify the components of CW quickly led to the development of a bioassay involving tiny explanted plugs of tissue from the phloem region of roots of cultivated carrot (see Steward and Krikorian 1971 for details).

The adoption of liquid rather than semisolid medium in this bioassay prompted almost immediate criticism from the establishment, especially P. R. White, but, as it turns out, the use of liquid also became, at Cornell University later, a key factor in the unexpected production of suspension cultures, and subsequently but indirectly, of somatic embryoids.

One part of the story has to do with the fortuitous production of carrot suspensions (Plate IV, 2). The other part of the story is that explants in liquid frequently produced roots (Plate IV, 3). I shall start with the suspensions. The need for large numbers of explants for biochemical studies led to the design of special 'nipple' flasks which were mounted on a horizontal axis and slowly (1 rpm) rotated at a 50 incline, thus allowing explants to be alternately submerged and exposed for aeration via a tumbling action (Plate IV, 1 and 2). Oftentimes some 100 explants were cultured in these distinctive flasks and not surprisingly, such large numbers of explants, the risk of microbial contamination increased considerably. Flasks were always warily watched and inspected. The fortuitous production of "free" cell suspensions occurred in such flasks as a result of a tumbling action and fragments being so released from the periphery of carrot plugs rapidly growing in the presence of CW-supplemented nutrient medium. Nowadays one would describe the suspensions as comprising very small chunks of tissues, still smaller clumps and progressively smaller groupings, even free-floating cells. Decanting or siphoning off aliquots of such material enabled serial culture—the flasks accommodated 250 ml of nutrient solution. [Parenthetically, some early attempts to culture absolutely free, or isolated cells "fished out" of suspensions were not successful despite fairly concerted efforts by one of Steward's technicians, Joan Smith (see Steward, Mapes and Smith 1958 for photographs of some characteristic cells)].

The other part of the story is that quite early in the use of plugs from carrot root phloem for bioassays it was observed that primary explants of some strains of carrot occasionally formed roots in culture. Wide-mouth pipetted or decanted "cell" cultures in liquid medium seemed to form "roots" much more frequently. Some cultures were so active in their "root"-forming capacity that its elimination was problematical. But these cultures never seemed to make shoots (Plate IV, 3).

An enterprising research assistant, Kathryn A. Mears, was the tissue culture technician at the time in Steward's laboratory and carried out the day-to-day operations involving the carrot bioassays (Plate IV, 1). She had an M.S. degree from the University of Vermont (coincidentally her thesis work utilized *in vitro* culture of excised embryos). On one occasion Miss Mears was challenged by F.C. Steward to demonstrate that a given flask culture, suspicious to him, was not contaminated with bacteria. Experience prompted her to believe it was not contaminated and she set out to convince Professor Steward. She transferred an aliquot containing granules and cells, most probably also including microscopic "rooty" forms to semisolid nutrient medium. There was no contamination; the opacity of the medium was due to suspended carrot cells! This was "turning point" number 1! It marked the beginning of suspension culture in Steward's laboratory. On another occasion, Miss Mears deliberately fished out a tiny greenish structure very much resembling an embryo from a large flask. She "planted" it on gelled medium to see whether it would grow. It yielded a leafy carrot plantlet. This was "turning point" number 2! The embryo-like structure that attracted her attention apparently derived or was dislodged from the fragmenting embryogenic suspension which, for the most part, was in retrospect, prevented from advancement by the suboptimum, even adverse culture environment. [Parenthetically, removal from the liquid environment also of conspicuous rooty forms, some of which seemed to have "buds" soon fostered shoot development and growth of the "rooty" structures into tiny plantlets ensued. We now know that these "rooty" forms were, for the most part, imperfect embryoids—some with minimally developed or even lacking shoot apices. Transfer from liquid to semisolid fostered structures to develop properly— see Krikorian 2000 for further details].

Thus, more than one chance event comprised a key "turning point" and led to the production and recognition of the embryoids. Certainly there were no carefully planned experiments devised to test the hypothesis that an isolated or "free" "totipotent" cell could give rise to a somatic embryo and eventually a mature plant!

The circumstances which led to the embryoids from liquid-grown "cell colonies" (to quote F.C. Steward) represented a marked departure from earlier methods used by others employing isolates made directly from their tissue of origin or procedures that depended on the disruption of proliferated callus cultures (initially grown on agar-stiffened media) through vigorous agitation on rotary shakers. By enabling carrot explants to proliferate in a gently tumbling liquid medium the environment was provided for the cells to divide and grow, and cell masses to become friable and yield structures (Plate IV, 2 through 11).

The providential testing of a liquid culture to prove lack of contamination by Kathryn Mears in F.C. Steward's laboratory opened the door for more systematic work with suspensions and observations of live cells with phase contrast under the inverted microscope. Miss Mears left Cornell at the end of November 1957, so her role in the work ended. Mrs. Marion Mapes (née Okimoto) (1913-1981), another "technician" who overlapped a bit with Kathryn Mears (who very soon after became Mrs. Joel S. Trupin), thereafter assumed responsibility for the aseptic culture operations in Steward's laboratory. Mrs Mapes had M.S. in Botany and had studied under the distinguished morphologist Arthur J. Eames at Cornell. She also had considerable experience in plant pathology (Plate IV, 9). Her technical skills were considerable and all would-be tissue culture workers in Steward's laboratory from the late 1950s onwards were inevitably trained by her. I was one of them.

The paper that brought together the observations on organization from carrot cells was published in the *American Journal of Botany*. It bears the names Steward, Mapes and Mears. (Incidentally, a preview of the development "from carrot cells to carrot plants" by Steward and Pollard appeared first in the September 27, 1958 issue of the British science weekly *Nature*— see Krikorian 1975 for citation). Following the usual convention of literary citation, reference to the classic paper soon became reduced to Steward "et al." 1958. It will perhaps be appreciated that Kathryn A. Mears is not the first laboratory research assistant responsible for making a pivotal observation in a laboratory headed by a distinguished scientist, and certainly she will not represent the last. A key point, one might argue, is that the significance of the initial preliminary observation was fully appreciated, and indeed Steward pursued the finding with vigor. He even stated in a report to the National Cancer Institute for work covering February 1957 to June 1958, that with the carrot "Haberlandt's dream of growth of whole plants from cells" had been achieved (see Krikorian 1975, p. 86). Marion Mapes served as F.C. Steward's "hands", as he termed it, for the more focused studies on morphogenesis from cell cultures. Kathryn Mears' mini-experiment leading to the observation that an embryo-like structure from a carrot culture could develop into a plantlet several inches high *in vitro* was not documented in detail as might well be appreciated. But that important finding was confirmed and extended and worked at on a larger scale—for example, all of the photographs in Steward et al., 1958 were from the cultures generated by Marion Mapes.

F.C. Steward was a charismatic personality and spellbinding lecturer and the "carrot and coconut [milk] work" gained widespread notice and acclaim due to his extensive lecturing and publishing activities (Krikorian 1995a). That the seminal role of Kathryn Mears soon dropped from view was not deliberate. But what she did as part of the trio was never delineated either in seminars, lectures or in print. It is important, however, to say that Kathryn Mears held F.C. Steward in high esteem, and still does, and he thought highly of her as well. In an analogous manner Marion Mapes' role in the culture work, that important task of finding out what 'really' had happened in the cultures all over again, from scratch as it were, became subsumed under Steward's strong personality and authority as Professor and director of the laboratory. Again, Marion Mapes regarded F.C. Steward with great admiration and affection, and vice versa. Even so, by 1963 one investigator summarized the carrot embryoid work as follows. "The work of Steward has shown that small groups of cells far removed in time and space from an initial fertilized egg, still capable of unrestricted meristematic activity, may elaborate tissues which culminate in maturity in a normal plant of the species in question. Support has thus been lent to the contention that any 2n cell of appropriate vital capacity might, at any stage remote from a fertilized egg, or zygote be totipotent in reconstituting a mature individual plant" (Ward 1963). This quote, although it represents the views of a single scientist, really reflects what was happening in the minds of people everywhere; the transition from "et al." to "Steward" was complete.

I also think it noteworthy that no mention is made of a *single* cell having been reared into a plant (more on that below).

Around the same time the carrot work became publicized from Steward's laboratory, support came for the possible single cell origin of embryoids when Armin Braun (1911-1986) of Rockefeller Institute, New York City, reported that isolated cultured cells of tobacco crown gall teratoma could recover and produce normal healthy plants (Braun 1959) and (Plate IV, 12–19). Braun concluded that somatic mutation at the gene level was an unlikely explanation for crown-gall tumorization since "the progeny of a single somatic parenchymatous cell of tobacco may possess all the potentialities necessary to reconstitute an entire tobacco plant" [as an aside, Braun seemed to prefer the word **pluripotent** (personal knowledge)]. Braun, a brilliant "hands-on" scientist, is perhaps best remembered today for his pioneering studies on crown gall tumor induction. That work paved the way for transformation of plant cells and genetic engineering (Kahl and Schell 1982; Khachatourians et al. 2002).

Photographs depicting embryoids in the early period show that most of the structures were not particularly elegant or "clean", meaning that morphological fidelity was not always particularly high. Pictures in the account of Steward "et al." (1958) showing how suspensions derived from root explants from the secondary phloem of cultivated carrot were capable of forming unorganized cell clusters, which could yield first roots and then shoots and ultimately whole plants, are incongruous, even contradictory to the modern observer (Plate IV, 9 and 11). In retrospect, it is all the more amazing that Steward was able to reconcile the observations into a coherent, if not experimentally defensible, consistent whole. But uncanny insight into a problem was one of his many talents (Krikorian 1995a).

Examination of the literature of the time shows that the sporadic development of root and shoot primordia from callus masses and even bipolar embryo-like structures in carrot callus grown on semisolid media had been published by others. Pierre Nobécourt (1895-1961), working in Strasbourg, France, was another pioneer in the development of plant tissue culture techniques (Plate III, 5). He published in 1946 that on occasion aging callus cultures derived from carrot root spontaneously formed tiny shoots and leafy structures. Roots subcultured from these as organ cultures formed minuscule stems, each with two oval or slender opposed leaves resembling cotyledons. In some root cultures, the phenomenon was such that the leafy structures were reminiscent of bryophyte thalli. Nobécourt emphasized that the production of these leafy stems was very unpredictable but that it only occurred in old cultures (cultures âgées). The phenomenon did seem worth describing, however, since those working with root organ cultures apparently had never observed anything like it (see Krikorian 1982 for exact citations).

Shortly thereafter Michael Levine (1886-1952), a plant physiologist working on the "cancer problem" at Montefiore Hospital in the Bronx, New York, demonstrated that cultivated carrot tap root tissue grown as callus on a semisolid medium could produce readily recognizable plantlets with well-differentiated roots, stems, and leaves (Krikorian 1982).

Few seem to have made much of Nobécourt's paper or to have made connections between Levine's work and the expression of totipotency by single cells. This may have been because Nobécourt pointed out that it had not been possible to determine the nature of the factors that induced the organized structures. Levine, too, pointed out that his experiments failed "to reveal any conditions necessary to the differentiation of this tissue..." For Steward, the Nobécourt and Levine observations would merit no special attention since he had witnessed much work on the anatomical basis of vegetative propagation when he was a graduate student in the laboratory of Joseph H. Priestley (1883-1944) at the University of Leeds in England in the 1920s. Buds routinely emerged from calluses on isolated slices of beet, kohlrabi and carrot roots (Priestley and Swingle 1929). The works of Anton Kerner von Marilaun (1831-1898) and Karl von Goebel, well known to many botanists since they were encyclopedic and had been translated from German into English, were full of examples of this sort. Kerner's *Pflanzenleben* [The Life of Plants] (1896-1897) and Goebel's *Einleitung in die experimental Morphologie der Pflanzen* [Introduction to Experimental Plant Morphology, 1908] and *Organographie der Pflanzen* [Organography of Plants, 1928-1933] are particularly noteworthy in this regard. One can hardly fault plant scientists in the early period for not viewing any of the aforesaid morphogenetic responses from aseptic callus cultures as surprising. The responses even seemed too uncontrollable to be of value in answering theoretically interesting questions, much less to be of practical use.

Had it not been for Steward's forceful and widespread promulgation of the "carrot embryoid story," the connection between "free cell" culture and totipotency and somatic embryogenesis of plants would not have gained the widespread attention and recognition that it did, or as early as it did. One of Steward's contemporaries asked in print: "Is there anyone anywhere who has not heard a lecture by Steward on 'Carrots and Coconuts'?"

Single-cell Origin of Plantlets?

When details of the totipotency of cultured carrot cells were first published there was considerable and justifiable skepticism, especially among botanists, as to whether Steward, Mapes and Mears had demonstrated that single cells had, in fact, given rise to whole plants. If one studies the papers, one will see that single cells were never invoked in unambiguous terminology, but one can in retrospect certainly read between the lines and come away with the impression that is what the author(s) (specifically, FCS) believed. F.C. Steward did little in his lectures and seminars to discourage any interpretation that single cells were not involved. His interpretive schemata (Plate IV, 11) make that stance very clear as well (Steward *et al.* 1961). There was no direct evidence to support the claim. His precipitous conclusion was based on indirect evidence, namely the repeated observation that large numbers of somatic embryos could develop from subcultured or plated cell suspensions which were relatively dilute. To Professor Steward's way of thinking, the sheer numbers could only be accounted for by invoking the totipotency of single cells. It would take a chapter in itself to analyze this in depth. Suffice it to say here that he was wrong in the specifics for the carrot suspensions in question, albeit correct in the generality that somatic embryos derive from single cells (Krikorian 2000). Somatic embryogenesis in carrot is an excellent example of how one can be right in the broad conclusion that embryoids can be produced, and from these plantlets, but quite incorrect in the particulars of the process.

To this day F.C. Steward has retained a high visibility in the history of somatic embryogenesis and will probably continue to do so although there has been considerable diminution in citation of the 1958 papers. More will be said below about his contemporaries Harry Waris and Jakob Reinert and their place in any comprehensive history of somatic embryogenesis. But first attention will be directed to single cell culture before returning to organized development *in vitro*.

Muir, Hildebrandt, and Riker—Single Cell Culture

In 1954, William H. Muir (1928-1985) a doctoral candidate at the University of Wisconsin working in the Department of Plant Pathology under the direction of Albert C. Hildebrandt (1916-2001) and A.[lbert] Joyce Riker (1894-1982), published the first successful attempt to culture individual higher plant cells derived from agar-grown callus cultures. These cells derived from callus that was disrupted in liquid medium. Single cells so isolated could be induced to grow by placing them on a "nurse" tissue provided in the form of a vigorously growing agar culture, from which it was separated by a piece of filter paper. This ingenious technique provided the milieu in which single cells could proliferate and form large calluses. But it was neither easy to follow their development, nor convenient for growing particularly large numbers of cells (Plate III, 6–9). R.S. de Ropp who himself did not succeed in inducing cell division in isolated cells of carrot, parthenocissus, and sunflower crown gall incubated in small capillary tube "microchambers", argued that the growth obtained by Muir was merely derived from callus that had grown through the filter papers. This criticism was proven unjustified, however (Muir et al. 1958). [Significantly, the potential for cloning plants from single cells was clearly enunciated by Riker and Hildebrandt in a comprehensive review article as early as 1958 (Riker and Hildebrandt 1958, p. 487)].

Louis G. Nickell (1921–) pioneered batch culture of plant cells in submerged liquid media using "fermentation" methods in the mid-1950s. Not much later, he with his colleague Walter Tulecke even sought to exploit plant cells for their biosynthetic potential in an industrial research laboratory setting (see Nickell and Tulecke 1960 for a perspective on the submerged cell culture work, and Krikorian and Steward 1969 for a perspective on production of chemical substances *in vitro*). Around this time, Lvdwig Bergmann (1927–) at the Botanical Institute in Cologne, Germany separated single cells from liquid-grown cultures by filtration and plated them onto agar medium. This should be viewed as the first successful cloning of plant cells in the sense one now uses the word. Approximately 40% of the cells (*Phaseolus*) divided and 20% grew into colonies that could be isolated (Bergmann 1960). John G. Torrey (1921-1993) was another of the early pioneers in free cell culture (Konar 1966; Vasil and Vasil 1972 for a summary of single cell cultures and morphogenesis).

From such points of departure, techniques rapidly developed and today there is little doubt that free cells are capable of dividing when the conditions are right. Even so, there are still many claims of single cells dividing when this has not been conclusively demonstrated (Krikorian 2000). A key point about the earliest cultures that grew from single cells was that they did not organize. We shall see below that this seeming limitation to organization was eventually overcome.

Folke Skoog et al. and the Identification of Kinetin: Its Significance for Morphogenesis

A point that merits reemphasis is that much significant and gradually convergent experimentation took place in the laboratories of a number of the investigators who have long been recognized as the 'founding fathers' of modern tissue culture. Folke Skoog (1908-2001) at the University of Wisconsin was one of these. In 1939, P.R. White published the observation that by submerging callus from a tumor-forming stationary liquid tobacco hybrid (*Nicotiana glauca* x *N. langsdorffii*) in liquid medium, "leafy buds" and shoot-like structures could be induced to form even though the callus had been maintained for long periods in an unorganized state. He interpreted this as a result of limiting the air supply (White 1939). By the late 1940s, Skoog and his coworkers had shown that growth of, and bud formation on the cultured callus of White's strain of this tumor, as well as bud formation on internodal segments and callus from *N. tabacum* cv. *"Havana" Wisconsin 38*, could be more or less fostered by adjusting the culture medium with respect to certain additives such as naphthalene acetic acid (NAA), adenine phosphate, and sucrose. Shoot initiation could be induced from callus and explants on media supplemented with high levels of adenine. On the other hand, media containing relatively high auxin with respect to adenine led to root formation. It was also established that further development of the shoots depended on the presence of roots.

A major development emerged in 1957 when Skoog and one of his research associates, Carlos O. Miller (1923-), announced chemically controlled induction of adventitious shoots and roots on proliferated callus initiated from tobacco pith (see Miller 1961). This observation arose from their prior discovery and isolation of a plant cell division-promoting substance from autoclaved preparations of herring sperm DNA. Skoog and his coworkers had been involved for quite some time in investigations dealing with growth and organogenesis in explanted tobacco tissues and segments (Skoog 1944, 1950; Sterling 1951—see Krikorian 1982 for citations).

As a result of the work of F.C. Steward, Samuel M. Caplin, and Edgar M. Shantz (1916-1995) on the effects of coconut water on carrot explants, the Wisconsin group tested coconut water and found that it also promoted the growth of explanted tobacco tissue. They soon discovered that the same type of continuous growth in tobacco brought about by coconut water could likewise be obtained by using certain preparations from yeast. The activity correlated with the presence of purines in yeast extract and a purine N^6-furfuryl-aminopurine was soon isolated and given the trivial name of kinetin (Skoog 1994). Although kinetin does not occur in plants, it is the prototype of several related adenyl compounds which do occur naturally. Since these substances in combination with auxins are active in promoting cell division (as contrasted with enlargement), they are called cytokinins. Several synthetic cytokinins, e.g. kinetin, N^6-benzylaminopurine (BAP), N^6-dimethylaminopurine, and naturally occurring ones, e.g. zeatin riboside, dihydrozeatin, N^6-$[)^2$-isopentenyladenosine (2-iP) are now readily available for use by tissue culture workers (Krikorian 1995b).

The discovery of kinetin gave an impetus to tissue culture technology that cannot be exaggerated. For the first time investigators could add to culture media known chemical substances, auxins (known since the 1930s) and (now) cytokinins, which could stimulate cell division. Moreover, by manipulating the exogenous levels of kinetin and auxin (e.g. IAA or NAA), one could attempt to foster shoot or root development from callus cultures. This work was of great interest to those working in experimental plant morphogenesis and development (Plate III, 11–13).

This may be an appropriate place to comment on the term "meristemoid". Using terminology earlier suggested by Erwin Bünning (1906-1990) of the University of Tübingen (Bünning 1952), John G. Torrey of Harvard University wrote in the course of an essay on the initiation of organized development, "that every living plant cell capable of being stimulated to divide is a **meristemoid**, potentially able and genetically capable of differentiating or developing into any cell type or of forming any one of a variety of multicellular structures" (Torrey 1966, p. 42). (Hopefully, it will occur to readers that this definition has a problem in that he says every cell is a meristemoid even when some are not. Perhaps, had he said "potential meristemoid" the statement would be more accurate.) Another definition that has been tendered is that meristemoids are "small embryonal regions within an already partially differentiated tissue that gives rise to specific cellular structures..." (Fink 1999, p. 218.) Whether this is any more edifying remains to be seen since the term "partially differentiated" is not very specific. Nevertheless, **meristemoid** has been frequently used by many investigators in a tissue culture context who wish to designate totipotent cells or groups of cells. (Interestingly the German word **Anlagen** was used to describe "dormant or latent rudiments" by workers such as Goebel long ago for the same purpose and in the same sense. In the 'Wisconsin 38' strain of tobacco, **meristemoids** are said to form in callus masses 3 × 3 × 2 mm in size between 9 and 11 days after the tissue is placed on an inductive medium and shoots begin to appear about 14 days later (Maeda and Thorpe 1979– see Krikorian 1982 for exact citation). Finally, but by no means least in this context, reference must be made to the insightful analyses of meristems and regenerative potential, pluripotency and the concept of stem cells in plants (a topic more frequently considered in animal cell systems) by Barlow (1978, 1997).

Jakob Reinert and Somatic Embryogenesis

There always was some sensitivity on F.C. Steward's part as to who was "the" first to discover somatic embryogenesis. Some little-known particulars about the discovery of carrot embryoids in F.C. Steward's laboratory have been recounted. The special feature of the Cornell work was that free cells as well as very small cell clusters sloughed off from mature root explants had grown and multiplied vigorously in liquid through successive passages. Moreover, despite the fact that relatively few cells were used to initiate each new subculture, they readily divided and formed structures suggestive of sexually derived pro- and later-stage embryos and many of these in turn gave rise to plantlets–especially when they were placed on semisolid media. Mention has been made that Steward quickly seized the opportunity to emphasize the theoretical aspects of the discovery. He stressed that throughout the course of development in higher plants, the genetic information in the nucleus and the ability to transcribe it in the cytoplasm could remain unimpaired as a plant matures.

In the early work, Steward characteristically overinterpreted the findings, to the extent not only of drawing special attention to "free" cells, but also invoking the supposedly special nurturing, even inductive role of coconut water on the somatic embryogenic process. In fact, coconut water was not necessary for any part of the process but its use had (and still has) significant value in many instances (Steward et al. 1975; Krikorian 1989).

Since Steward wrote and lectured in English, it is no surprise perhaps that he is most often credited as being the first to have discovered somatic embryogenesis. No doubt he was an independent thinker and a forceful, talented scientist. On the other hand, Jacob Reinert (1912-2002) then working at the University of Tübingen in West Germany, is generally credited with the discovery by those whose primary language is German. In general, neither Reinert, nor Steward cited each other's work, and although their relations were cordial in each other's presence, Reinert openly claimed priority (Reinert 1959, 1968 and personal knowledge; (see Plate V, 1–10).

There has been as yet no grand inquest as to who actually was the very "first" to discover somatic embryogenesis. I have been asked from time to time to write a few critiques of the topic. These have emphasized various aspects of the history and have generally addressed the conceptual aspects of the work (Krikorian 1989). An attempt was made to document how one underappreciated participant deserves a place in the history of somatic embryogenesis. Indeed, as in all science, there have been real flesh and blood people who conducted sideshows not even realized at the time to be paramount to the story. They reveal their strengths and weaknesses at each unfolding of the major event. One such person was the Finnish scientist Harry Waris (Krikorian and Simola1999; Simola 2000).

Harry Waris and Somatic Embryogenesis

Harry Waris (1893-1973) was Professor of Botany at the Botanical Institute of the University of Helsinki. Waris first drew attention in 1957 and thereafter to the production *in vitro* of unusual growth forms (later referred to by him as **neomorphs**) by cultured but intact umbellifer seedlings (*Oenanthe*, carrot etc.) stressed with high levels of certain amino acids such as glycine. He observed that the seedlings in liquid initially "presented a normal appearance", "but within 3 – 4 months they became morbid and ceased to grow." He added: "In fact, to judge from their brown colour, they were on the point of dying. It was, therefore, a great surprise to find that later on a number of fresh green plants had grown in the flask containing the original, almost dead, plant. The new plants presented a strange, *Vallisneria*-like appearance, and reproduced vigorously until the bottom of the 3/4 liter Erlenmeyer flask became filled with fresh green "seedlings". He then continues: "It could be established that the new type had originated from minute grains formed by some root tips of the original, morbid plant. Further reproduction of the new type took place from cell groups that were spontaneously detached from colourless outgrowths formed by the green leaves, which indicates that the epidermis was responsible for the reproduction. After being freed, or even earlier, the cell groups developed colourless 'embryos' which later became green and assumed a shape resembling normal seedlings, though more than two cotyledons were sometimes formed. In the subsequent development, tape-formed leaves, a few centimeters in length and a few millimeters in width appeared, but the final size of the individuals did not exceed a few centimeters....When transferred to new nutrient solutions containing sucrose but no glycine, the new type was maintained for several months....After about four months, however, a part of the plants began to develop normal leaves. Transfer to inorganic solution was not successful" (Plate V, 11–17).

All the important points are clear – strikingly different ("new"-*neo*) morphological structures (**neomorphs**) appeared in/from a place where they are not normally expected. They could sustain themselves and apparently no permanent genetic change had taken place in their formation.

Harry Waris also makes it clear that *groups of cells from the root, possibly from the epidermis, developed into embryos.* Clearly Waris deserves to be credited with being among the earliest to recognize somatic embryogenesis under sterile culture conditions. In a paper communicated in 1957 there is a photograph of carrot neomorphs produced through use of excess glycine that are quite similar to some of the carrot "pro-embryo-like growth unit(s)" [later named "embryoids" by Steward *et al.* (1961, p. 205)] first figured by Steward, Mapes and Mears (1958)— (compare Plate V, 12 and Plate IV, 4). Waris' work is undeniably contemporary with the work of Steward or Reinert. Waris was certainly very aware of the botanical literature and he mentions the work of Steward, Mapes and Mears in his 1959 paper. The reasons for his somatic embryo work being underappreciated have been discussed by Krikorian and Simola (1999).

THE FIRST PLANTS REGENERATED FROM SINGLE CELLS

Direct and unequivocal support for the totipotency of single cultured cells first came in 1965 when Dr. Vimla Vasil was working in A.C. Hildebrandt's laboratory at the University of Wisconsin (Plate V, 18 and 19). She reared a mature tobacco plant from a single cell which was initially grown in a microchamber in "microculture" (Plate V, 20–29). The cell she started with first divided progressively into a cluster of cells and then into a callus mass. This was then induced on agar to yield shoots and roots (Plate V, 30–33). The theoretically achievable vegetative reproductive cycle from isolated cell to mature flowering plant with fertile seed finally became a reality (see Plate V, 34 and Vasil and Hildebrandt 1967 for a full account). However, there was no suggestion whatever that somatic embryogenesis or an adventive embrogeny similar to that encountered in carrot and other umbellifers had occurred. Development was via meristemoids (Plate V, 30–34). The matter of the single cell origin of embryoids was laid to rest for most people when an apparently single cell was followed by Backs-Hüsemann and Reinert (1970).

EPILOGUE

After cultured carrot and tobacco cells were shown to be capable of growing into plants botanists finally relinquished the view that the genetic material was progressively permanently altered or even lost during the process of differentiation and development. Space limitations prevent delving here into the contributions of those many investigators who fostered a broader and deeper understanding of the events associated with somatic embryogenesis and expression of totipotency (Ammirato 1983; Thorpe 1995; Bhojwani and Razdan 1996; Raghavan 1997; Krikorian 2000 and references cited therein). For example, Walter Halperin (1932- ; now retired from the Botany Department, University of Washington) working as a doctoral candidate in the laboratory of Donald F. Wetherell at the University of Connecticut showed that CW was not necessary for somatic embryogenesis, and that the synthetic auxin 2,4-D served very well to foster initiation of embryogenic cultures of "Queen Anne's Lace", the wild carrot (Halperin 1995). Others showed that exogenous growth regulators need not be added to culture media to initiate embryogenic cultures (Krikorian and Smith 1992; Krikorian 2000 for details). Production of haploids from miscrospores was first reported by Sipra Guha (now Mukherjee) and Satish C. Maheshwari, University of Delhi, in 1964 (see Touraev et al. 2001 for a comprehensive modern review). The first report of a plant (tobacco) successfully regenerated from protoplasts was published from the laboratory of Georg Melchers (1906-1997) at the Max Planck Institute in Tübingen, Germany by Itaru Takebe *et al.*, in 1971. (Dr. Takebe's home base was the Institute for Plant Virus Research in Chiba, Japan.) The production of interspecific hybrid tobacco through protoplast fusion was reported from the laboratory of Harold H. Smith (1910-1994) at Brookhaven National Laboratory, Upton, New York by Peter S. Carlson in 1972 (see Smith et al. 1976 for details). The rest is history!

ACKNOWLEDGMENTS

I thank the Cambridge University Press for permission to base some of this chapter on a review published in the *Biological Reviews* in 1982 (Krikorian, 1982). I thank Mrs. Kathryn M. Trupin for scrutinizing the section that relates to the Cornell work in which she participated. I also thank Dr. John K. Pollard, Jr. (Ph.D. Cornell 1955), a Graduate student and long-time Research Associate of F.C. Steward for reading the manuscript. Over the years I have collected photographs of scientists and maintained files of classic works in experimental botany and plant tissue culture. Many investigators have indulged me in this hobby by providing photographs and reprints. I also acknowledge NASA for having aided and abetted me in my somatic embryogenesis investigations over many years. The late Professor F.C. Steward, himself very much interested in history, nurtured my interest in the history of our discipline. He often emphasized that a scientist who had no appreciation of the history of his/her subject could be likened to a mule that had "neither pride of descent nor hope for posterity"!

REFERENCES

Allen CC (1923) The potentialities of a cell. Amer. J. Bot. 10:387-398.

Ammirato PV (1983) Embryogenesis. *In:* DA Evans, WR Sharp, PV Ammirato, Y Yamada, (eds.). Handbook of Plant Cell Culture. Macmillan, New York, NY, vol.1, pp. 82-123.

Arditti J, Krikorian AD (1996) Orchid micropropagation: the path from laboratory to commercialization and an account of several unappreciated investigators. Bot. J. Linn. Soc. 122:185-241.

Asker SE, Jerling L (1992) Apomixis in Plants. CRC Press, Boca Raton, FL, USA.

Backs-Hüsemann D, Reinert J (1970) Embryobildung durch isolierte Einzelzellen aus Gewebekulturen von *Daucus carota*. Protoplasma 70:49-60.

Bailey IW (1943) Some misleading terminologies in the literature of 'Plant Tissue Culture'. Science 93:539.

Barlow PW (1978) The concept of the stem cell in the context of plant growth and development. *In:* BI Lord, CS Potten, RJ Cole, (eds.). Stem Cells and Tissue Homeostasis. Second Symp. British Society for Cell Biology. Cambridge Univ. Press, Cambridge, UK, pp. 87-113.

Barlow PW (1997) Stem cells and founder zones in plants, particularly their roots. *In:* CS Potten, (ed.). Stem Cells, Acad. Press, London, UK. pp. 29-57.

Bergmann L (1960) Growth and division of single cells of higher plants *in vitro*. J. Gen. Physiol. 43:841-851 .

Bhojwani SS, Razdan MK (1996) Plant Tissue Culture. Theory and Practice. Elsevier, Amsterdam.

Braun AC (1959) A demonstration of the recovery of the crown-gall tumor cell with the use of complex tumors of single cell origin. Proc. Natl. Acad. Sci. (USA) 45:932-938.

Bünning E (1952) Morphogenesis in plants. Surveys Biol. Prog. 2:105-140.

Denny AD III (1952) Erwin Frank Smith: A Story of North American Plant Pathology. Mem. Amer. Phil. Soci., vol. 31. Ameri. Phil. Soc. Philadelphia, PA.

Fink S (1999) Pathological and Regenerative Plant Pathology. Encyclopedia Plant Anatomy/Handbuch der Pflanzenanatomie, Vol. 14, pt. 6. Gebrüder Borntraeger, Berlin.

Gautheret R J (1982) Plant tissue culture: The history. *In:* A Fujiwara, (ed.). Plant Tissue Culture 1982. Proc. 5th Inter. Cong. Plant Tissue Cell Culture, Jpn. Assoc. Plant Tissue Culture, Maruzen Co., Tokyo, pp. 7-12.

Gautheret R J (1983) Plant tissue culture: A history. Bot. Mag. (Tokyo) 96:393-410.

Gautheret R J (1985) History of plant tissue and cell culture: A personal account. *In:* IK Vasil, (ed.). Cell Culture and Somatic Cell Genetics of Plants. Acad. Press, New York, NY, Vol 2. pp.1-59.

Haberlandt G (1902) Kulturversuche mit isolierten Pflanzenzellen. Sitzungsberichte der Koniglichen Kaiserlichen Akademie der Wissenschaften Mathematisch-Naturwissenschaftliche Klasse 111:69-92.

Haberer, G, Kieber JJ (2002) Cytokinins: new insights into a classic phytohormone. Plant Physiol. 128: 354-362.

Halperin W (1995) *In vitro* embryogenesis: some historical issues and unresolved problems. *In:* TA Thorpe, (ed.). *In vitro* Embryogenesis in Plants. Kluwer Acad. Publi., Dordrecht, Netherlands, pp. 1-16.

Harrison RG (1928) On the status and significance of tissue culture. Archiv experim. Zellforschung 6:4-27.

Hartmann HT, Kester DE, Davies, Jr. FT, Geneve RL (2002) Hartmann and Kester's Plant Propagation: Principles and Practices. Prentice-Hall, Inc., Englewood Cliffs, NJ. (7th ed.).

Jones LE, Hildebrandt AC, Riker AJ (1960) Growth of somatic tobacco cells in microculture. Amer. J. Bot. 47:468-475.

Kahl G, Schell JS (eds) (1982) Molecular Biology of Plant Tumors. Acad. Press, New York, NY.

Kerner von Marilaun A (1895-1896) The Natural History of Plants: Their Growth, Reproduction, and Distribution. Henry Holt, New York, NY (2 Vols.)

Khachatourians GG, McHughen A, Scorza R, Nip W-K , Hui YH (2002) Transgenic Plants and Crops. Marcel Dekker, New York, NY.

Konar KN (1966) Single cell culture and morphogenesis (review). Portugaliae Acta Biologicae (Serie 4) 10:1-32.

Korschelt E (1927,1990) Regeneration and Transplantation. – Translated from German by Sabine Lichtner Ayed, with a new Preface by Bruce M. Carlson. Science History Publ., Canton, MA (USA). (3 Vols.)

Krikorian AD (1975) Excerpts from the History of Plant Physiology and Development. *In:* P Davies, (ed.). Historical and Current Aspects of Plant Physiology: A Symposium Honoring F.C. Steward. Cornell Univ., College Agricu. Life Sciences, Ithaca, NY, pp. 9-97.

Krikorian AD (1982) Cloning higher plants from aseptically cultured tissues and cells. Biol. Rev. 57:151-218.

Krikorian AD (1989) Introductory essay to: Growth and organized development of cultured cells. II. Organization in cultures grown from freely suspended cells by Steward, F.C., Mapes, M.O. and Mears, K. *In:* J Janick, (ed.). Classic Papers in Horticultural Science. W.H. Freeman and Co., pp 40-47.

Krikorian AD (1995a) Frederick Campion Steward 16 June 1904-13 September 1993. Biog. Mem. Fellows Royal Soc. 41:420-437

Krikorian AD (l995b) Hormones in tissue culture and micropropagation. *In:* P Davies (ed.). Plant Hormones: Physiology, Biochemistry and Molecular Biology. Kluwer, Dordrecht, Netherlands, pp.774-796.

Krikorian AD (2000) Historical insights into some contemporary problems in somatic embryogenesis. *In:* SM Jain, PK, Gupta, RJ Newton, (eds.), Somatic Embryogenesis in Plants. Kluwer Acad. Pub., Dordrecht, Netherlands, pp. 17-49.

Krikorian AD, Berquam DL (1969) Plant cell and tissue cultures: The role of Haberlandt. Bot. Rev. 35:58-88.

Krikorian AD, Steward FC (1969) Biochemical differentiation: The biosynthetic potentialities of growing and quiescent tissue. *In:* F.C. Steward (ed.). Plant Physiology: A Treatise. Acad. Press, New York, NY, Vol. 5 B, pp. 227-326.

Krikorian AD, Smith DL (l992) Somatic embryogenesis in carrot (*Daucus carota*). *In:* K Lindsey (ed.). Plant Tissue Culture Manual: Fundamentals and Application. Kluwer Acad. Publi., Dordrecht, Netherlands, pp. PTCM-A9 1-32.

Krikorian AD, Simola LK (1999) Totipotency, somatic embryogenesis and Harry Waris (1893-1973). Physiologia Plantarum 105:348-355.

Laimer M, Rücker W, (eds.) (2002) Plant Tissue Culture:100 Years since Gottlieb Haberlandt. Springer-Verlag, Wien.
Loeb J (1924) Regeneration from a Physico-Chemical Viewpoint. McGraw-Hill, New York, NY.
McKinnel RG, Di Berardino MA (1999) The Biology of Cloning: History and rationale. Bio Science 49: 875-885
Miller CO (1961) Kinetin and related compounds in plant growth. Ann. Rev. Plant Physiol. 12:395-408.
Morgan TH (1901) Regeneration. Macmillan, London, UK.
Muir WH, Hildebrandt AV, Riker AJ (1958) The preparation, isolation and growth in culture of single cells from higher plants. Amer. J. Botany 45:589-597.
Murbeck Sv. (1902) Über Anomalien im Baue des Nucellus und des Embryosackes bei Parthenogenetische Arten der Gatttung *Alchemilla*. Lunds Universiteit Årskrift 38:2, 2, 1-12.
Nickell LG, Tulecke W (1960) Submerged growth of cells of higher plants. J. Biochem. Microbiol. Tech. Eng. 2:287-297.
Nogler GA (1984) Gametophytic apomixis. *In:* BM Johri, (ed.). Embryology of Angiosperms. Springer-Verlag, Berlin, pp 475-518.
Pratt H (1948) Histo-physiological gradients and plant organogenesis. Bot. Rev. 14:603-643.
Pratt H (1951) Histo-physiological gradients and plant organogenesis (Part II) Bot. Rev. 17:693-746.
Priestley JH, Swingle CF (1929) Vegetative propagation from the standpoint of plant anatomy. U.S. Dep. Agri. Tech. Bull. 151:1-98.
Raghavan V (1997) Molecular Embryology of Flowering Plants. Cambridge Univ. Press, New York, NY.
Reinert J (1958) Morphogenese und ihre Kontrolle an Gewebekulturen aus Carotten. Die Naturwissenschaften 45:344-345.
Reinert J (1959) Über die Kontrolle der Morphogenese und die Induktion von Adventive-embryonen an Gewebekulturen aus Karotten. Planta 53:318-333.
Reinert J (1968) Morphogenese in Gewebe-und Zellkulturen. Naturwissenschaften 170-175.
Riker AJ, Hildebrandt AC (1958) Plant tissue cultures open a botanical frontier. Ann. Rev. Microbiol. 12:469-490.
Robbins WJ (1957) The influence of Jacques Loeb on the development of plant tissue culture. Bull. Jardin Botanique de l'Etat Bruxelles 27:189-1970.
Roux W ed (1912) Terminologie der Entwicklungsmechanik der Tiere und Pflanzen, in Verbindung mit Carl Correns, Alfred Fischel, Ernst Küster. W. Engelmann, Leipzig.
Savidan Y, Carman JG, Dresselhaus T, (eds.) (2001) The Flowering of Apomixis: From Mechanisms to Genetic Engineering. CIMMYT Publications, Institut de Recherche pour le Développement, [Paris]: IRD [Brussels]: European Commission, Mexico D.F.
Schel JHN (1989) Plant cell biology 150 years after Matthias Schleiden. Sexual Plant Reprod. 2:59-64.
Schleiden MJ (1852) Die Pflanze und ihr Leben. Wilhelm Engelmann. Leipzig, Germany (3rd ed.).

Schober A (1904) Matthias Jacob Schleiden. Nach der Gedenkrede im Naturwissenschaftlichen Verein am 13. April 1904. Lütcke & Wulff, Hamburg.

Simola LK (2000) Harry Waris, a pioneer in somatic embryogenesis. *In:* SM Jain, PK Gupta, RJ Newton, (eds.), Somatic Embryogenesis in Plants. Kluwer Academic Pub. Dordrecht, Netherlands, Vol 6. pp. 1-16.

Sinnott EW (1960) Cell and Psyche. The Biology of Purpose. Univ. North Carolina Press, Chapel Hill, NC.

Skoog F (1994) A personal history of cytokinin and plant hormone research. *In:* DWS Mok, MC Mok (eds.). Cytokinins. Chemistry, Activity, and Function. CRC Press, Boca Raton, FL, pp. 1-14.

Skoog F, Miller CO (1957) Chemical regulation of growth and organ formation in plant tissues cultured in vitro. Symp. Soci. Experim. Biol. 11:118-131.

Smith EF (1920) An Introduction to Bacterial Diseases of Plants. W.B. Saunders, Philadelphia, PA.

Smith HH, Kao KN, Combatti NC (1976) Interspecific hybridization by protoplast fusion in *Nicotiana*. J. Heredity 67:123-128.

Steward FC, Shantz EM (1956) The chemical induction of growth in plant tissue cultures. *In:* RL Wain, F Wightman (eds.). The Chemistry and Mode of Action of Plant Growth Substances. Proc. Symp. Wye College (Univ. London) July 1955. Butterworths, London, pp 165-186.

Steward FC, Krikorian AD (1971) Plants, Chemicals and Growth. Acad. New York, NY.

Steward FC, Mapes MO, Mears K (1958) Growth and organized development of cultured cells. II. Organization in cultures grown from freely suspended cells. Amer. J. Bot. 45:705-708.

Steward FC, Mapes MO, Smith J (1958) Growth and organized development of cultured cells. I. Growth and division of freely suspended cells. Amer. J. Bot. 45:693-703.

Steward FC, Shantz EM, Pollard J K, Mapes MO, Mitra J (1961) Growth induction in explanted cells and tissues: Metabolic and morphogenetic manifestations. *In:* D Rudnick, (ed.). Molecular and Cellular Synthesis. Soc. Study of Develop. Growth. 19th Growth Symp. Ronald Press, New York, NY, pp. 193-246.

Steward FC, Israel HW, Mott RL, Wilson HJ, Krikorian AD (1975) Observations of growth and morphogenesis using cultured cells of carrot. Phil. Trans. Roy. Soc. London, 273:33-53.

Takebe I, Labib G, Melchers G (1971) Regeneration of whole plants from isolated mesophyll protoplasts of tobacco. Naturwissenschaften 58:318-320.

Thorpe TA, (ed.) (1995) *In Vitro* Embryogenesis in Plants. Kluwer Acad. Pub., Dordrecht, Netherlands.

Torrey JG (1966) The initiation of organized development in plants. Advances in Morphogenesis 5:39-91.

Touraev A, Prosser M, Heberle-Bors E (2001) The miscrospore: a haploid multipurpose cell. Adv. Bot. Res. 35:53-109.

Trân Than Vân K, Bui Vân Lê (2000) Current status of thin cell layer method for the induction of organogenesis or somatic embryogenesis. *In:* SM Jain, PK Gupta, RJ Newton (eds.), Somatic Embryogenesis in Plants. Kluwer Acad. Pub., Dordrecht, Netherlands, Vol 6. pp 51-92.

Van Harten AM (1998) Mutation Breeding: Theory and Practical Applications. Cambridge Univ. Press, Cambridge, UK.
Vasil V, Hildebrandt AC (1967) Further studies on growth and differentiation of tobacco plants from single isolated cells of tobacco *in vitro*. Planta 75:139-151.
Vasil IK, Vasil V (1972) Totipotency and embryogenesis in plant cell and tissue cultures. *In vitro* 8:117-127.
Vöchting H (1878) Über Organbildung im Pflanzenreich. Physiologische Untersuchungen über Wachsthumsursachen und Lebenseinheiten. Max Cohen & Sohn, Bonn.
Von Goebel K (1900,1905) Organography of Plants. Oxford Univ. Press, Oxford, UK. (2 Vols.)
Von Goebel K (1908) Einleitung in die Experimentelle Morphologie der Pflanzen.
Ward M (1963) Developmental patterns of adventitious sporophytes in *Phlebodium aureum* J. Sm. J. Linn. Soc. (Botany) 58:377-380.
White PR (1936) Plant Tissue Cultures. Bot. Rev. 2:419-437.
White PR (1939) Controlled differentiation in a plant tissue culture. Bull. Torrey Botan. Club 66: 507-513.
White PR (1943) A Handbook of Plant Tissue Culture. Ronald Press, New York, NY.
Williams G (1960) Virus Hunters. Alfred A. Knopf, New York, NY.
Wilmut I, Schnieke AE, McWhir J, Kind AJ, Campbell KHS (1997) Viable offspring derived from fetal and adult mammalian cells. Nature 385:810-813.

Plate I

A.D. Krikorian examining flasks of cell suspensions in the laboratory, State University of New York, Stony Brook. Note the distinctive 'nipple' culture flasks positioned by means of spring clamps on Plexiglas wheels which rotate at 1 rpm around the horizontal axis (see text page 32 Plate IV, 2 for a close-up)

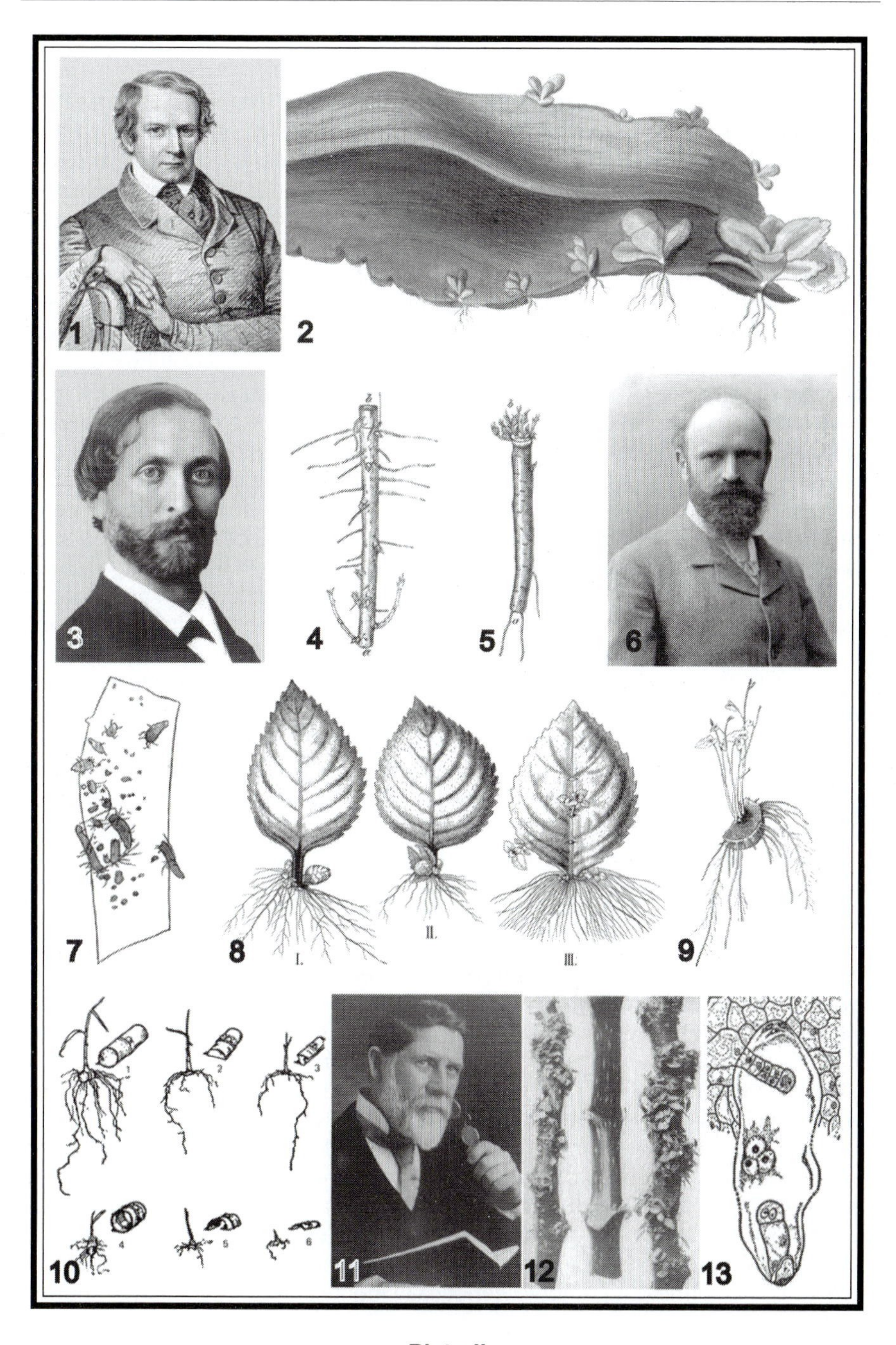

Plate II

Plate II

Some pioneers in the study of plant development and examples of natural (*in vivo*) regeneration. 1-Matthias J. Schleiden, a cofounder of the cell theory, as he appeared in 1845 (Schober, 1904). 2-*Bryophyllum calycinum* leaf showing plantlets at varying stages of development emerged from leaf notches (Schleiden, 1852). 3-Early photograph of Hermann Vöchting. Courtesy Botanisches Institut der Universität Bonn. 4-Regeneration from an inverted stem segment of *Salix viminalis* shoots from the 'shoot end' (a) and with roots emerged from the 'root end' (b). This species shows strong polarity (Vöchting, 1878). 5-Regeneration from an inverted stem segment of *Populus pyramidalis* with roots emerged from the 'shoot end' (a) and shoots from the 'root end' (b). This species shows virtually no polarity for regeneration (Vöchting, 1878). 6-Karl Goebel, student of plant form and experimental morphologist. Undated but showing him probably in his early 50s. Courtesy Botanische Stattssammlung, Munich. 7-Portion of the thallus of a liverwort (*Blyttia lyellii*) with adventitious shoots at various stages of development emerged from the surface (Goebel, 1908). 8-Variations on the theme of regeneration from leaf bases using an *Achimenes* hybrid (Gesneriaceae): I, tiny bulbous tubers formed at the petiolar base; II, adventitious shoot with three foliage leaves to be followed by the formation of a bulbous tuber; III, leaf from which adventitious shoots have emerged at the site of cuts on the main vein both on the adaxial and abaxial surfaces. Also, shoots have formed at the basal end (Goebel, 1908). 9-Portion of a *Dioscorea sinuata* tuber with adventitious shoots at the cut edge surface and roots emerged from the periphery (Goebel, 1908). 10-Experiment showing "irreducible minimum" needed for sprouting of a sugarcane cutting. Presence of a bud guarantees vigorous "germination" but the smaller the unit, the less vigorous the growth:1, single-budded "sett" with one-inch internode on either side; 2, as in 1 except it is longitudinally sliced in two—the bud half being planted; 3, like 2 but a quadrant; 4, like1 but minus the pith; 5, like 2 but pith removed; 6, similar to 3 but pith removed (Venketraman, 1926). 11-Erwin F. Smith in the early 1920s. Smith, "Father of Bacterial Plant Pathology", was Chief of Plant Pathology in the Bureau of Plant Industry, US Department of Agriculture from 1889 to 1927. From Denny (1952) with permission of the American Philosophical Society. 12-*Begonia phyllomaniaca* stems showing degrees of tumor- or parasite-free teratosis. The center axis is nearly free from leafy adventitious proliferations. The left bore 425 shoots and the right one 591. **Phyllomania** is the production of leaves in unusual numbers or in unusual places. Such plants are also known as epiphyllous or viviparous. Smith likened these to "young plants bedded in the tissues of the mother plant" (Smith 1920, p. 615). 13-Example of both unreduced parthenogenesis and sporophytic budding in the embryo sac (*Alchemilla pastoralis,* Rosaceae). The developing gamete (egg) at "6 o'clock" has the zygotic number of chromosomes since the embryo sac cells did not derive from a reduction division followed by mitosis. It is forming one embryo. At "10 o'clock" a nucellar cell is forming an embryo genetically identical to the sporophytic parent (Murbeck, 1902).

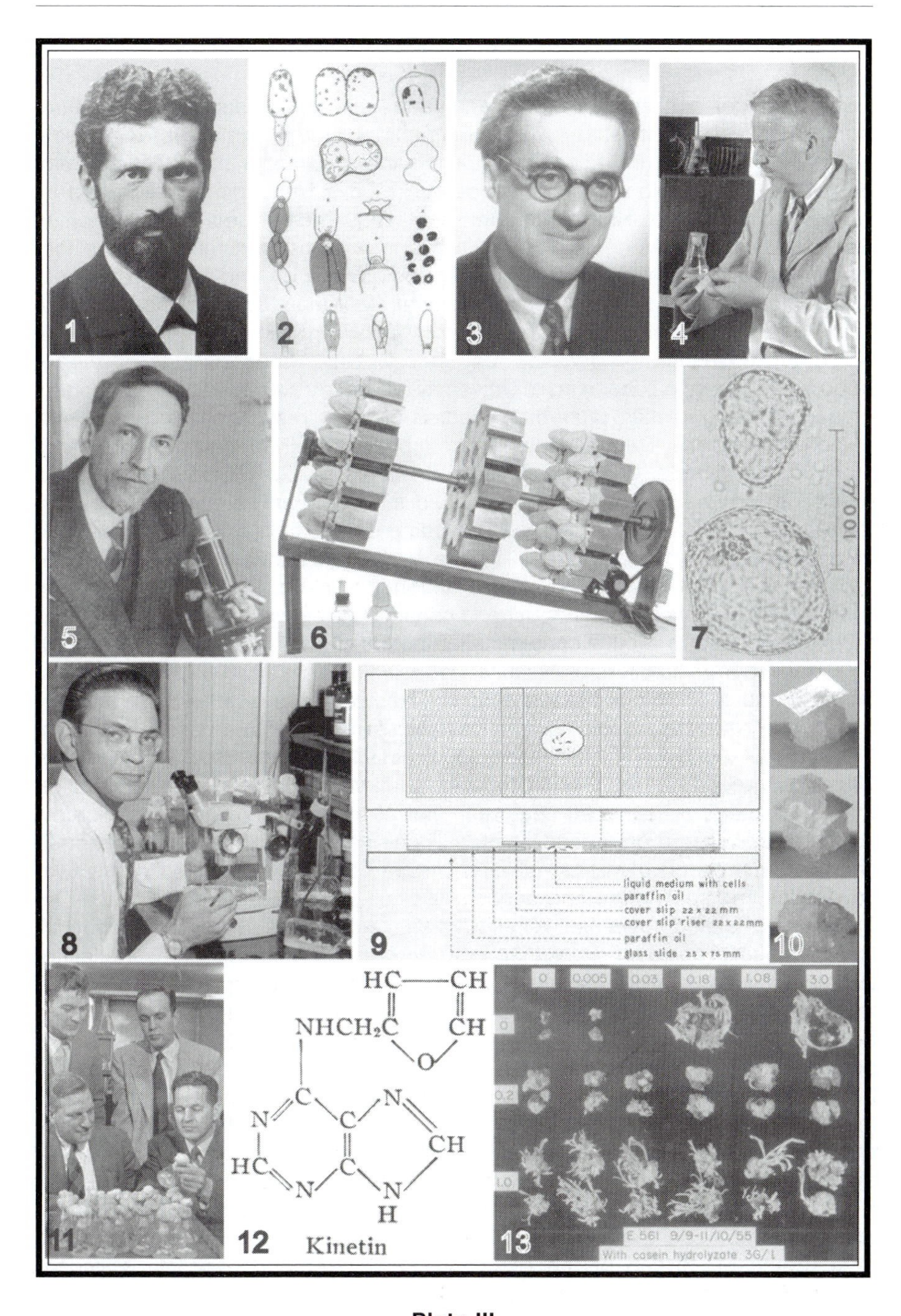
1
2
3
4
5
6
7
100 μ
8
9
liquid medium with cells
paraffin oil
cover slip 22 x 22 mm
cover slip riser 22 x 22 mm
paraffin oil
glass slide 25 x 75 mm
10
11
12 Kinetin
13
0
0.005
0.03
0.18
1.08
3.0
0.2
1.0
E 561 9/9-11/10/55
With casein hydrolyzate 3G/l

Plate III

Plate III

Some early cell and tissue culture personages and landmarks in tissue culture work. 1-Gottlieb Haberlandt in Graz (sometime during 1880-1909). Courtesy Institut für Anatomie und Physiologie der Pflanzen, Graz, Austria. 2-Plate from Haberlandt (1902). No cell division was observed but enlargement occurred in certain cases (see Krikorian and Berquam 1969 for full details.) 3-Roger J. Gautheret in 1952. Along with P.R. White and P. Nobécourt he pioneered the modern period of plant tissue culture. 4-Philip R. White, around 1953, measuring cultured tomato roots which, by weekly transfer, were maintained for more than 30 years. The initial root tip was excised on March 1, 1933! Photo courtesy late Dr. Charity Waymouth, Jackson Laboratory, Bar Harbor, Maine. 5-Pierre Nobécourt at the microscope in his University laboratory around 1950 in Grenoble, France. Nobécourt also pioneered studies of plant resistance to fungal and bacterial pathogens. Courtesy late Mme. Nobécourt. 6-Photograph of a rotating drum aerator used by William H. Muir as a student of A.C. Hildebrandt and A.J. Riker. Tissues were grown in 40 ml liquid in largemouth culture bottles and rotated at 1 rpm. Friable cultures, including single cells, were obtained by fragmentation of callus. Single cells were fished out with a micropipette or by dragging friable masses over an agar-stiffened surface. At the lower left are culture bottles with a rubber stopper and cotton-plugged wide glass tube; at right, a 2-ply filter used to reduce contamination potential (Muir et al., 1959). 7-Single cells isolated from a culture of *Nicotiana tabacum* (Muir et al., 1959). 8-Muir in his laboratory in the Biology Department of Carleton College in 1959. Note the 'prescription medicine bottle' culture vessels. A hot wire was used as a glass breaking tool to cut the bottle open for full access to its contents. Courtesy of the Carleton College Archives. 9-Microculture chamber used for single cells. A droplet of heavy mineral oil is placed near each end of a standard microscope slide. A cover slip is lowered onto each droplet. These serve to form a shallow central chamber. A rectangle of mineral oil is then placed on the slide connecting the two cover slips and sealing the end of each. A small droplet of medium is placed in the center of a third cover slip. A cell from shake-culture is placed in the drop of liquid medium. The cover slip is then inverted onto the rectangle of mineral oil to surround the medium with its cell and to seal the ends of the top cover slip. There is no liquid/air interface along the optical axis and hence good observations can be made (Courtesy late Prof. A.C. Hildebrandt; Jones *et al.* 1960; 6, 7, and 9 with permission of the American Journal of Botany). 10-Examples of callus growth from single cells on filter paper over established "nurse" cultures. Upper, 8 x 8 mm^2 filter paper on a sunflower culture. Center, marigold culture of single cell origin grown on filter paper over sunflower; Lower, free-growing culture removed from filter. Photograph provided by the late Professor A.C. Hildebrandt. 11-Folke Skoog and co-investigators of kinetin, University of Wisconsin. Photograph taken in 1955. Left front, Folke Skoog of the Botany Department and right front, Frank M. Strong of the Biochemistry Department; rear left, Malcolm H. von Saltza; rear right, Carlos O. Miller, Ph.D. candidate and postdoctoral researcher respectively. Photograph provided by Prof. Carlos Miller. 12-structure

Contd.

Contd.

of kinetin. 13-Latin Square display of the effect of kinetin and indole-3-acetic acid on growth and organ formation in tobacco callus (*Nicotiana tabacum* var. 'Havana' Wisconsin 38) after 62 days on a semisolid White's basal nutrient medium supplemented with casein hydrolyzate (CH, 3 gm/L^{-1}, a source of reduced Nitrogen). Kinetin level (left): 0, 0.2 and 1.0 mg/L^{-1}; IAA level (top) 0, 0.005, 0.03, 0.18, 1.08 and 3.0 mg/L^{-1}. The control, lacking kinetin or IAA shows virtually no growth. In the presence of high auxin, roots were favored whereas in the presence of kinetin, bud development was enhanced. With the isolation and synthesis of kinetin it became possible for the first time to add two identified growth substances (IAA and kinetin) to a culture medium and control organ development. Two cultures were lost to contamination (Skoog and Miller, 1957). Courtesy late Professor F. Skoog and permission of the Society for Experimental Biology, London.

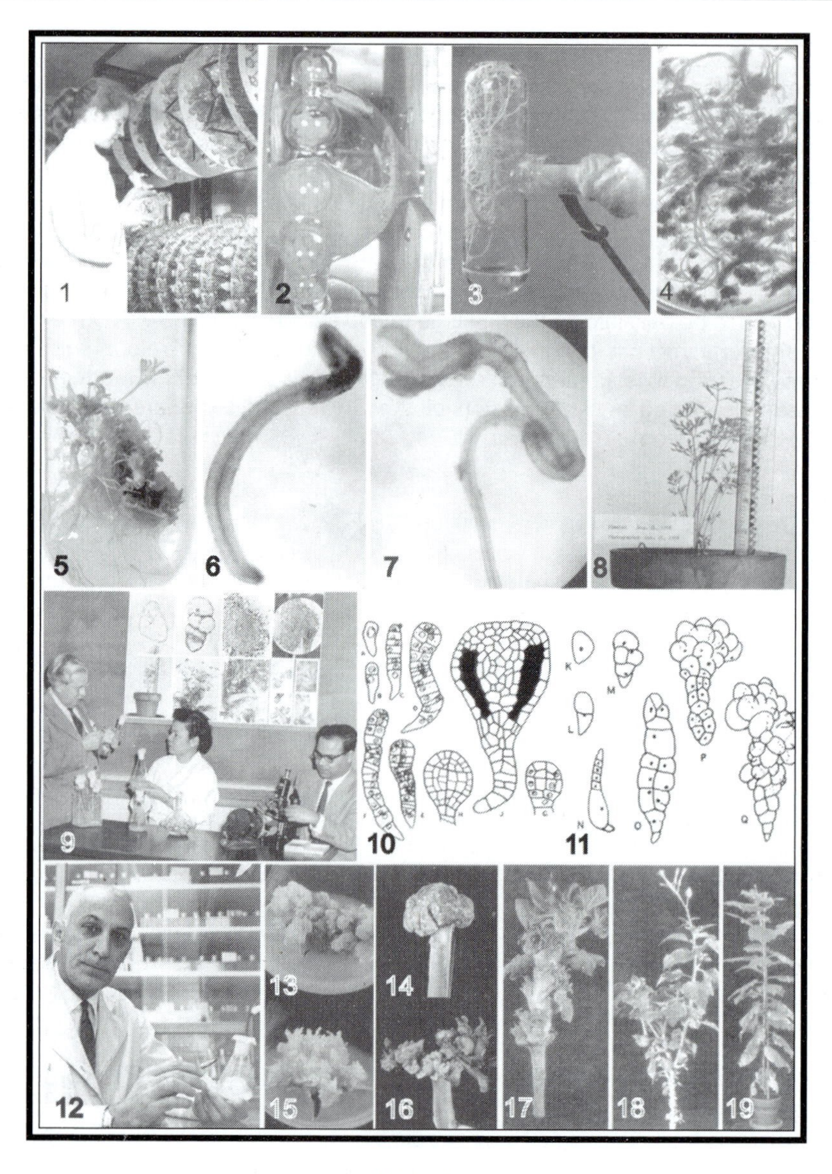

Plate IV

Organized development from cultured cells: some highlights. 1- Kathryn A. Mears at the "auxophyton" apparatus designed to rotate "nipple" flasks and "T-tubes". The rotation exposes tissues to a gentle tumbling action. Photograph around 1957. 2-Culture flask on a rotating "wheel" of the auxophyton. The upper two "nipples" show carrot root phloem explants at a nonsubmerged stage of rotation, whereas explants at the lower part of the photograph are submerged. Alternate

Contd.

Contd.

exposure to air and liquid aerates explants and exposes them to nutrient. The medium is opaque from detached fragments and cells (Steward and Shantz 1956). 3-"T-tube" and contents with extensive root development from an inoculum of suspended carrot cells. This kind of culture can yield embryoids when placed on a semi-solid medium (Steward and Shantz 1956). 4-Organized forms from a carrot suspension. Range of embryoids as well as neomorphic, long-rooted structures without well-formed shoot apices. 5-Carrot embryoids grown in a culture tube from tiny 'rooty' structures 'fished out' of liquid medium. Asynchronous development and root growth has lifted some culture off the medium. Photograph taken in the late 1950s. 6 and 7-Carrot embryoids (Steward, Mapes and Mears 1958; with permission of the American Journal of Botany). 8-Potted carrot plant from an embryoid. The *in vitro*-generated plantlet (embryoid) was grown to a point where it could tolerate transplanting into vermiculite on August 21 1958; it was photographed on October 21 1958. The above-ground portion of the plant is about 6 inches high (Steward *et al.*, 1961). 9-F.C. Steward (at left) holding a culture tube with a carrot plantlet, Marion O. Mapes with an Erlenmeyer flask of embryoids and plantlets, Dr. Jyotirmay Mitra prepares cultures for cytological analysis. The collage shows "a sequence in the development from carrot cell to carrot plant". Upper left: a "freely suspended carrot cell in liquid medium"; "similar cell after divisions in two planes at right angle"; colony with "organized growth centers" or nodules, and the next shows a "proliferated mass" of cells with a root emerging from a "nodule." Potted plant (see number 8), followed by a photograph of a mature carrot plant in the second cycle of biennial growth. Storage roots with normal carotene content were also presented as well as images of stained chromosomes. Although cell cultures could show tetraploidy, the regenerated plants were diploid with the usual diploid complement of 18 chromosomes. Photographed in early1960. 10 and 11-Zygotic embryogenesis and embryo-like cell clusters (Steward et al. 1961) show similarities between normal embryogeny and "growth forms in the outgrowth of single carrot phloem cells in liquid medium." Left, a zygote and its subsequent development into a precotyledonary stage embryo with vascular development apparent (in black). Complexes encountered in cultures initiated from "free cells " drawn in 11 are suggestive but in fact, are at odds with the biological facts. 12-Armin C. Braun in his laboratory at Rockefeller Institute, New York City in1960. He holds a flask of tobacco tumor tissue reared in microculture from a single cell. 13 through 19; Reversibility of the tumor state using tobacco crown-gall of single cell origin. 13-Culture of unorganized tumor tissue; 14, Result when sterile tumor (13) was grafted to the cut stem (distal end) of tobacco plant; 15-Aseptic culture of teratoma of single cell origin; 16-Result when sterile teratoma (see 15) was similarly grafted; 17, 18, 19 show three stages to the full recovery of a crown-gall teratoma:17-Growth from a tumor bud (like 16) grafted to the cut stem end of a normal tobacco plant. Note the abnormal growth; 18-Tumor shoot like that in17 similarly grafted. The growth is more normal. One branch has even flowered; 19-Seeds obtained from a flowering scion like the one shown in 18 and germinated yielded normal plants. Thus autonomous teratoma cells can lose their tumor properties in a gradual process and develop normally once again. Photographs supplied by the late Professor Braun.

1
2
3
4
5
6
7
8
9
10
11
12
13
14
15
16
17
18
19
20
21
22
23
24
25
26
27
28
29
30
31
32
33
34
PLANTLET
YOUNG PLANT
MATURE PLANT
STEM PITH IN
CALLUS FROM PITH
NICOTIANA GLUTINOSA
X
N. TABACUM
ROOT FORMATION
SHOOT FORMATION
SUSPENSION CULTURE
CELL MASS FROM SINGLE CELL
CALLUS FROM CELL MASS
CELL MASS FROM MICROCULTURE
MICROCULTURE

Plate V

Plate V

Embryogenic development from carrot callus cultures, "embryos" and neomorphic embryos from stressed seedlings of carrot and *Oenanthe* (both umbellifers) in liquid and a single cell of tobacco microcultured in liquid to a mature plant in soil. 1-Jakob Reinert photographed around 1963. Presented to the author by the late Professor Reinert. 2-Organization from a carrot callus in culture for several years on a medium supplemented with coconut water (7% by volume) and IAA (10 mg/ L^{-1}). The culture was characterized as essentially undifferentiated. Occasionally roots were produced when the medium was presumed to be depleted. After transferral to a synthetic medium, supplemented with various organics and an auxin (either IAA at 10 mg/L^{-1} or 2,4-D at .05 mg/L^{-1}) the growth rate changed as well as the growth habit. It became harder and nodular structures formed from the surface. At that point cultures were transferred to auxin-free medium of otherwise unchanged composition. After a while the culture went through a period of root formation and then increasing numbers of shoot buds were observed. Reinert interpreted all this as controlled morphogenesis. A seemingly monopolar root is the lower part of the photograph whereas the upper culture shows roots as well as shoot buds (Reinert, 1958). 3-8: Presumed stages in adventive embryogenesis from carrot. Photographs provided by the late Professor J. Reinert. 3, 4, 5-single to several celled forms–yielding a proembryo. 6-Precotyledonary stage. 7 and 8-Cotyledonary stages (7 in section). 9 and 10-Plantlets: 9 with thickened primary root. 11-Harry Waris in 1954. 12-Aberrant adventive embryo structures (neomorphs) of carrot showing elongated roots and varied forms. From a paper published by Professor Waris in 1957. 13-17, Neomorphs of the umbellifer *Oenanthe aquatica*. Various deviations from normal phenotype are evident. Small arrow in 13 indicates the original seedling challenged with 13.3 mM glycine; the flask was filled with neomorphs some 10 months after the initial exposure to nonphysiological levels of the amino acid. 14-Right, the original seedling and left, neomorphs produced from the seedling challenged with 53.3 mM glycine. 15-Original "mother plantlet" from which neomorphs in 13 derived. 16-Flattened, thalloid neomorph selected from the population shown in 14. 17-"Seedling-like" neomorphs from Krikorian and Simola 1999; photographs reproduced courtesy of Physiologia Plantarum). 18-Vimla Vasil in the laboratory at the University of Florida. Photograph taken around 1990. 19-A.C. Hildebrandt in the laboratory January 1965 examining a parsley culture grown by Indra K. Vasil. 20-through 33-"From a clone of single cell origin to flowering tobacco plant". 20 to 29-Variations in cell division pattern of single cells in microculture. 30-Newly transferred three-week-old callus. 31 and 32-Leafy shoot production on a medium supplemented with kinetin and IAA. 33-Production of numerous shoots in another culture with different proportions of IAA and kinetin after growth in light for 6 weeks. 34-Schematic life cycle of hybrid tobacco (*N. glutinosa* X *N. tabacum*) showing stages starting with a single cell grown in microculture in a microchamber shown at "5 o' clock" (see Plate III, 9 for microchamber details) and ending with a flowering tobacco plant ("12 o'clock") and, continuing yet again with reestablishment of a new culture from which a single cell clone can derive (Vasil and Hildebrandt 1967). Photographs 18 and 19 courtesy Drs. Indra and Vimla Vasil. Permission to reproduce photographs 2, and 20 through 34 from Springer-Verlag, Berlin.

2

Isolation and Culture of Plant Protoplasts: Historical Perspectives and Current Status

Michael R. Davey[1*], Paul Anthony[1], J. Brian Power[1] and Kenneth C. Lowe[2]
[1]*Plant Sciences Division, School of Biosciences, University of Nottingham, Sutton Bonington Campus, Loughborough LE12 5RD, UK*
[2]*School of Biology, University Park, University of Nottingham, Nottingham NG7 2RD, UK*
**corresponding author:*
e-mail: mike.davey@nottingham.ac.uk

INTRODUCTION

Plant cells are unique in their totipotency since each living cell carries the genetic information that theoretically enables it to develop into one or more fertile plants under the correct culture conditions. The living contents of plant cells consist of the cytoplasm bound by the plasma membrane that collectively constitute the protoplast, a term first coined by von Hanstein (1880). A pectin-rich matrix, the middle lamella, joins the walls of adjacent cells. The mechanical isolation of plant protoplasts also dates back to the 1880s, when it was demonstrated that protoplasts could be released from the cut ends of plasmolyzed cells following slicing of tissues in a hypertonic solution. The most significant advance was made during the early 1960s with the enzymatic isolation of protoplasts from root tips of tomato seedlings. However, it was not until the early 1970s that intact tobacco plants were regenerated from isolated leaf protoplasts. Currently, large populations of protoplasts provide the starting material both for physiological and genetic manipulation/genomic investigations, particularly somatic hybridization/cybridization involving protoplast fusion, exploitation of genetic variation and aspects of transformation.

This chapter presents an historical overview of the methodology for the isolation and culture of plant protoplasts. Emphasis is given to key milestones in the evolution of protoplast technologies and identification of crucial variables influencing the behavior of protoplasts in culture. Some current and predicted basic and applied applications of isolated protoplasts in plant biotechnology are also discussed.

ISOLATION OF PLANT PROTOPLASTS

Mechanical and Enzymatic Isolation Procedures

Early attempts to isolate protoplasts from higher plants involved mechanical methods. In the late 19th century, Klercker (1892) described a procedure for the isolation of protoplasts from plasmolyzed leaf material of *Stratiotes aloides.* This study involved slicing tissues through cell walls at a right angle to the cell axes, permitting the extrusion of intact protoplasts. However, such a mechanical method, while providing suitable material for investigating the osmotic behavior of isolated protoplasts, was limited by the restricted number of protoplasts that were released and the fact that they failed to undergo further development. In the 1960s, e.g. with the moss *Funaria hygrometrica,* mechanically isolated protoplasts were cultured through to regenerated plants (Binding 1966), although examples of plant regeneration from mechanically isolated protoplasts are rare.

Demonstrations that the cell walls of bacteria and fungi could be degraded to release protoplasts and the realization that protoplasts are released by natural enzymatic digestion of cell walls during fruit ripening, prompted investigations in the early 1960s to isolate protoplasts from root tips of tomato seedlings (Cocking 1960). These early experiments represented a significant advance in terms of numbers of protoplasts released, but were constrained by the fact that only limited quantities of cell wall degrading enzymes were available for experimentation. Consequently, studies were restricted to small-scale isolations and some preliminary investigations of cultural responses. Indeed, it was not until 1968 that commercially available cellulase from the fungus *Trichoderma viride,* allowed Japanese workers to carry out large-scale protoplast isolations from plant tissues other than those of root tips. In this respect, Takebe et al. (1968) pretreated tobacco leaves with a pectinase to separate the cells, followed by a cellulase treatment to digest the walls and release protoplasts. Subsequently, this two-step approach was modified to develop a single-step treatment using a mixture of cellulases and pectinases for protoplast isolation (Power and Cocking 1970). This latter approach simplified the procedure for protoplast isolation from leaf tissues, since the previous sequential method of isolation involved repeated centrifugation of cells and isolated protoplasts, resulting in a progressive reduction in yield. It was clear from these initial studies using mixed enzyme formulations that cell walls could be removed completely, making it feasible to isolate large populations of protoplasts for studies on auxin responses and uptake of macromolecules. Additionally, studies on cell wall regeneration by tobacco leaf protoplasts would have been difficult to interpret if parts of the original walls had remained after treatment (Nagata and Takebe 1970). The addition of a low concentration of pectinase, such as pectolyase, in combination with cellulase, improved the yields of protoplasts from tobacco leaves (Nagata and Ishii 1979) and cell suspensions (Nagata et al. 1981).

A range of wall-degrading enzymes is currently available commercially, many of which are mixtures of several enzymes, although one component generally predominates. The effectiveness of crude and hence less costly enzyme preparations was improved by gel filtration (Kao et al. 1971), while the passage of enzymes prior to use through Sephadex G-25 or G-75 resins or Biogel P-6 columns (Santos et al. 1980) reduced contaminants and improved activity with concomitant increases in protoplast yield and viability. The use of purified enzyme preparations was also shown to be beneficial, especially for maximizing the plating efficiency and plant regeneration from single protoplasts during subsequent culture, as in *Petunia parodii* (Patnaik et al. 1982). Interestingly, many workers subsequently found that such enzyme purification became unnecessary as the range of commercially available enzymes increased.

Factors have been recognized which optimize protoplast release. These include tissue/cell type, preplasmolysis, duration and temperature of enzyme incubation, and purification by filtration or flotation (Davey and Kumar 1983). An important observation was that preplasmolysis of source tissues in a suitable salts solution (Frearson et al. 1973), containing the same osmoticum as the enzyme mixture [e.g. 13% (w/v) mannitol], reduced protoplast damage and spontaneous fusion during subsequent enzyme incubation (Gamborg et al. 1981). This procedure seals plasmodesmata between adjacent cells and reduces enzyme uptake by endocytosis that normally occurs during the incubation period when cells are immersed directly in the enzyme mixture. Generally, sugar alcohols, such as mannitol or sorbitol, are used as osmotica in the enzyme mixtures, rather than carbohydrates such as sucrose and glucose. This is because the latter are taken up and metabolized during enzyme incubation, resulting in starch accumulation and protoplast instability.

Enzymatic digestion of source tissues is usually performed at 25–28ºC and may be of short duration (usually up to 6 hours) or, for convenience, overnight (12-20 hours). Incubation, initially at a temperature of 14°C, followed by a short period at 30°C, appeared to be beneficial in cereals (Vasil and Vasil 1980). The efficiency of protoplast release can be further improved by gentle agitation at *ca.* 40 rpm (Uchimiya and Murashige 1974). Overall, procedures for the isolation of protoplasts during the subsequent 3 decades have remained relatively unchanged with only minor modifications to protocols. Indeed, the simplest and still the most efficient method for purification of highly vacuolated protoplasts involves their flotation on sucrose solutions (Gregory and Cocking 1965). Removal of the lower epidermis of leaves with fine forceps, or the dissection of leaves into thin strips, facilitates cell wall digestion, thereby maximizing protoplast release. Preconditioning of donor plants or explants by exposure to low light intensity, reduced photoperiods (including darkness), or preculture on suitable media, may increase protoplast yield and viability.

Source Material for Protoplast Isolation

While it is possible to release protoplasts from virtually any plant tissue containing cells with primary walls, leaves have been the most frequently employed source of tissues for protoplast isolation. In this respect, much of the success of isolating totipotent protoplasts can be attributed to recognition of the importance of optimizing the growth conditions of donor plants. Uniformity in the growth of intact plants is often difficult to achieve even under the most stringent glasshouse conditions. For example, parameters influencing protoplast yield and viability were studied in detail in tobacco (Watts et al. 1974) and potato (Shepard and Totten 1977).

The use of controlled-environment chambers for the growth of donor plants eliminates the seasonal variation in light intensity, temperature, and humidity experienced under glasshouse conditions. Indeed, use of such controlled environments was found to be essential for ensuring reproducible yields and sustained mitotic division in *Medicago* mesophyll protoplasts (Santos et al. 1980). Variation in the physical and chemical parameters to which source material is exposed can be reduced through the use of *in-vitro* grown shoots and seedlings (Binding 1974). The advantage of the latter as source material is that protoplasts can be isolated from radicles, hypocotyls and cotyledons within a few days of seed germination. A further benefit of this approach is that the donor material is axenic which circumvents the requirement for surface sterilization.

Protoplasts with a haploid genome have been isolated from pollen tetrads and, with more difficulty, from mature pollen (Zhou 1989). Callus and cell suspensions are useful sources of large populations of protoplasts, although embryogenic suspensions must be employed, as in the case of cereals, if the released protoplasts are required to express their totipotency. One disadvantage of using such donor material for protoplast isolation is that cultures may be genetically unstable. Chromosome variation in cell suspensions was reported as early as the first sub-culture (Evans and Bravo 1983), although it is more often associated with cells that have undergone long-term culture. To minimize such ploidy changes, a strategy of selection pressure, based on frequent subculture, proved beneficial in maintaining cells in the diploid state (Evans and Bravo 1983). Protoplasts isolated from such cultures are more likely to regenerate into plants and, as a consequence, are of more use in genetic manipulation programs, such as those involving somatic hybridization.

CULTURE OF ISOLATED PLANT PROTOPLASTS

Nutritional Requirements of Protoplasts and Culture Media

Over the years, an extensive range of media has been developed for culturing isolated protoplasts, many of which have been based on the well-tested MS (Murashige and Skoog 1962) and B5 (Gamborg et al. 1968) formulations, often with modifications to the concentrations of components compared to the original composition. For example, a deleterious effect of ammonium ions on protoplasts of potato (Shepard and Totten 1977) and *Salpiglossis sinuata* (Boyes and Sink 1981) necessitated replacing ammonium nitrate of the MS formulation with organic nitrogen in the form of glutamine and serine. For ease of reproducibility, media for protoplast culture have tended to be as simple as possible and defined. An exception is the complex and undefined medium developed by Kao and Michayluk (1975) for the culture of single protoplasts or populations of protoplasts at low densities. Sucrose is the most common carbon source, although glucose may act as both a carbon source and osmotic stabilizer. Sugar alcohols, in particular mannitol, were introduced as nonmetabolizable osmotica. Maltose as a carbon source may be beneficial in stimulating shoot regeneration during the latter stages of culture of protoplast-derived cells, as in the case of cereals (Jain et al. 1995).

Most protoplasts have been found to require one or more auxins or cytokinins as growth regulators in the culture medium, to sustain mitotic division. Exceptions to this rule are carrot, where only auxin appears to be necessary (Grambow et al. 1972) and *Citrus*, in which both auxin and cytokinin are detrimental to protoplast growth (Vardi et al. 1982).

Protoplasts may change their hormonal requirements as culture proceeds. For example, a tenfold reduction of naphthaleneacetic acid concentration after 4 days of growth stimulated colony formation from haploid tobacco protoplasts (Caboche 1980). Similarly, rapid growth of protoplasts in culture followed by prolific shoot regeneration in *Hyoscyamus muticus* and *Nicotiana tabacum* necessitated a reduction of auxin concentration in the culture medium (Wernike and Thomas 1980). The time from protoplast isolation to plant regeneration in potato can be reduced by changing the balance of growth regulators during the culture period (Shepard 1982). An important advance in this area has been the incorporation into culture media of phenylurea-type derivatives, with potent cytokinin activity to enhance protoplast growth. Notable examples of such compounds are thidiazuron (Chupeau et al. 1993) and N-(2-chloro-4-pyridyl)-N′-phenylurea (Sasamoto et al. 2002). Indeed, in the former case, thidiazuron as the sole cytokinin was essential to sustain division of isolated protoplasts of white poplar. Interestingly, related studies have demonstrated that some naturally occurring brassinosteroids, with structural similarity to animal steroid hormones, act as growth regulators and promote division of protoplasts in culture (Oh and Clouse 1998).

Experimental Systems for Protoplast Culture

Several procedures have been developed and reported for the culture of isolated protoplasts. Incubation in liquid medium was the most commonly employed method of the emerging technology. Thus, protoplasts were suspended in small volumes of liquid medium and plated in Petri dishes (Binding 1974). An important observation from these early investigations was that it was necessary to culture protoplasts in thin layers (approx. 1 mm in depth) of culture medium to ensure an adequate oxygen supply. Recent advances in protoplast culture technology have focused on the applications and benefits of novel approaches for enhancing gas supply, involving the use of artificial oxygen carriers based on hemoglobin or chemically inert fluorocarbon liquids. Such procedures are described later. Protoplasts have also been cultured successfully in liquid medium overlying nylon mesh supports (Russell and McCown 1986), filter papers (Partanen 1981), or in hanging droplets (Kao et al. 1970). The latter technique involved placing droplets (each 100-150 μL in volume) of protoplast suspension on the inverted lids of Petri dishes. When the lids were reapplied to the dishes, the droplets became suspended from the lids. A variation of this technique was the multiple drop array, in which the volume of the droplets was reduced to *ca.* 50 μL, making it possible to suspend up to 50 drops from the lid of a 9 cm diameter dish (Potrykus et al. 1979). The advantage of this approach was that it facilitated rapid evaluation of culture media, particularly during assessment of several growth regulator combinations.

Media semisolidified with agar, of different qualities, have been used extensively for protoplast culture. This technique was originally developed for the plating of cells from suspension culture (Bergmann 1960) and was subsequently modified and exploited for isolated tobacco mesophyll protoplasts (Nagata and Takebe 1971). A protoplast suspension, at twice the required plating density, is mixed with molten medium to give a final agar concentration of 0.6% (w/v), either before (Cella and Galun 1980) or after (Coutts and Wood 1977) protoplasts have regenerated new cell walls. A benefit of this approach is that embedded protoplasts remain fixed and physically separated in the semisolid medium. This reduces polyphenol production that normally inhibits cell growth and also facilitates monitoring individual cell colonies. Other gelling agents, such as agarose, gelatin and K-carrageenan, have also been evaluated as media gelling agents. Agarose has consistently given the best results in terms of the retention of protoplast viability, with several studies reporting improved protoplast plating efficiencies compared to the use of agar (Lörz et al. 1983). A further advantage of agarose is that the gelling temperature is lower than that of agar, enabling protoplasts to be plated without a heat shock. Alginate was found to be a useful alternative to agar or agarose (Adaohambanasco and Roscoe 1982), particularly with protoplasts that are heat sensitive. The alginate is gelled by pouring the warm alginate-containing medium in which the protoplasts are suspended over an agar layer containing calcium ions, or by gently dropping the molten medium into a solution containing calcium ions. Subsequently, the release of protoplast-derived colonies is readily affected by depolymerizing the alginate through exposure to sodium citrate to remove calcium ions. The medium containing the suspended protoplasts may be dispensed as layers or droplets (usually up to 250 µl in volume) in Petri dishes. Cutting the semisolid layers of medium into sections and bathing the sections or droplets in liquid medium of the same composition was observed to stimulate the growth of embedded protoplasts. The osmotic pressure of the semisolid phase containing the embedded protoplasts/cells can be reduced by changing the bathing medium (Shillito et al. 1983). A further refinement of this technology was the plating of protoplasts in a liquid medium over the same medium semisolidified with a gelling agent, such as agar or agarose (Power et al. 1976).

Plating Density and the Use of Nurse Cells

The plating density at which protoplasts are cultured has been known for many years to be crucial in sustaining mitotic division and cell colony formation. Generally, the optimum plating density is 5×10^4–1.0×10^6 ml^{-1}, depending on the species. For those species in which protoplast-to-protoplast relationships influence the cultural response, a high initial plating density, typically > 1.0×10^5 ml^{-1}, was shown to be required for example, for cereals (Vasil 1987) and woody plants (Ochatt 1990). It is known that dividing cells stimulate division of their neighbors, probably as a result of the release of growth factors, in particular amino acids, into the surrounding medium. This approach was extended to include "feeder" protoplasts following their X-irradiation to prevent cell division (Raveh et al. 1973). In these early experiments, the irradiated protoplasts were mixed with a population of viable protoplasts and plated in agar medium. Alternatively, viable protoplasts were plated over a thin layer of agar medium containing the irradiated protoplasts (Cella and Galun 1980). "Nurse" cells or tissues may likewise be employed to promote division in cultured protoplasts. Interestingly, in this respect it is not essential for the nurse cells to be from the same species as the isolated protoplasts. For example, the growth of protoplasts isolated from embryogenic cell suspensions of Japonica and Indica rice varieties was stimulated by fast-growing nurse cells of wild rice (*Oryza ridleyi*) or Italian ryegrass (*Lolium multiflorum*), used either alone or in combination (Jain et al. 1995). The isolated protoplasts can be cultured in a layer of liquid medium, embedded in semisolidified medium, or spread in a liquid layer on a cellulose nitrate membrane overlying the semisolid layer of nurse cells (Jain et al. 1995).

Protoplasts should be assessed for their viability and for removal of their cell walls immediately following isolation and prior to experimentation. Various methods have been developed to assess protoplast viability. For example, Evans Blue and Tryphan Blue are excluded by intact plasma membranes (Glimelius et al. 1974), but viable protoplasts take up fluorescein diacetate (Widholm 1972). Any residual wall material remaining after protoplast isolation can be visualized following staining with Calcofluor White (Nagata and Takebe 1970) or the fluorescent brightener, Tinopal (Cocking 1985).

INNOVATIVE APPROACHES TO PROTOPLAST CULTURE

Significant advances were made in protoplast and cell culture technologies during the three decades 1960s, 1970s, 1980s, particularly in terms of refining media and optimizing the physical parameters for protoplasts of a range of dicotyledonous and monocotyledonous species. However, more recently several innovative approaches have been evaluated that address specific chemical and physical parameters, often in combination. It is important to note that some of these approaches have been modified from animal cell culture technologies (Lowe et al. 1998), emphasizing the requirement for dialogue between workers in different disciplines of plant and animal sciences.

Chemical Supplements for Protoplast Culture Media

The nonionic surfactant *Pluronic*® F-68, a copolymer of polyoxyethylene and polyoxypropylene, is often used in the culture of animal cells to reduce membrane damage during growth under forced aeration. In evaluating the application of this surfactant to plant protoplast systems, *Pluronic*® F-68 increased the plating efficiency of protoplasts of *Solanum dulcamara* when added to the culture medium, with a maximum response at 0.1% (w/v) of the surfactant (Kumar et al. 1992). The precise mode of action of surfactants, such as *Pluronic*® F-68, is still not clear, although such compounds may increase the permeability of plasma membranes to growth regulators in the culture medium (Lowe et al. 1993). In future, a range of surfactants is worthy of assessment as medium supplements, although the optimum growth conditions at which they exert their physiological effect may be influenced by their hydrophilic-lipophilic balance (HLB number). The HLB number is an indicator of membrane-permeabilizing properties of surfactants.

Certain antibiotics also stimulate protoplast division, this probably being a casual observation arising from the incorporation of antibiotics into culture media to control bacterial infection. For example, shoot differentiation in tobacco and carrot was enhanced by low concentrations of the aminoglycoside antibiotics, kanamycin and streptomycin (Owens 1979). Furthermore, the cephalosporin antibiotic, cefotaxime, stimulated mitotic division and cell colony formation from protoplasts isolated from seedling leaves of the woody plant passionfruit (d'Utra Vaz et al. 1993), when added to culture medium at 250 $\mu g\ ml^{-1}$. Cefotaxime may be metabolized to a growth regulator-like compound(s) (Mathias and Boyd 1986), although the manner in which this antibiotic exerts its effect on cultured plant cells remains obscure.

Manipulation of Respiratory Gases

An adequate and sustainable gaseous exchange is essential to maximize protoplast development. Gassing of culture vessels with oxygen soon after plating of protoplasts was demonstrated to increase the plating efficiency of those of jute and rice (d'Utra Vaz et al. 1992), with oxygen in the culture dishes gradually reverting to the normal atmospheric concentration after the first few days of growth. Similarly, perfluorocarbon (PFC) liquids were evaluated in higher plant protoplast systems, following their use in animal (including human) cell cultures (Lowe et al. 1998).

Such inert, linear, cyclic or polycyclic organic compounds are capable of dissolving large volumes of respiratory gases. PFC liquids, typically about twice as dense as water, form a stable layer beneath aqueous media, permitting the culture of protoplasts at the interface between the lower PFC layer and the overlying liquid medium. Such use of oxygen-gassed perfluorodecalin significantly stimulated mitotic division in protoplasts of several species, including *Oryza sativa, Passiflora giberti, Petunia hybrida,* and *Salpiglossis sinuata* (Lowe et al. 1998). Evidence has also been obtained that surfactants and PFCs act synergistically, since supplementation of the culture medium with 0.1% (w/v) *Pluronic*® F-68 overlying oxygenated PFC further increased the plating efficiency of *P. hybrida* suspension cell protoplasts (Anthony et al. 1994).

A further novel approach for enhancing oxygen supply to cultured protoplasts has involved supplementation of media with a commercial bovine hemoglobin solution. For example, the mean initial plating efficiency of protoplasts of *P. hybrida* and *O. sativa* following culture with 1:50-1:100 (w:v) *Erythrogen*™ was significantly enhanced over control (Anthony et al. 1997; Azhakanandam et al. 1997). Again, *Pluronic*® F-68 at 0.01% (w/v) also exerted a synergistic effect when added to culture medium containing *Erythrogen*™ (Anthony et al. 1997).

Physical Procedures to Stimulate Protoplast Growth in Culture

Physical parameters have been investigated which stimulate protoplast development in culture. A simple experimental system, which probably acts by increasing aeration, involves the insertion of glass rods into a layer of agarose-solidified culture medium in Petri dishes. This approach increased mitotic division and cell colony formation from leaf protoplasts of cassava (*Manihot esculenta*) (Anthony et al. 1995). The division of protoplasts in culture is also stimulated by electrical currents (Davey et al. 1996). For example, prolonged, low-voltage currents enhanced division in protoplasts of *Trifolium* (Li et al. 1990) and stimulated the development of protoplasts into somatic embryos in *Medicago* species (Dijak et al. 1986). High-voltage (about 1,250 V), short-duration (10–15 μsec) electrical pulses had the most marked effect in stimulating mitotic division and DNA synthesis in protoplasts of, for example, *Prunus, Pyrus, Solanum* and *Glycine* (Davey et al. 1996). Cell colony formation and shoot regeneration were also promoted, with shoots from electropulsed protoplasts of *Prunus* and *Solanum* developing more vigorous and more extensive root systems than those from tissues derived from untreated protoplasts. There is evidence that these effects of electrostimulation of protoplasts are long term, since they can be recognized after many cell generations. Clearly, considerable scope still exists for assessing the application of both chemical and physical parameters to those protoplast systems that are recalcitrant in culture.

PROTOPLAST-TO-PLANT SYSTEMS

Since the first report of whole plant regeneration from tissues derived from leaf protoplasts of tobacco (Takebe et al. 1971), it has been recognized that the early stages of culture are crucial in the development of protoplast-to-plant systems. Consequently, most attention has focused on these stages of culture. Procedures already established for plant regeneration from explant-derived tissues, through shoot regeneration and/or somatic embryogenesis, have also been applied to protoplast-derived material. Indeed, during the past 20 years there has been a dramatic increase in the number of plant species for which shoot regeneration from protoplasts has been reported (Bajaj 1996; Xu and Xue 1999). Presently, protoplast-to-plant systems exist for almost 400 species, most notably in the case of members of the Solanaceae. The latter is followed, in decreasing order, by representatives of the Leguminosae, Gramineae, Compositae, Cruciferae, Umbelliferae and Rosaceae. Surprisingly, in the Orchidaceae, one of the largest families of higher plants, there has been no reproducible demonstration of the successful regeneration of plants from protoplast-derived tissues (Xu and Xue 1999). It is tempting to speculate that some of the more innovative culture approaches already discussed, including supplementation of culture medium with surfactants or PFCs and electrostimulation, may be beneficial in maximizing shoot development from those protoplast-derived tissues that, at present, respond poorly or are recalcitrant to regeneration. In this respect, the plant genotype can influence shoot formation as, for example, in the case of *Lycopersicon esculentum* where, of 14 genotypes evaluated, protoplast-derived plants were obtained only from the cultivar Lukullus (Morgan and Cocking 1982).

EXPLOITATION OF PROTOPLAST-TO-PLANT SYSTEMS: SOMATIC AND GAMETOSOMATIC HYBRIDIZATION

The ability to regenerate plants routinely from protoplast-derived cells and tissues is fundamental to genetic manipulation approaches involving somatic hybridization and, to a lesser extent, transformation. As reviewed in detail elsewhere in this volume, fusion of protoplasts isolated from various plant genera, species, and varieties presents few difficulties. Hybrid cells that result may generate plants with balanced or asymmetric nuclear genomes. Cell-cell interactions following fusion may result in chloroplast segregation, or more rarely, retention of a mixed plastid population. Additionally, elimination of the nuclear genome of one protoplast parent may result in the production of cytoplasmic hybrids (cybrids). Generation of such cybrids has been exploited to transfer organellar traits, such as mitochondrially encoded cytoplasmic male sterility (CMS) or chloroplast DNA-encoded herbicide (e.g. atrazine) resistance into target plants. Fusion is not restricted to diploid protoplasts; the latter can be fused with protoplasts isolated from cells of haploid plants. In this respect, fusion of diploid with haploid protoplasts was used to generate fertile triploid plants through gametosomatic hybridization (Desprez et al. 1995). Collectively, somatic and gametosomatic hybridization provide opportunities to create novel nuclear and cytoplasmic combinations, increasing the genetic diversity available to plant breeders. An excellent example is provided by the studies of Helgeson et al. (1998) in which protoplasts of *Solanum tuberosum* were fused with protoplasts of wild potato, producing novel hybrids having improved resistance to potato late blight.

TRANSFORMATION OF PROTOPLASTS BY ISOLATED DNA

Reports during the 1970s that cultured animal cells were capable of taking up and expressing cloned genes or isolated genomic DNA provided the impetus for assessing the feasibility of adopting this approach for isolated plant protoplasts. Interaction of the Ti plasmid from an octopine strain of *Agrobacterium tumefaciens* with protoplasts from cell suspensions of *P. hybrida* in the presence of the membrane-stimulating agent, poly-L-ornithine, resulted in protoplast-derived colonies having the ability to grow on medium lacking growth regulators and to synthesize octopine (Davey et al. 1980). Other studies confirmed polyethylene glycol (PEG)-induced transformation of tobacco leaf protoplasts with Ti plasmids and the production of fertile, transgenic plants. An interesting and significant observation in such studies was that the T-DNA border sequences were not recognized during DNA uptake and integration into protoplasts (Krens et al. 1982). Such experiments provided "proof of concept" that isolated DNA could transform freshly isolated protoplasts. Consequently, the natural progression in this technology was to clone genes on small plasmids derived from *Escherichia coli*, to isolate the plasmids, and to use this DNA in protoplast transformation experiments (Paszkowski et al. 1984).

Undoubtedly, the most important application of the uptake of DNA into isolated protoplasts has been in the transformation of those plants that are not readily amenable to *Agrobacterium*-mediated gene transfer.

Primarily, such studies have been directed to the transformation of cereals and grasses, particularly rice (Hayashimoto et al. 1989), once protoplast-to-plant systems became available for these target crops. It seems likely, following the recent success in the transformation of the major cereals with *A. tumefaciens* (Repellin et al. 2001), that DNA uptake into isolated protoplasts will assume less relevance in the context of cereal transformation. However, this may not be the case in certain crop plants, in which the uptake of DNA into isolated protoplasts will continue to be a routine approach for generating transgenic plants. An excellent example is sugarbeet, in which introduction of DNA into guard cell protoplasts provides a reproducible transformation system for this important root crop (Hall et al. 1996).

In addition to the introduction of foreign DNA into the nuclear genome of recipient protoplasts, the targeting of genes to organelles has also been demonstrated by PEG-mediated DNA uptake into isolated protoplasts, representing a major advance in plant cell engineering technology. Plastid transformation was first demonstrated in tobacco with plant clones bearing transformed plastomes being selected by their spectinomycin resistance encoded by a mutant 16S ribosomal RNA gene carried by the plasmid used for transformation (Golds et al. 1993). Plastome engineering is likely to assume considerable importance in future, particularly in commercial terms (Maliga 2003), because of the ability of transplastomic plants for high level transgene expression (Daniell et al. 2002). Currently, the main limitation of this approach is in the development of a robust transformation technology applicable to a range of plants. Additionally, gene insertion into mitochondrial genomes could make a significant contribution to modifying plant cell genetic diversity, once this technology has also been developed.

GENETIC ENGINEERING THROUGH PROTOCLONAL VARIATION

Since 1981, when the term "somaclonal variation" was adopted to describe variability in plants regenerated from cultured cells (Larkin and Scowcroft 1981), it became established that such variation might affect a range of traits, including those for plant morphology and vigor, flower color, yield, nutritional value, production of secondary products, tolerance to environmental conditions, and resistance to pathogens. The combination and concentration of growth regulators may affect DNA structure and expression, while imbalances in the nucleotide pool in the medium may influence DNA base structure. In general, the longer cells are in culture, particularly at the callus stage, the greater the variation to be expected. This applies to both explant-derived tissues and protoplast-derived cells.

Consequently, unless a conscious effort is made to induce and identify variation associated with protoplast culture (protoclonal variation), the period from protoplast isolation to plant regeneration should be kept to a minimum. In contrast, where genetic variability is required, protoclonal variation has several important attributes. For example, unlike transformation, it requires no knowledge of the genetic basis of the trait(s) and does not necessitate gene isolation and cloning. It negates the use of mutagenic agents, specialized apparatus, or containment procedures, is inexpensive, and can be exploited as a spin-off of routine culture. Exposure of protoclonal variation also has the potential to increase genetic diversity and it bypasses the sexual cycle. It may be considered a simple, or probably the simplest, form of genetic manipulation and, as such, is a useful adjunct to conventional plant breeding alongside more technological approaches, such as somatic hybridization and transformation.

STUDIES WITH PLANT PROTOPLASTS

While the principal application of isolated protoplasts has been undoubtedly in plant genetic manipulation, large populations of protoplasts have provided ideal experimental material for physiological, ultrastructural and genetic studies for more than four decades. The fact that isolated protoplasts develop in culture into colonies of single cell origin, has been exploited to isolate clonal lines of cells and plants, especially those for increased secondary product synthesis. Exposure of protoplasts to mutagenic agents or irradiation permits the induction and selection during subsequent culture of mutant cells and plants (Dix 1994). Protoplasts take up macromolecules other than DNA and this property has been exploited in studies of endocytosis at the plasma membrane, using compounds such as ferritin as molecular markers, which can be visualized by electron microscopy. In early experiments, protoplasts also featured as systems for studying virus uptake and replication in plant cells (Cocking 1969). The osmotic fragility of protoplasts released from the confines of their cell walls permits their controlled lysis for the isolation of cellular fractions, including membranes, intact vacuoles and organelles. In physiological studies, isolated vacuoles have been used to monitor the accumulation of compounds, such as sucrose and its associated transporter, as in sugarcane (Thom et al. 1992). Protoplasts have provided ideal systems for studies of ion transport through the plasma membrane and regulation of the osmotic balance of cells, enabling patch-clamp studies to be performed (Colombo and Cerana 1991), comparable to those with animal cells. Other notable investigations include the detection of elicitor binding sites and the binding of fungal phytotoxins to the plasma membrane (Diekmann et al. 1994), auxin accumulation and metabolism (Delbarre et al. 1994), and light-induced proton pumping in guard cell protoplasts (Herscovich et al. 1992).

Protoplasts have also provided unique material for studying the early stages of cell wall synthesis, focusing on the role of the plasma membrane and cellular organelles in this process, as revealed by transmission electron microscopy. Such experiments remain classic ultrastructural studies in the literature (Fowke et al. 1983). Related investigations have focused on the role of microtubules during cell development from isolated protoplasts (Meijer et al. 1988). Protoplasts from fern prothalli have also been used to study the influence of gravity and light in the development of cell polarity (Edwards and Roux 1998). Generally, there have been relatively few investigations with protoplasts of lower plants, although those of mosses have been exploited in genetic studies, including investigations of the fate of mutant macrochloroplasts following somatic hybridization (Rother et al. 1994).

CONCLUSIONS AND FUTURE PROSPECTS

The ever-increasing literature reflects the continuing interest in protoplasts as experimental systems for studying many aspects of plant physiology and development, together with genetic manipulation involving simple exposure of genetic variation, to more complex cell fusion and molecular technologies. Coupled with these developments has been the progressive introduction of sophisticated molecular assays to assess gene transfer and expression in plants arising from the genetic manipulation of isolated protoplasts. Fundamental to many of the investigations using isolated protoplasts is the need to develop protoplast-to-plant systems for those species that, to date, remain recalcitrant in culture. In this respect, it is interesting to note that a protoplast-to-plant system was finally achieved, after about 20 years of research, for the cereal sorghum (Sairam et al. 1999). Currently there is renewed interest in exploiting simple, plant-based systems for toxicological assays, including the screening of pharmaceuticals, food additives, cosmetics and agrochemicals, as well as radiation-related interactions. Isolated protoplasts, especially if totipotent, are useful for assessing both short-term and, more important, long-term effects of such agents on cells (Lowe et al. 1995). However, the effects of chemical and environmental factors on plant cells may be apparent only after several seed generations. Consequently, rapid cycling plants, such as *Arabidopsis thaliana* and *Brassica napus*, are particularly useful systems for generating several seed generations in a relatively short period (e.g. twelve months). Comparable studies of the long-term effects of chemical and environmental factors are not possible with animal cells, since the latter do not express the unique feature of totipotency characteristic of cultured plant cells. Several of the recent innovative culture approaches applied to isolated protoplasts described in this chapter were, in fact, developed for the culture of animal cells. Moreover, procedures such as the uptake of isolated DNA were developed and evaluated in animal cells before being assessed in plant systems. Such observations provide the clear message that plant and animal cells have certain similarities, emphasizing the fact already discussed, that plant biologists should be aware of, and exploit advances reported in the animal cell literature.

REFERENCES

Adaohambanasco EN, Roscoe DH (1982) Alginate—an alternative to agar in plant protoplast culture. Plant Sci. Lett. 25:61-66.

Anthony P, Davey MR, Power JB, Lowe KC (1995) An improved protocol for the culture of cassava leaf protoplasts. Plant Cell, Tiss. Org. Cult. 42:299-302.

Anthony P, Lowe KC, Davey MR, Power JB (1997) Strategies for promoting division of cultured plant protoplasts: synergistic beneficial effects of haemoglobin (*Erythrogen*™) and Pluronic F-68. Plant Cell Rep. 17:13-16.

Anthony P, Davey MR, Power JB, Washington C, Lowe KC (1994) Synergistic enhancement of protoplast growth by oxygenated perfluorocarbon and *Pluronic*® F-68. Plant Cell Rep. 13:251-255.

Azhakanandam K, Lowe KC, Power JB, Davey MR (1997) Hemoglobin (*Erythrogen*™)-enhanced mitotic division and plant regeneration from cultured rice protoplasts (*Oryza sativa* L.). Enzyme Microb. Tech. 21:572-577.

Bajaj YPS. (1996) (ed.), Biotechnology in Agriculture and Forestry. Springer-Verlag, Berlin, Vol. 38, pt VII.

Bergmann L (1960) Growth and division of single cells of higher plant *in vitro*. J. Gen. Physiol. 43:841-845.

Binding H (1966) Regeneration un Verschmelzung nackter Laubmoosprotoplasten. Z. Pflanzenphysiol. 55:305-321.

Binding H (1974) Cell cluster formation by leaf protoplasts from axenic cultures of haploid *Petunia hybrida* L. Plant Sci. Lett. 2:185-188.

Boyes CJ, Sink KC (1981) Regeneration of plants from callus-derived protoplasts of *Salpiglossis*. J. Amer. Soc. Hort. Sci. 106:42-46.

Caboche M (1980) Nutritional requirements of protoplast-derived haploid tobacco cells grown at low densities in liquid medium. Planta 149:7-18.

Cella R, Galun E (1980) Utilization of irradiated carrot suspensions as feeder layer for cultured *Nicotiana* cells and protoplasts. Plant Sci. Lett. 19:243-252.

Chupeau MC, Lemoine M, Chupeau Y (1993) Requirement of thidiazuron for healthy protoplast development to efficient tree regeneration of a hybrid poplar (*Populus tremula* x *Populus alba*). J. Plant Physiol. 141:601-609.

Cocking EC (1960) A method for the isolation of plant protoplasts and vacuoles. Nature (London) 187:927-929.

Cocking EC (1969) An electron microscopic study of the initial stages of infection of isolated tomato fruit protoplasts by tobacco mosaic virus. Planta 68:206-214.

Cocking EC (1985) Protoplasts from root hairs of crop plants. Bio/Tech. 3:1104-1106.

Colombo R, Cerana R (1991) Inward rectifying K+ channels in the plasma-membrane of *Arabidopsis thaliana*. Plant Physiol. 97:1130-1135.

Coutts RHA, Wood KR (1977) Improved isolation and culture methods for cucumber mesophyll protoplasts. Plant Sci. Lett. 9:45-51.

d'Utra Vaz FB, Slamet IH, Khatun A, Cocking EC, Power JB (1992) Protoplast culture in high molecular oxygen atmospheres. Plant Cell Rep. 11:416-418.

d'Utra Vaz FB, Santos AVP dos, Manders G, Cocking EC, Davey MR, Power JB (1993). Plant regeneration from leaf mesophyll protoplasts of the tropical woody plant, passionfruit (*Passiflora edulis* fv. Flavicarpa Degener)—the importance of the antibiotic cefotaxime in the medium. Plant Cell Rep. 12: 220-225.

Daniell H, Khan MS, Allison L (2002) Milestones in chloroplast genetic engineering: an environmentally friendly era in plant biotechnology. Trends Plant Sci. 7:84-91.

Davey MR, Blackhall NW, Lowe KC, Power JB (1996) Stimulation of plant cell division and organogenesis by short-term, high voltage electrical pulses. *In:* PT Lynch, MR Davey, (eds.), Electrical Manipulation of Cells. Chapman and Hall, New York, NY, pp. 273-286.

Davey MR, Cocking EC, Freeman JP, Pearce N, Tudor I (1980) Transformation of *Petunia* protoplasts by isolated *Agrobacterium* plasmids. Plant Sci. Lett. 18:307-313.

Davey MR, Kumar A (1983) Higher plant protoplasts—retrospect and prospect. Int. Rev. Cytol. Suppl. 16:219-299.

Delbarre A, Muller P, Imhoff V, Morgat JL, Barbierbrygoo H (1994) Uptake, accumulation and metabolism of auxins in tobacco leaf protoplasts. Planta 195:159-167.

Desprez B, Chupeau MC, Vermeulen A, Delbreil B, Chupeau Y, Bourgin JP (1995) Regeneration and characterization of plants produced from mature tobacco pollen protoplasts. Plant Cell Rep. 14:204-209.

Diekmann W, Herkt B, Low PS, Nurnberger T, Scheel D, Terschuren C, Robinson DG (1994) Visualization of elicitor-binding loci at the plant-cell surface. Planta 195:126-137.

Dijak M, Smith DL, Wilson TJ, Brown DCW (1986) Stimulation of direct embryogenesis from mesophyll protoplasts of *Medicago sativa*. Plant Cell Rep. 5:468-470.

Dix PJ (1994) Isolation and characterisation of mutant cell lines. In IK Vasil, TA Thorpe, (eds.), Plant Cell and Tissue Culture, Kluwer Acad. Pub., Dordrecht, pp. 119-138.

Edwards ES, Roux SJ (1998) Gravity and light control of the developmental polarity of regenerating protoplasts isolated from prothallial cells of the fern *Ceratopteris richardii*. Plant Cell Rep. 17:711-716.

Evans DA, Bravo JE (1983) Plant protoplast isolation and culture. Int. Rev. Cytol. Suppl.16:33-53.

Fowke LC, Griffing LR, Mersey BG, Van der Valk P (1983) Protoplasts for studies of the plasma membrane and associated organelles. Experientia Suppl. 46: 101-110.

Frearson EM, Power JB, Cocking EC (1973) The isolation, culture and regeneration of *Petunia* leaf protoplasts. Dev. Biol. 33:130-137.

Gamborg OL, Miller RA, Ojima K (1968) Nutrient requirements of suspension cultures of soybean root cells. Experim. Cell Res. 50:151-158.

Gamborg OL, Shyluk JP, Shahin EA (1981) Isolation, fusion and culture of plant protoplasts. *In:* TA Thorpe,(ed.), Plant Tissue Culture: Methods and Application in Agriculture. Acad. Press, New York, NY, pp. 115-153.

Glimelius K, Wallin A, Eriksson TC (1974) Agglutinating effects of Concanavalin A on isolated protoplasts of *Daucus carota*. Physiol. Plant. 31:225-230.

Golds TJ, Maliga P, Koop H-U (1993) Stable plastid transformation in PEG-treated protoplasts of *Nicotiana tabacum*. Bio/Tech. 11:95-97.

Grambow HJ, Kao KN, Miller RA, Gamborg OL (1972) Cell division and plant development from protoplasts of carrot cell suspension cultures. Planta 103:348-355.

Gregory DW, Cocking EC (1965) The large scale isolation of protoplasts from immature tomato fruits. J. Cell Biol. 24:143-146.

Hall RD, Riksen-Bruinsma T, Weyens GJ, Rosquin IJ, et al. (1996) A high efficiency technique for the generation of transgenic sugar beets from stomatal guard cells. Nature Biotech. 14:1133-1138.

Hayashimoto A, Li ZJ, Murai N (1989) A polyethylene glycol-mediated protoplast transformation system for the production of fertile transgenic rice plants. Plant Physiol. 93:857-863.

Helgeson JP, Pohlman JD, Austin S, Haberlach GT, et al. (1998) Somatic hybrids between *Solanum bulbocastanum* and potato: a new source of resistance to late blight. Theor. Appl. Genet. 96:738-742.

Herscovich S, Tallman G, Zeiger E (1992) Long-term survival of *Vicia* guard-cell protoplasts in cell-culture. Plant Sci. 81:237-244.

Jain RK, Khehra GS, Lee, S-H, Blackhall NW, et al. (1995) An improved procedure for plant regeneration from indica and japonica rice protoplasts. Plant Cell Rep. 14:515-519.

Kao KN, Michayluk MR (1975) Nutritional requirements for growth of *Vicia hajastana* cells and protoplasts at a very low population density in liquid media. Planta 126:105-110.

Kao KN, Keller WA, Miller RA (1970) Cell division in newly formed cells from protoplasts of soybean. Experim. Cell Res. 62:338-340.

Kao KN, Gamborg OL, Miller RA, Keller WA (1971) Cell division in cells regenerated from protoplasts of soybean and *Haplopappus gracilis*. Nature (New Biol.) 232:124.

Klercker JAF (1892) Eine methode zur isolierung lebender protoplasten. Ofvers. Vetensk. Akad. Forh. Stokh. 9:463-475.

Krens FA, Molendijk L, Wullems GJ, Schilperoort RA (1982) *In vitro* transformation of plant protoplasts with Ti plasmid DNA. Nature 296:72-74.

Kumar V, Laouar L, Davey MR, Mulligan BJ, Lowe KC (1992) Pluronic F-68 stimulates growth of *Solanum dulcamara* in culture. J. Exp. Bot. 43:487-493.

Larkin PJ, Scowcroft WR (1981) Somaclonal variation- a novel source of variability from cell-cultures for plant improvement. Theor. Appl. Genet. 60:197-214.

Li ZY, Tanner GJ, Larkin PJ (1990) Callus regeneration from *Trifolium subterraneum* protoplasts and enhanced protoplast division by low-voltage treatment and nurse cells. Plant Cell, Tiss. Org. Cult. 21:67-73.

Lörz H, Larkin PI, Thomson I, Scowcroft WR (1983) Improved protoplast culture and agarose media. Plant Cell, Tiss. Org. Cult. 2:217-226.

Lowe KC, Davey MR, Power JB (1998) Perfluorochemicals: their applications and benefits to cell culture. TIBTECH. 16:272-277.

Lowe KC, Davey MR, Power JB, Mulligan BJ (1993) Surfactant supplements in plant culture systems. Agro-Food-Ind. Hi-Tech. 4:9-13.

Lowe KC, Davey MR, Power JB, Clothier RH (1995) Plants as toxicity screens. Pharm. News 2:17-22.

Maliga P (2003) Progress towards commercialization of plastid transformation technology. TIBTECH. 21:20-28.

Mathias RJ, Boyd LA (1986) Cefotaxime stimulates callus growth, embryogenesis and regeneration in hexaploid bread wheat (*Triticum aestivum* L. Em Thell). Plant Sci. 46:217-223.

Meijer EGM, Keller KA, Simmonds DH (1988) Cytological abnormalities and aberrant microtubule organization during early divisions in mesophyll protoplast cultures of *Medicago sativa* and *Nicotiana tabacum*. Physiol. Plant. 74:233-29.

Morgan A, Cocking EC (1982) Plant regeneration from protoplasts of *Lycopersicon esculentum* Mill. Z. Pflanzenphysiol. 106:97-104.

Murashige T, Skoog F (1962) A revised medium for rapid growth and bioassays with tobacco tissue cultures. Physiol. Plant. 15:473-497.

Nagata T, Ishii S (1979) A rapid method for the isolation of mesophyll protoplasts. Can. J. Bot. 57:1820-1823.

Nagata T, Okada K, Takebe I, Matsui C (1981) Delivery of tobacco mosaic virus RNA into plant protoplasts mediated by reverse-phase evaporation vesicles (liposomes). Mol. Gen. Genet. 184:161-165.

Nagata T, Takebe I (1970) Cell wall regeneration and cell division in isolated tobacco mesophyll protoplasts. Planta 92:301-308.

Nagata T, Takebe I (1971) Plating of isolated tobacco mesophyll protoplasts on agar medium. Planta 99:12-20.

Ochatt SJ (1990) Protoplast technology and top-fruit tree breeding. Acta Hort. 280:215-226.

Oh MH, Clouse SD (1998) Brassinolide affects the rate of cell division in isolated leaf protoplasts of *Petunia hybrida*. Plant Cell Rep. 17:921-924.

Owens LD (1979) Binding of ColE1-kan plasmid DNA by tobacco protoplasts. Plant Physiol. 63:683-696.

Partanen CR (1981) Filter papers as a support and carrier for plant protoplast cultures. *In Vitro* 17:77-80.

Paszkowski J, Shillito RD, Saul MW, Mandak V, et al. (1984) Direct gene transfer to plants. EMBO J. 3:2717-2722.

Patnaik G, Wilson D, Cocking EC (1982) Importance of enzyme purification for increased plating efficiency and plant regeneration from single protoplasts of *Petunia parodii*. Z. Pflanzenphysiol. 102:199-203.

Potrykus I, Harms CT, Lörz H (1979) Multiple-drop array (MDA) technique for the large-scale testing of culture media variations in hanging drop microdrop cultures of single cell systems. I. The technique. Plant Sci. Lett. 14:231-235.

Power JB, Cocking EC (1970) Isolation of leaf protoplasts: macromolecular uptake and growth substance response. J. Exp. Bot. 21:64-70.

Power JB, Frearson EM, George D, Evans PK, et al (1976) The isolation, culture and regeneration of leaf protoplasts in the genus *Petunia*. Plant Sci. Lett. 7: 51-55.

Raveh D, Huberman E, Galun E (1973) *In vitro* culture of tobacco protoplasts: use of feeder techniques to support division of cells plated at low densities. *In Vitro* 9:216-222.

Repellin A, Baga M, Jauhar PP, Chibbar RN (2001) Genetic enrichment of cereal crops via alien gene transfer: new challenges. Plant Cell, Tiss. Org. Cult. 64:159-183.

Rother S, Hadeler B, Orsini JM, Abel WO, Reski R (1994) Fate of a mutant macrochloroplast in somatic hybrids. J. Plant Physiol. 143:72-77.

Russell JA, McCown BH (1986) Culture and regeneration of *Populus* leaf protoplasts isolated from non-seedling tissue. Plant Sci. 46:133-142.

Sairam RV, Seetharama N, Devi PS, Verma A, Murthy UR, Potrykus I (1999) Culture and regeneration of mesophyll-derived protoplasts of sorghum (*Sorghum bicolour* L. Moench). Plant Cell Rep. 18:927-977.

Santos AVP dos, Outka DE, Cocking EC, Davey MR (1980) Organogenesis and somatic embryogenesis in tissues derived from leaf protoplasts and leaf explants of *Medicago sativa*. Z. Pflanzenphysiol. 99:261-270.

Sasamoto H, Ogita S, Wakita Y, Fukui M (2002) Endogenous levels of abscissic acid and gibberellins in leaf protoplasts competent for plant regeneration in *Betula platyphylla* and *Populus alba*. Plant Growth Regul. 38:195-201.

Shepard JF (1982) Cultivar dependent cultural refinements in potato protoplast regeneration. Plant Sci. Lett. 26:127-132.

Shepard JF, Totten RE (1977) Mesophyll cell protoplasts of potato: isolation, proliferation and plant regeneration. Plant Physiol. 60:313-316.

Shillito RD, Paszkowski J, Potrykus I (1983) Agarose plating and bead type culture technique enable and stimulate development of protoplast-derived colonies in a number of plant species. Plant Cell Rep. 2:244-247.

Takebe I, Labib G, Melchers G (1971) Regeneration of whole plants from isolated mesophyll protoplasts of tobacco. Naturwissenschaften 58:318-320.

Takebe I, Otsuki Y, Aoki S (1968) Isolation of tobacco mesophyll cells in intact and active state. Plant Cell Physiol. 9:115-124.

Thom M, Getz HP, Maretzki A (1992) Purification of a tonoplast polypeptide with sucrose transport properties. Physiol. Plant. 86:104-114.

Uchimiya H, Murashige T (1974) Evaluation of parameters in the isolation of viable protoplasts from cultured tobacco cells. Plant Physiol. 54:936-944.

Vardi A, Spiegel-Roy P, Galun E (1982) Plant regeneration from *Citrus* protoplasts: variability in methodological requirements among cultivars and species. Theor. Appl. Genet. 62:171-176.

Vasil IK (1987) Developing cell and tissue culture systems for the improvement of cereal and grasses crops. J. Plant Physiol. 128:193-218.

Vasil IK, Vasil V (1980) Clonal propagation. Int. Rev. Cytol. Suppl.11A 145-173.
Von Hanstein J.(1880) Biologie des protoplasmas. Bot. Abh. 4:1-56.
Watts JW, Motoyoshi F, King JM (1974) Problems associated with the production of stable protoplasts of cells of tobacco. Ann. Bot. 38:667-671.
Wernicke W, Thomas E (1980) Studies on morphogenesis from isolated protoplasts: shoot formation from mesophyll protoplasts of *Hyoscyamus muticus* and *Nicotiana tabacum*. Plant Sci. Lett. 17:401-407.
Widholm J (1972) The use of FDA and phenosafranine for determining viability of cultured plant cells. Stain Tech. 47:186-194.
Xu Z-H, Xue H-W (1999) Plant regeneration from cultured protoplasts. *In:* W-Y Soh, SS Bhojwani, (eds.), Morphogenesis in Plant Tissue Culture. Kluwer Acad. Publ., Dordrecht, pp. 37-70.
Zhou C (1989) Cell division in pollen protoplast culture of *Hemerocallis fulva* L. Plant Sci. 62:229-235.

3

Plant Protoplasts: Consequences of Lost Cell Walls

Attila Fehér*, Taras P. Pasternak, Krisztina Ötuüs and Dénes Dudits
Laboratory of Functional Cell Biology, Institute of Plant Biology, Biological Research Centre, H.A.S., Temesvári krt. 62, H-6726, Szeged, Hungary
**corresponding author: e-mail: fehera@nucleus.szbk.u-szeged.hu*

INTRODUCTION

A plant protoplast can be defined as the content of a plant cell enclosed by the plasmalemma without the surrounding cell wall. Experimentally, plant protoplasts can be obtained from cells of various plant tissues by either mechanical or enzymatic removal of their wall. The protoplasts, formed as a result of these manipulations, can be cultured in suitable media, provided that the appropriate osmotic pressure is maintained to preserve cellular integrity. Cultured protoplasts represent single, separated entities capable of reforming cell walls, dividing and initiating defined developmental pathways. That is why protoplast cultures represent the ideal "free cell" systems in the case of plants. The culture of protoplasts not only allows large scale, clonal propagation of a given plant, but as a population of discrete plant cells, protoplasts can be handled like microbes and are amenable to techniques used in microbial and somatic cell genetics. *In vitro* cultured protoplast-derived cells are freed from the effects of growth regulator gradients and positional information functioning *in planta,* which is why they are the most suitable experimental objects for studies of basic cellular responses of plant cells.

The elaboration of experimental techniques for the isolation and culture of plant protoplasts dates back for several decades. Although these techniques were already adapted and routinely applied for a huge number of species as well as various explants, the basic procedures are still the same and have not changed markedly from the early times of protoplast research. That is why this review does not deal with technical issues. Several comprehensive overviews and handbooks have already been published on this subject (e.g. Maheshwari et al. 1986; Fehér and Dudits 1994). Here we attempt to summarize the experimental information gained mainly during the last few years, on the physiology and development of plant protoplast-derived cells originating from differentiated tissues, in light of the emerging possibilities for using these single cells as useful experimental models in modern plant biology.

ISOLATION AND CULTURE: A STRESSFUL START

Without doubt, isolation of plant protoplasts is a stressful procedure. During isolation and purification, plant cells are exposed to wounding, mechanical forces (during filtration and centrifugation), and osmotic stress as well as elicitation by impure fungal enzymes and cell wall debris. In addition to these obvious stress factors, due to the removal of the cell wall matrix, plasmodesmatal connections and cellular communications, including the supply of growth regulators and metabolites from adjacent cells, are disrupted. Furthermore, during culture, the cells become exposed to nonphysiological concentrations of chemicals and growth regulators. Wounding represents a significant stress to differentiated plant tissues (e.g. leaf, hypocotyl, etc.) used to isolate protoplasts. In tobacco, it was shown that neither removal of the leaf from the plant, nor surface sterilization or enzymatic maceration, but the mechanical slicing of the leaf in order to facilitate enzyme penetration was the main factor inducing the expression of stress-related genes (Grosset et al. 1990). Even the composition of the culture medium had little effect on the overall gene expression changes induced during protoplast isolation (disappearance of leaf mRNAs and appearance of callus mRNAs; Grosset et al. 1990).

Further, commercial cell wall-degrading enzyme preparations contain impurities (proteinaceous as well as low molecular weight compounds) that may also be harmful for plant cells. Gel filtration of the enzyme solution frequently yields a higher percentage of viable protoplasts. Nonpurified cellulase, induced the formation of active oxygen species (AOS) such as superoxide radical or hydrogen peroxide (Papadakis and Roubelakis-Angelakis 1999). Purified cellulase, on the other hand, was inactive as an inducer of oxidative burst in protoplasts. The response was rapid: AOS accumulation started 5 min after application. Incubation of maize coleoptiles in solutions containing cell wall-degrading enzymes caused a rapid depolarization of the membrane voltage (Brudern and Thiel 1999). Denaturation of the macerozyme (0.5%) and cellulase (1.5%) enzymes did not affect membrane depolarization; contrarily, active pectolyase (0.1%) was significantly more potent in this respect than the denatured enzyme. Active pectolyase treatment was also followed by an oxidative burst and enhanced K^+ efflux. This observation indicates that protoplast isolation-derived pectic cell wall fragments may act as elicitor-like signals.

The complex metabolic modifications of cells during protoplast isolation may also include alterations in oxidative processes and the accumulation of active forms of oxygen. AOS may affect morphogenic response of plant cells and their accumulation during protoplast isolation was associated with decreased regeneration potential of protoplasts (Siminis et al. 1993). Papadakis and Roubelakis-Angelakis (1999) found two different AOS synthase activities in tobacco protoplasts and plasmamembranes: one showed specificity to NAD(P)H and sensitivity to diphenyleneiodonium (DPI, a well-known inhibitor of the neutrophyl NAD(P)H oxidase), and a second NAD(P)H oxidase-peroxidase was sensitive to KCN and NaN_3. Protoplast-derived cells of tobacco capable of cell divisions possessed both enzyme activities, while those of grape exhibiting no cell division activity showed only the second activity. Net H_2O_2 accumulation was reported to be associated with cell division and the addition of catalase in order to remove extracellular H_2O_2 prevented cell division in the case of tobacco protoplast-derived cells (De Marco and Roubelakis-Angelakis 1996a,b). *Nicotiana tabacum* protoplast-derived cells with and without the potential for cell division (the difference was due to the different length of maceration in cell wall-digesting enzyme solution) have been isolated and H_2O_2-dependent activities were recorded during 14 days of culture by De Marco and Roubelakis-Angelakis (1996a). The authors found that higher H_2O_2-scavenging capability of protoplasts ensured better division potential. In addition, differences in the activities of enzymes related to cell wall reconstitution were also observed in a comparison of regenerating and non-regenerating cultures. It was hypothesized that peroxidase-mediated cell wall reconstruction requires apoplastic H_2O_2 and its removal blocks cell wall formation and cell division (De Marco and Roubelakis-Angelakis 1996a,b; Papadakis and Roubelakis-Angelakis 1999, Papadakis et al. 2001).

Further evidence for the link between oxidative stress response and cell formation from protoplasts was provided by Kato and Esaka (2000) who found that ascorbate oxidase is involved in the elongation of tobacco protoplast-derived cells. However, based on their observations they hypothesized that this enzyme rather influenced transport processes and not the cell wall regeneration (Kato and Esaka 2000). In alfalfa, it was demonstrated that the decrease in the level of apoplastic H_2O_2, due to either DPI or dimethylthiourea (DMTU, H_2O_2-scavenger) application, had a dual effect on leaf protoplasts: it delayed cell elongation, which accords with the role of apoplastic H_2O_2 in cell wall formation but, parallelly, cell activation was also prevented, as indicated by the compact nuclei of the cells, inhibition of DNA synthesis, and low abundance of the signaling MAP-kinase protein MMK1 (Pasternak TP and Feher A, Unpublished results). This observation suggests that DPI/DMTU interferred with cellular signaling in addition to cell wall formation. The signaling role of H_2O_2 is well established in plant and animal species as well (for review, see Gamaley and Klyubin 1999, Adler et al. 1999; Vranová et al. 2002).

Surprisingly enough, plant cells can not only survive the serious stresses caused by protoplast isolation procedures, but rapidly accommodate to the changed environmental conditions. Plant protoplasts can develop a specific strategy against oxidative stress, which differs from the one that operates in the donor tissue. During culture, Ascorbate Peroxidase (APX) activity is enhanced and substitutes for decreasing activity of catalase (CAT) as H_2O_2 scavenger in the protoplast-derived cells (Siminis et al. 1993; De Marco and Roubelakis-Angelakis 1996a; Fig. 3.1). As shown by Northern analysis, the mRNA of the cytoplasmic form of APX was actively trans-cribed in dividing protoplasts whereas no APX mRNA was detectable in nonregenerating protoplasts (De Marco and Roubelakis-Angelakis 1996b). Cytoplasmic APX is suggested to be the predominant isozyme expressed in cells derived from protoplasts. Protoplast-derived cells can not survive if H_2O_2 accumulates in the cytoplasm due to APX inhibition, and fail to divide if H_2O_2 is removed from the outer space by catalase (De Marco and Roubelakis-Angelakis 1996b) or DMTU/DPI (Pasternak TP and Feher A, Unpublished results). In addition, chloroplastic Fe SOD activity decreases while cyto-plasmic Cu/Zn SOD activity increases during protoplast culture parallel with chloroplast degeneration (Papadakis et al. 1999; Pasternak TP and Feher A, Unpublished results). It was suggested that the localization of H_2O_2 rather than its absolute concentration is responsible for oxidative stress and controlled amounts of apoplastic H_2O_2 are necessary to allow proper cell wall reconstitution while low H_2O_2 concentrations are highly toxic in the chloroplasts and the cytoplasm (De Marco and Roubelakis-Angelakis 1996a).

A beneficial effect of mild oxidative stress on protoplast-derived cell development has also been reported in the case of alfalfa mesophyll protoplast-derived cells (Pasternak et al. 2002). Well-known oxidative stress-inducing compounds such as iron menadione (Pasternak et al. 2002), were demonstrated to enhance cell division at low, sublethal concent-rations. The earlier division of the treated cells may result from a

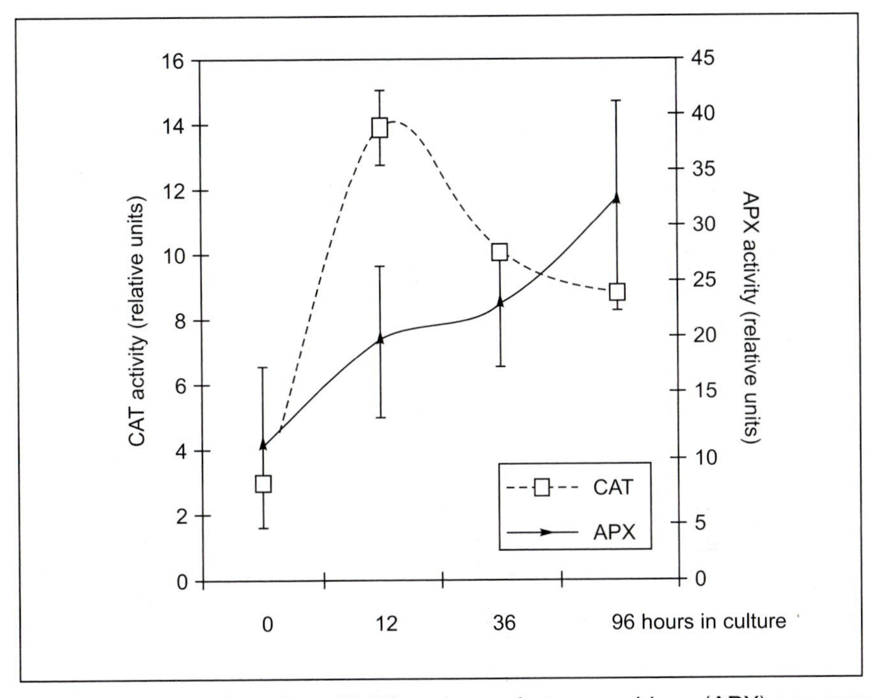

Fig. 3.1 Activity of catalase (CAT) and ascorbate peroxidase (APX) enzymes during culture of leaf protoplast-derived alfalfa cells. Increasing activities of the enzymes, especially CAT, within the first 12 h of culture may be ascribed to isolation stress. During further culture CAT activity continuously decreased and APX activity increasesd. APX (EC 1.11.1.11) activity was determined by monitoring the decrease in absorbance at 290 nm in 1 ml reaction mixture containing 50 mmol potassium phosphate buffer (pH 7.0), 0.5 mmol ascorbate, 0.1 mmol H_2O_2, 0.1 mmol EDTA and 10 μg protein (Nakano and Asada 1987). CAT (EC 1.11.1.6) activity was determined by monitoring the decomposition (consumption) of H_2O_2 in 1 ml reaction mixture containing 66 mM potassium phosphate buffer (pH 7.0), 12.5 mM H_2O_2, and 10 μg protein. Activities are expressed by using the absorbance readings as relative units/μg protein/min. (Gosset et al. 1994).

shortened dedifferentiation period due to stress-induced cellular changes (see below).

Before cell division, however, protoplasts have to accomplish another significant process: the resynthesis of their removed cell wall.

CELL WALL SYNTHESIS: NO WALL, NO PROBLEM?

The cell wall is an indispensable element of the plant cell that, among other things, determines cell shape. It not only provides a rigid extracellular matrix but can be considered a functional cellular "organelle" undergoing dynamic changes due to a plethora of different stimuli during plant development (Braam 1999). Studies on the structure and function of plant cell walls are inevitably required to understand how plants and plant cells function. Protoplasts provide a unique opportunity to investigate the general mechanisms of cell wall synthesis. The carefully isolated protoplasts have no or very limited preexisting cellulosic material or other microfibrils; hence cell wall deposition can be well studied in these experimental systems.

A number of early investigations on cell wall formation relied on staining or electron microscopic procedures (for review, see e.g. Maheshwari et al. 1986). Studies on incorporation of radioactive precursors showed that synthesis starts within a few minutes after removal of cell wall-degrading enzymes (e.g. Klein et al. 1981). The initial phase of cell wall reformation is characterized by callose β-(1-3)-glucan) secretion that precedes cellulose deposition, although the original cells *in planta* do not synthesize this compound (Klein et al. 1981, vanAmstel et al. 1996, Caumont et al. 1997). This accords with the view of callose deposition as a wounding response of cells (e.g. Levina et al. 2000), although, in unwounded suspension cultured cells, the transient appearance of callose was hypothesized to be related to cellulose microfibril deposition (vanAmstel et al. 1996).

Several other observations also indicate that, especially in liquid culture, a normal cell wall may not always be made by protoplasts (e.g. Blaschek et al. 1982). However, it was also demonstrated that protoplasts isolated from specific cell types can retain their ability to synthesize specific cell wall constituents. For example, endosperm-derived protoplasts were shown to produce galactomannans (Sotiriou and Spyropoulos 2002). Pollen tube protoplasts resynthesize new cell wall material that like the subapical region of actively growing pollen tubes, is rich in callose polymers and pectin epitopes recognized by the JIM7 antibody (Majewska-Sawka et al. 2002). These observations further support the usefulness of protoplasts as models in cell wall synthesis research.

The cell wall is not only a product of the protoplast, but one of the critical factors determining cellular fate and development (see later, and for more detailed reviews Brownlee et al. 1998; Wojtaszek 2001; Malinowski and Filipecki 2002). For example, cell division and cell elongation, the two basic processes of plant morphogenesis, both rely on the remodeling and *de novo* synthesis of the cell wall. Futhermore, there is increasing evidence that the cell wall is more than an "exoskeleton" (Kohorn 2000). It is rather a carbohydrate matrix which provides a dynamic scaffold for a variety of other carbohydrates and proteins. Recently, the use of antibodies against carbohydrate and protein epitopes became a very useful approach to investigating the dynamic organization and function of plant cell walls (Knox 1997). Many plant cell wall associated proteins, such as endo-1-4-β-D-glucanases (del Campillo 1999), expansins (Cosgrove 1996), wall-associated kinases (WAKs, Kohorn 2001), and arabinogalactan proteins (AGPs) (Showalter 2001), have been shown to be involved in diverse processes of plant growth and development.

AGPs, for example, although their exact functions are not clear, are hypothesized to be involved in molecular interactions and cellular signaling at the cell surface. They exhibit organ-, tissue-, and cell-specific expression and are markers of cellular differentiation. The JIM8 AGP epitope was demonstrated to represent a soluble signal that stimulates embryo development in carrot cell cultures (Pennell et al. 1995; McCabe et al. 1997). AGP epitopes have also been detected in newly formed cell walls of protoplasts (Butowt et al. 1999; Majewska-Sawka et al. 2002).

Physical connections between the cell wall and the plasma membrane can be observed in many ways in plant cells; however, the existence of "adhesion sites" containing integrins, receptors, kinases, their ligands and the cytoskeleton as described in metazoan cells has not been proven in plant cells so far. Integrins were hypothesized as present in plant cells based on studies with plant protoplasts (e.g. Barthou et al. 1999; Blackman et al. 2000, 2001) but no direct molecular or functional evidence has been adduced to now. Considering the importance of cell-to-cell communication in plant development and cell differentiation, the removal of cell walls and the separation of plant cells necessarily results in significant developmental consequences.

DEDIFFERENTIATION: LOSING THE WAY

Stress, if not exceeding the level of tolerance, induces a general reprogramming of gene expression and reorganization of metabolism in plant cells, providing better flexibility for them to survive unfavorable conditions. Removal of plant cells from their tissue environment during protoplast isolation is not only a stress. It also means that the cells no longer experience hormonal gradients and developmental signals provided by surrounding cells and tissues. This situation results in two major consequences. On the one hand, differentiated plant cells as a result of protoplast isolation lose their differentiated functions; in other words, they start to dedifferentiate. On the other hand, to ensure the viability of these dedifferentiated cells, hormonal (developmental) signals have to be artificially provided in the culture medium. While nuclear RNA synthesis of freshly isolated protoplasts is not dependent upon exogenously supplied hormones during the first 18 h of culture (Cook and Meyer 1981; Bergounioux et al. 1988), cell division of protoplast-derived cells requires the exogenous application of auxin and cyokinin (Pasternak et al. 2000). Despite the absolute requirement for exogenous auxin, cultured plant cells produce substantial amounts of the native auxin, indoleacetic acid (IAA) (Bartel 1997). In alfalfa mesophyll protoplasts, a transient increase in endogenous auxin and cytokinin levels was observed prior to the first cell division (Fig. 3.2). Modification of the pH (by 2-morpholinoethanesulfonic acid, MES), auxin (2,4-D) or iron (Fe-EDTA) concentration in the medium resulted in a shift in the timing of the endogenous auxin peak (Pasternak et al. 2002). The earlier appearance of the cellular IAA peak correlated well with a shorter time period required before the entry of the protoplast-derived cells into the cell division cycle. Measurement and immunocytochemical localization of IAA in immature zygotic embryos of sunflower before, during and after induction of somatic embryo development provided another experimental evidence that an endogenous auxin pulse may be associated with a developmental switch (Charriére et al. 1999; Thomas et al. 2002).

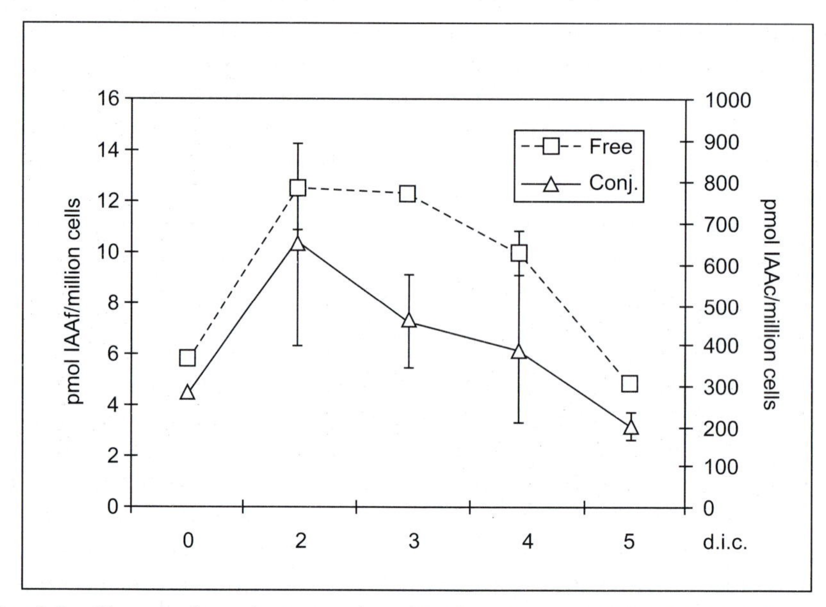

Fig. 3.2 Changes in endogenous free (IAAf) and conjugated (IAAc) indoleacetic acid levels in alfalfa leaf protoplasts during the first four days of culture. A transient increase in the cellular level of both IAA forms can be well recognized during the first few days of protoplast culture. Measurements were made by micro LC/LC ESI MS/MS after solid phase extraction in the library of Harry Van Onchelen (University of Antwerp) as described in Pasternak et al. (2002). d.i.c. – days in culture

Reactivation of cell division in differentiated cells is necessarily linked with changes in gene expression profiles. Existing gene expression patterns ensuring differentiated cell functions have to be erased and in parallelly, new genes are expected to be activated in order to allow adaptation and initiation of new developmental pathways. As dedifferentiation of plant cells is a basic phenomenon underlying *in vitro* cell and tissue culture responses, there have been many attempts to isolate potential key genes with altered transcriptional profile in this process. The synthesis of proteins involved in photosynthesis is completely inhibited while other proteins absent from differentiated mesophyll cells are actively synthesized in protoplasts as soon as they are isolated from leaves (Fleck and Durr 1980; Vernet et al. 1982). Among those genes responsible for the synthesis of these proteins, many are associated with stress response, primary metabolism, cell wall synthesis or cell division (Grosset et al. 1990; Marty et al. 1993; Nagata et al. 1994; Yu et al. 1999; Criqui et al. 1992). During dedifferentiation, all these functions participate in the reorganization of cellular states (e.g. metabolism, physiology) associated with the altered developmental program. However, no master genes responsible for the regulation of the dedifferentiation process have been identified up to now (reviewed recently in Fehér et al. 2003). One can hypothesize, however, that such master genes, if they exist, should be involved in the general reprogramming of gene expression (chromatin remodeling, transcription machinery).

Chromatin structure and the mode of DNA organization around basic nuclear proteins, the histones, change dynamically and are continuously remodeled during development because intrinsically involved in the regulation of nuclear processes such as DNA repair, replication and especially transcription (for reviews, see Varga-Weisz and Becker 1998; Tsukiyama and Wu 1997). Chromatin-dependent gene silencing is a common mechanism for maintaining the differentiated state of cells (Gregory and Horz 1998). As a consequence, chromatin remodeling has necessarily to be linked with cellular dedifferentiation when change of cell fate requires general reorganization of gene expression patterns. In tobacco protoplast-derived cells, probing the chromatin structure by nuclease digestion as well as accessibility of DNA to the intercalating fluorescent dye, propidium iodide, revealed two separated periods of chromatin decondensation: the first took place during protoplast isolation and was linked to dedifferentiation, while the second was induced by plant hormones, auxin and cytokinin, and linked to the reactivation of cell division (Zhao et al. 2001). In this model, dedifferentiation (first phase of chromatin decondensation) represents a transitory stage when cells become competent to alter their fate. However, their further development is dependent on hormonal signals. In the absence of auxin, the cells under went apoptosis. Auxin treatment alone resulted in differentiation (elongation), while auxin and cytokinin were both required for cell division

(Zhao et al. 2001). Similar observations were made with leaf protoplast-derived cells of alfalfa: the structure and size of the nucleus and especially the ratio of the volumes of the nucleus and nucleolus, as well as the stainability of the chromatin with fluorescent dyes, could be correlated with auxin/cytokinin-dependent phases of cell reactivation and division (Pasternak et al. 2000). Differentiated leaf cells of alfalfa have compact nuclei with a small nucleolus probably as a consequence of the moderate transcription limited to the few genes necessary for the well-defined functions of a leaf cell. In the absence of any growth regulators, small but obvious increase in the size of the nuclei/nucleoli could be observed; however, for full dedifferentiation both auxin and cytokinin were exogenously required (Pasternak et al. 2000; Fig. 3.3). At the beginning of the *in vitro* culture of alfalfa leaf protoplasts, dedifferentiation could be

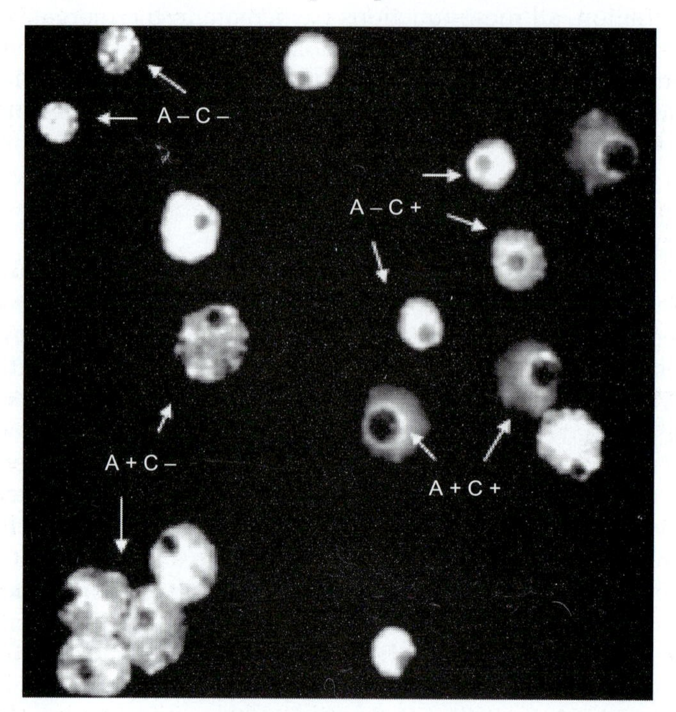

Fig. 3.3 Morphology of nuclei isolated from alfalfa protoplast-derived cells cultured in the presence of various auxin/cytokinin combinations for three days. Protoplasts were cultured separately in different media with (+) or without (-) 2 mg/L^{-1} 2,4-D as an auxin (A) and 0.5 mg/L^{-1} zeatin as a cytokinin (C). Nuclei were isolated and stained by 4',6-diamidino-2-phenylindole (DAPI) as described elsewhere (Pasternak et al. 2000). The images of fluorescent nuclei isolated from the cells grown in the presence of the various hormone combinations were taken separately by a CCD camera and overlayed artificially to make a composite picture.

characterized with increase in size of the nucleolus. The nucleolus is the compartment for the synthesis of ribosomal genes (rDNA) and for the assembly of pre-ribosomes within the nucleus (Hernandez-Verdun 1991). It is obvious, that the increased need for protein synthesis due to stress responses, the shift from photoautotrophic to heterotrophic metabolism, cell wall synthesis and cell cycle reactivation require new ribosomes, which have to be reflected in the increased size of the nucleolus. Similar changes were observed in animal cells reentering the cell division cycle in regenerating liver (e.g. Muramatsu and Busch 1965). Nucleolar size in alfalfa leaf protoplasts was significantly influenced by the presence of exogenous cytokinin (Pasternak et al. 2000). This accorded well with the fact that cytokinin induces the expression of ribosomal genes (Tepfer and Fosket 1978; Gaudino and Pikaard 1997).

Establishment of a new cellular state is not only governed at the gene expression level, but requires modification and/or removal of unnecessary polypeptides and protein complexes. Regulated protein degradation attracts more and more attention as an important mechanism controlling cell cycle progression and differentiation in eukaryotic cells including plants (Udvardy 1996; Genschik et al. 1998; Criqui et al. 2000). The ubiquitin-proteasome pathway is responsible for the degradation of many cellular proteins, not just regulators and short-lived proteins but also those that serve structural roles and are long lived (Rock et al. 1994). Ubiquitin, a small peptide of 76 amino acids, can be covalently linked to a variety of proteins and serves as a biochemical tag that marks proteins for degradation (Ingvardsen and Veierskov 2001). Dedifferentiation of mature tobacco cells was shown to be accompanied by a sharp increase in ubiquitin gene expression (Jamet et al. 1990). This elevation in ubiquitin gene activity may be related to the cellular reorganization during dedifferentiation and required for the selective destruction of proteins associated with the previous, differentiated cell state. In line with this observation, it has recently been reported that in the case of reactivated tobacco protoplasts, entry into the DNA replication phase could be prevented by MG132, a specific compound inhibiting the ubiquitin-proteasome pathway, but not by leupeptin, a protease inhibitor (Zhao et al. 2001). Furthermore, it has been demonstrated that auxin signaling is based, at least in part, on the degradation of specific proteins (Leyser 2001; Rogg and Bartel 2001).

Reorganization of metabolism, gene expression, and developmental events is necessarily reflected by physiological parameters of the cells. For example, reactivation of quiescent cells is linked with characteristic changes in cellular pH gradients. In animal as well as yeast cells, modification of cytoplasmic pH (pHc) is important for control of the cell cycle, cell division and growth (Anand and Prasad 1989; Swann and Whitaker 1990; Whitaker 1990). Acidic pHc conditions are usually associated with a quiescent or dormant cellular state while an increase in pHc often accompanies cellular activation (for review, see Frelin et al. 1988). The speed of medium acidification correlated with the appearance of dividing cells in alfalfa leaf protoplast cultures and manipulation of the medium pH sufficed to accelerate or delay the dedifferentiation process (Pasternak et al. 2002; Fig. 3.4). In the same culture system the intracellular distribution of fluorescein diacetate (FDA), a pH-indicator fluorescent dye, was found to be a useful marker of dedifferentiation and cell fate (Pasternak et al. 2002). FDA was not detectable in the chloroplasts, only in the cytoplasm, in small, dense, embryogenic-type cells in contrast to highly vacuolated callus-forming cells (Pasternak et al. 2002). Photosynthetic electron transport results in the establishment of a ΔpH across the thylakoid membrane of chloroplasts, significantly acidifying the thylakoid lumen (pH app. 5.0) versus the stroma (pH app. 8.0). It was suggested that FDA accumulation in the cytoplasm of protoplast-derived cells is related to the loss of photosynthetic ability as a differentiated function.

CELL DIVISION AND ELONGATION: WHICH WAY TO FOLLOW?

Cell elongation and division are two basic processes underlying plant morphogenesis, linked respectively to cellular differentiation and dedifferentiation. Cultured mesophyll protoplasts can be directly induced to follow different developmental pathways. In a medium containing only auxin (1 mg/L^{-1} naphthyleneacetic acid, NAA), tobacco protoplasts elongated and finally differentiated into very long tubular cells (Tao and Verbelen 1999). The presence of cytokinin induced cell division and cells could be switched from one developmental pathway to another by giving the appropriate hormonal signal. The auxin-induced elongation and differentiation was coupled to DNA endoreduplication (Valente et al. 1998).

Concomitant with the subsequent fate of protoplasts, microtubules and actin-filaments as well as cellulose microfibrils are randomly arranged in spherical cells and perpendicular to the long axis in elongated ones (Dijak and Simmonds 1988; Vissenberg et al. 2000). It is well accepted that the cellulose synthase complex is somehow associated with the cytoskeleton (e.g. Hasezawa and Nazaki 1999). Both cellulose microfibrils and cellulose fibers are traverse to the direction of cell elongation. However, elongation and cellulose deposition can be genetically separated as shown by mutant analysis (Kohorn 2000). This finding supports a model that a mechanism driving cell expansion determines the orientation of microtubules and cellulose fibrils independently, but in the same direction.

In the case of alfalfa, it was shown that cell elongation and the timing of cell cycle reactivation correlated inversely (Fig. 3.5). Small, cytoplasm-

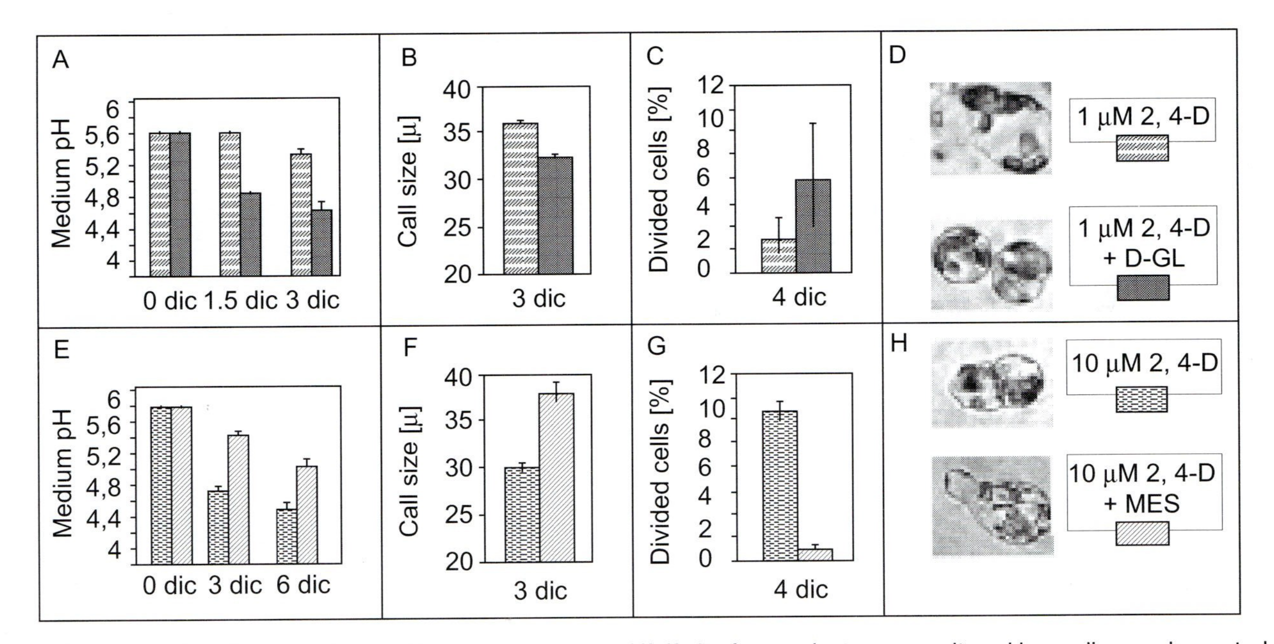

Fig. 3.4 Effect of the medium pH on protoplast development. Alfalfa leaf protoplasts were cultured in media supplemented with 10 mM freshly dissolved D-galctolactone (D-GL)(A,B,C,D) or 15 mM morpholinoethanesulphonic acid (MES) (E,F,G,H). As shown in panels A and E, MES delayed while D-GL enhanced medium acidification during the first four days of protoplast culture. Cell size (B,F) and cell division (C,G) as well as morphology of the cells (D,H) was altered due to the altered medium pH: a delay in medium acidification resulted in development of larger cells dividing at a lower rate; enhanced medium acidification enhanced cell divisions at a lower size. D-GL-mediated medium acidification in the presence of 1 µM 2,4-D resulted in the development of small, round, cytoplasmic-rich cells morphologically similar to those developed at a higher (10 µM) 2,4-D concentration (D), while MES-mediated inhibition of medium acidification in the presence of 10 µM 2,4-D resulted in a similar elongated, vacuolized cell type otherwise developed in the presence of 1 µM 2,4-D. d.i.c. – days in culture.

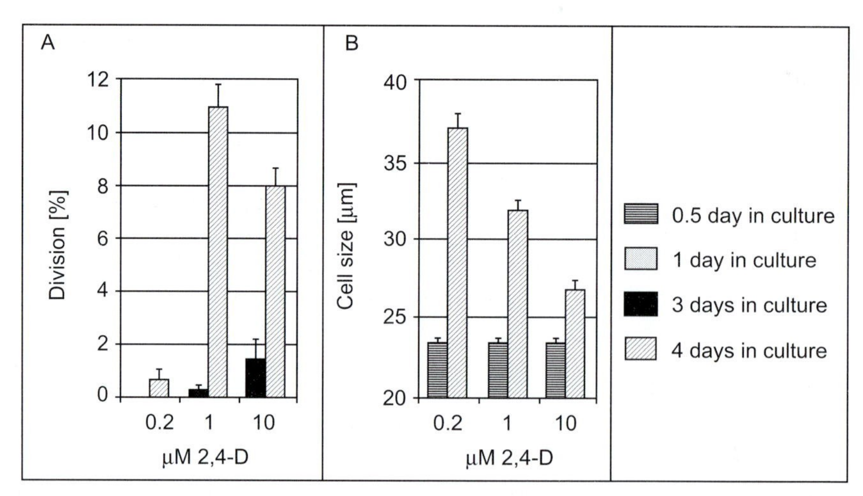

Fig. 3.5 Cell division and elongation correlate inversely with increase in exogenous auxin concentration in alfalfa leaf protoplast cultures. Alfalfa leaf protoplasts were isolated and cultured as described by Pasternak et al. (2000, 2002) in the presence of the indicated 2,4-D concentrations and the frequency of divided cells (500-1,000 cells counted) as well as the average cell length as a parameter of cell size (30 cells) were determined. d.i.c. – days in culture.

rich embryogenic cells divided earlier than elongated, vacuolized differentiating cells (Pasternak et al. 2002). Bögre et al. (1990) also reported that leaf protoplast-derived round-shaped small cells of an embryogenic alfalfa genotype activated earlier compared to elongated vacuolized cells of a nonembryogenic genotype.

The key plant growth regulators considered responsible for cell division activity *in vivo* as well as *in vitro* are auxin and cytokinin. There are no, or very few, protoplast systems in which no exogenous hormones are required to maintain cell viability and especially to initiate cell divisions (Vardi et al. 1987, Sim et al. 1988, Phillips and Darrel 1988). Cell expansion was found to be dependent on the auxin concentration in alfalfa (Fig. 3.5) and tobacco (Vissenberg et al. 2000) mesophyll protoplast cultures. Application of TIBA drastically reduced elongation of tobacco protoplast-derived cells, indicating the importance of auxin efflux in addition to the original auxin concentration in the medium (Vissenberg et al. 2000). Protoplasts of corn coleoptiles or *Arabidopsis* hypocotyls responded to auxin with rapid change in volume (Steffens et al. 2001). Antibodies raised against the C-terminus of corn or *Arabidopsis* auxin-binding protein 1 (ABP1) inhibited, while antibodies against the auxin-binding domain induced, auxin response, indicating that the auxin signal was perceived by extracellular ABP1 (Steffen et al. 2001). Both auxin and cytokinin were found to be indispensable for the S-phase in tobacco (Cook and Meyer 1981; Carle et al. 1998) and *Petunia* (Trehin et al. 1998) leaf protoplasts. In a specific, low density culture system of tobacco, auxin, (NAA) was able to induce DNA synthesis which finally resulted in endoreduplication in the absence of cytokinin, but cytokinin was required for mitosis (Valente et al. 1998). Similarly, in other culture systems as in tobacco suspension cell-derived protoplasts (Jouanneau and Tandeau de Marsac 1973) or hormone-deprived tobacco cell cultures (John et al. 1993, Zhang et al. 1996), cytokinin was required for mitosis while DNA synthesis could be accomplished in the presence of auxin alone. In cycling cells it was also shown that endogenous cytokinin concentration reaches a significant peak before mitosis suggesting its specific role at this cell cycle phase transition (Redig et al. 1996; Laureys et al. 1998). However, in alfalfa mesophyll protoplast-derived cells the requirement of cytokinin for S-phase entry was also reported (Pasternak et al. 2000).

It is a well-established fact that in plants, as in all eukaryotes, the core of cell cycle regulation includes homologues of the protein kinase and proteins such as cyclins, inhibitors and phosphatases regulating $p34^{cdc2}$ activity (for review, see Mészaros et al. 2000). At present, we have only limited information about the mode of action of auxin and cytokinin on the cell cycle machinery, especially on the reactivation of differentiated plant cells (G0-G1/S transition) (for reviews, see den Boer and Murray 2000, Stals and Inze 2001). The positive action of auxin on the expression of the gene encoding for the cognate plant $p34^{cdc2}$ kinase has been well demonstrated in different experimental systems (e.g. Hirt et al. 1991; Miao et al. 1993). However, there are contradictory results considering the requirement of auxin and cytokinin for the expression of the *cdc2* gene in leaf protoplast-derived cells of plants. While in the case of *Petunia* both auxin and cytokinin were absolutely required for *cdc2* expression (Trehin et al. 1998), in tobacco either auxin or cytokinin alone could induce *cdc2* transcription, although the combination of the two hormones was the most effective (Hemerly et al. 1993; Carle et al. 1998). In the case of alfalfa mesophyll protoplast-derived cells, auxin alone or together with a cytokinin resulted in the accumulation of similar amounts of *Cdc2* protein (Pasternak et al. 2000). It was also showed that cytokinin was required for the post-translational activation of the cell cycle regulatory protein kinase, Cdc2MsA/B, via unknown mechanisms, both at the G1/S and the G2/M cell cycle phase boundary (Pasternak et al. 2000). The central role of the *Cdc2*-related kinase in regulating the entry of protoplasts into the division cycle was further demonstrated by experiments, in which the transient overexpression of the alfalfa *cdc2* gene under the control of the 35S promoter could significantly accelerate S-phase entry of leaf protoplast-derived cells of tobacco (Damla Dedeoglu, Metin Bilgin, Attila Fehér and Dénes Dudits, unpubl. results; Fig. 3.6).

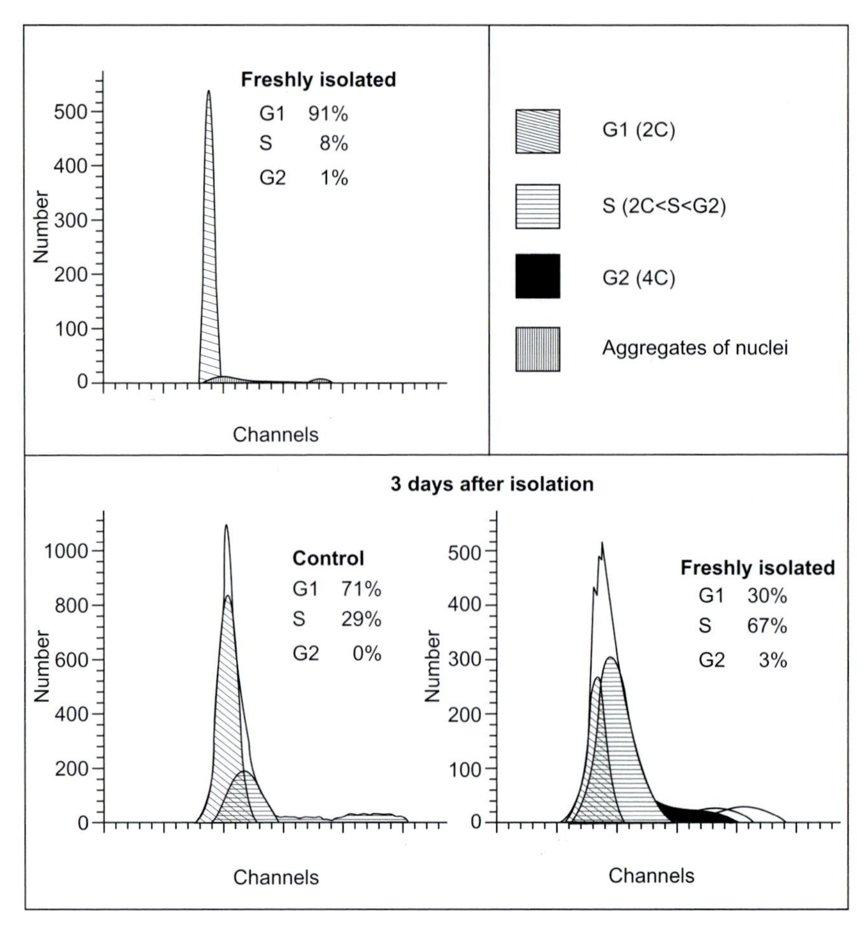

Fig. 3.6 Transient expression of the alfalfa *cdc2* gene in tobacco leaf protoplasts increases the frequency of S-phase cells at three days of culture. Tobacco (*Nicotiana tabacum* SR1) protoplasts were isolated and transformed by plasmid DNA as described by Negrutiu et al. (1997). Flow cytometric analysis on isolated nuclei was performed by a FACSCalibur flow cytometer following ethidium bromide staining as described elsewhere (Pasternak et al. 2000). The data obtained were analyzed by ModFit software.

DEVELOPMENTAL PATHWAYS: WHERE TO GO?

Although differentiated cells obviously lose many of their specific functions during *in-vitro* culture, an increasing number of experimental observations indicate that correctly isolated and cultured protoplasts can retain some of their functional characteristics (Sheen et al. 2001). Such was the case with guard cell protoplasts of *Nicotiana glauca* (Taylor et al. 1998). Dedifferentiation of guard cell protoplasts was found to be dependent on the culture temperature and the presence of 0.1 μM abscisic acid (ABA) and under the different culture conditions they developed into various cell types. The degree of dedifferentiation varied in these cell types as revealed by biochemical and molecular characterization (e.g. investigating the signal transduction from blue light photoreception to activation of the plasmalemma ATPase). Under optimal culture conditions, however, the protoplast-derived cells could preserve a high degree of morphological and functional similarity to freshly isolated guard cell protoplasts. As another example, the cultured aleuron protoplasts provided an excellent system to study the hormonal (abscisic acid, gibberellic acid) control of cell death (Bethke et al. 1999). Various protoplast cultures and transient gene expression analyses have also been successfully used in the dissection of diverse signal transduction pathways in plants (for a recent review, see Sheen et al. 2001).

However, cell fate can also be easily altered in protoplast cultures and the underlying mechanisms studied. Stress, exogenous hormones, and reactivation of cell division, all being associated with the isolation and culture of protoplasts, result in a flexible cellular state due to the general reorganization of chromatin structure, reprogramming of gene expression, and cellular metabolism (reviewed by Fehér et al. 2003). This cellular flexibility favors a developmental switch in response to an appropriate developmental signal. From alfalfa leaf protoplasts, embryogenic competent cells can be formed in response to 2,4-D (Dudits et al. 1991; Pasternak et al. 2002). While a relatively low 2,4-D concentration (1 μM) resulted in the formation of elongated vacuolated cells, the embryogenic competent cell type could be observed in the presence of 10 μM 2,4-D (Fig. 3.7). This latter type cell could develop into proembryogenic cell clusters accumulating the embryogenic AGL-15 protein in their nuclei (Perry et al. 1999) after removal of the exogenous auxin-analogue (Pasternak et al. 2002). In other experimental systems, direct somatic embryos were also produced from cells which were small in size, exhibited a dense cytoplasm, and were metabolically very active (for review, see Fehér et al. 2003). It can be hypothetised that the small, protoplast-derived alfalfa cells, resembling meristematic cells in morphology and division capability, represent a dedifferentiated cell state with the potency to initiate a new developmental program, and as such are the analogues of the "state 0" embryogenic competent cells defined by Nomura and Komamine in carrot cultures (Nomura and Komamine 1995).

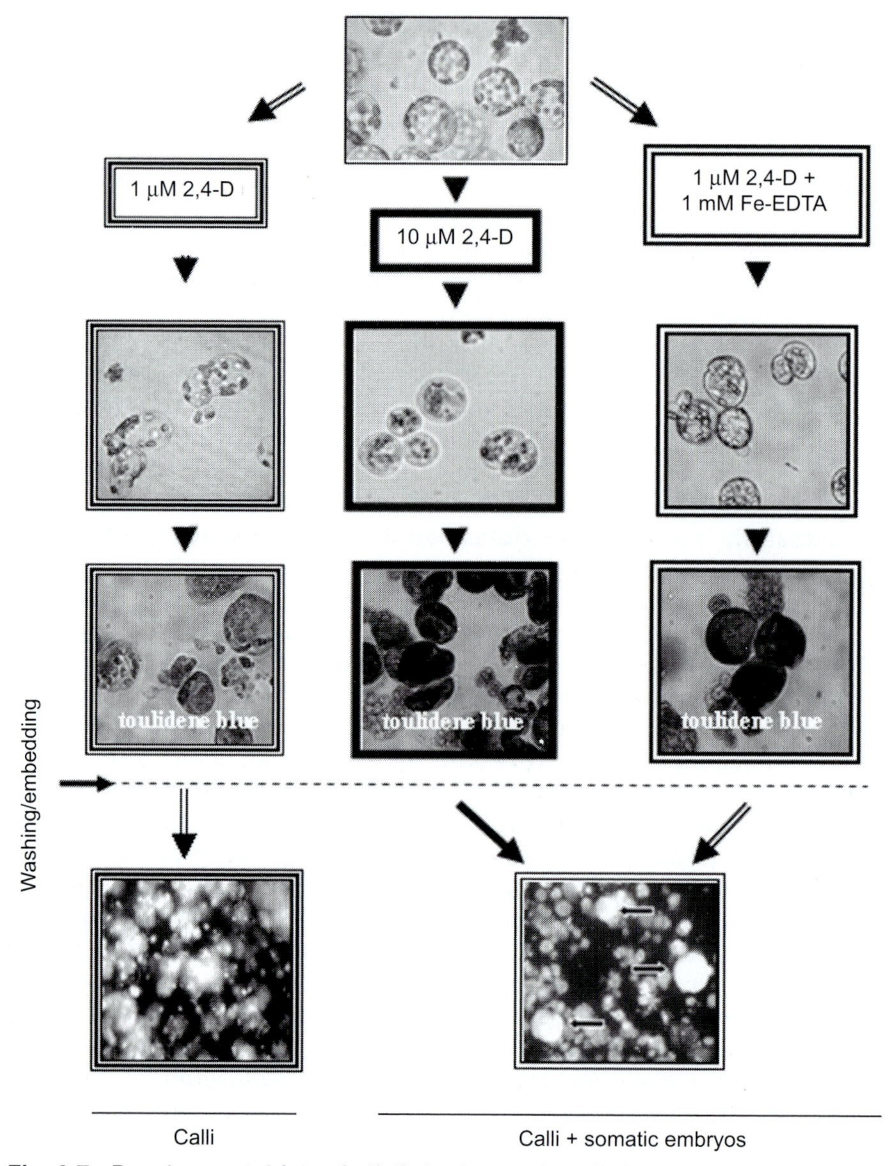

Fig. 3.7 Developmental fate of alfalfa leaf protoplast-derived cells is dependent on initial culture conditions. Leaf protoplasts from the embryogenic alfalfa (*Medicago sativa*) genotype A2 were isolated and cultured as described by Pasternak et al. (2000, 2002). Protoplast-derived cells cultured at 1 or 10 µM 2,4-dichlorophenoxyacetic acid (2,4-D) for three-five days underwent different fate. Cells grown in the presence of the higher 2,4-D concentration became embryogenic and could developed into somatic embryos if subcultured into fresh, agarose-

Contd.

It is well established that cellular polarity is an important determinant of cell fate. The initial zygotic division in higher plants is asymmetric (Scheres and Benfey 1999). This event establishes the basic polarity of the plant which probably determines subsequent pattern formation (Jurgens et al. 1997). Asymmetric cell divisions by definition generate daughter cells with different fates. The different developmental potentials of the cells may arise from the unequal distribution of cell fate determinants (intrinsic mechanism) or can be a result of differential interactions with the surrounding cells (extrinsic mechanism) (Scheres and Benfey 1999). The intrinsic mechanism can probaly be best studied in free-cell systems, such as protoplast cultures. Alfalfa leaf protoplasts could be induced to form asymetrically dividing embryogenic cells directly (Dijak and Simmonds 1988; Song et al. 1990; Dudits et al. 1991). 2,4-D stimulated type of cell divisions asymmetric in a concentration- (Pasternak et al. 2002) and genotype- (Bögre et al. 1990) dependent manner. Electric stimulation also induced the formation of embryogenic cells from alfalfa leaf protoplasts which were characterized by disordered microtubules compared to nonembryogenic control cells (Dijak and Simmonds 1988). The nucleus was positioned peripherally in these cells. This finding indicated early establishment of cellular polarity in response to the electric stimulation used to induce embryogenic development. Unequal distribution of cellular determinants can also be observed in polar egg cells: the nucleus is positioned at the cytoplasm-rich chalazal pole, while the micropylar pole is highly vacuolated (Russell 1993). The microtubular cytoskeleton is particularly dense near the nucleus and has no specific orientation, similar to the above mentioned embryogenic alfalfa cells. The actin microfila-ments have a similar conformation.

Nomura and Komamine (1985) showed that isolated small, cytoplasm-rich carrot cells developed into somatic embryos after an unequal first division and polarized synthesis of macromolecules. In certain *Arabidopsis* ecotypes, 2-5% of leaf protoplast-derived embryogenic cells were reported to exhibit a cell division pattern reminiscent of the first division of the zygote forming embryo proper and suspensor-like cells (Luo and Koop 1997). These experimental observations are clearly in favor of a hypothesis that intrinsic mechanisms play a central role in polarity determination.

Fig. 3.7 Contd.

solidified medium with the lower 2,4-D level. Cells cultured at the lower concentration from the beginning developed into elongated cells and could only form callus when subcultured in a similar manner. Iron, a known oxidative stress-inducing compound applied at nonlethal concentrations to the protoplasts grown in the presence of 1 μM 2,4-D resulted in the development of a small, round, densely cytoplasmed, embryogenic cell type strongly stained by toulidene blue, similar to that appeared in the presence of the higher 2,4-D concentration.

In other culture systems, e.g. *Cichorium* leaf cells (Blervacq et al. 1995) or hypocotyl-derived carrot cell cultures (Toonen et al. 1994), morphological asymmetry of the dividing embryogenic cells was not always observed. However, the asymmetric distribution of certain cellular determinants among the daughter cells formed by morphologically equal cell divisions cannot be excluded, especially since our knowledge is very limited about the possible determinants of polarity and cell fate in plants. Accumulating pieces of experimental evidence strengthen the importance of polarized secretion of cell wall materials as a determinant of axis fixation during embryogenesis (for reviews, see Scheres and Benfey 1999; Belanger and Quatrano 2000; Malinowski and Filipecki 2002; Fehér et al. 2003). The cell wall forms a continuum with the plasmamembrane and the cytoskeleton (reviewed e.g by Fowler and Quatrano 1997; Wojtaszek 2001). Plasmolysis of leaf cells of *Tradescantia* revealed the adhesion of the plasmamembrane to the cell wall at sites coinciding with cytoskeletal arrays involved in the polarization of the cells undergoing asymmetric division and in the establishment and maintenance of the division site (Cleary 2001). In sunflower, protoplasts derived from hypocotyls divided symmetrically in liquid culture medium whereas they underwent unequal division and gave rise to embryoids when embedded in agarose (Petitprez et al. 1995). Barthou et al. (1999) showed that adhesion sites between the plasmamembrane of protoplasts and the agarose matrix are involved in embryoid formation. The authors proposed a model in which anchorage of the membrane to the agarose matrix is mediated by integrin-like (arg-gly-asp (RGD)-binding) proteins connected with microtubules. The RGD tripeptide has been shown to inhibit carrot embryogenesis as well as cell wall regeneration of algal protoplasts (Blackman et al. 2000, 2001). Using the sunflower protoplast culture system, a QTL mapping of genomic regions involved in symmetric and asymmetric division of protoplasts was carried out (Berrios et al. 2000). Twelve and eleven putative loci were found to be associated with protoplast division activity and asymmetric division frequency, respectively. Clustering of QTLs for organogenesis, somatic embryogenesis, and protoplast division in certain chromosomal regions has also been suggested.

In parallel with the polarization of the extracellular matrix, intracellular changes have to occur to determine cell fate. In sunflower protoplast-derived cells, as indicated by the fluorescent probe DMBodipy-PAA, there was a translocation of Ca^{2+} channels, which depended on the division type (Xu et al. 1999; Vallee et al. 1997, 1999). In freshly isolated protoplasts, the probe was strictly localized to distinct points on the plasmamembrane. This compound also labeled cytoplasmic strands and a region around the nucleus in cultured protoplast-derived cells. When cells divided, staining was localized homogeneously in the plane of cell division in symmetrically dividing cells, but marked only a peripheral ring around the site of cell division during asymmetric division (Xu et al. 1999; Vallee et al. 1999). Later, after embryoids had developed from the asymmetrically divided cells, the Ca^{2+} channel probe was clearly localized on the basal part of the embryoids, as was the case with zygotic embryos (Xu et al. 1999; Vallee 1997, 1999). Unequal distribution of Ca^{2+} channels and the established Ca^{2+} gradient were important in early determination of the axis in *Pelvetia* and *Fucus* zygotes as well (Souter and Lindsey 2000; Cove 2000). In alfalfa leaf protoplasts cultured in the presence of 2,4-D, the endogenous IAA levels increased considerably during the first days of culture (Pasternak et al. 2002; Fig. 3.2). This auxin wave was transient and comparable under both embryogenic and non-embryogenic conditions; however, the timing of the synthesis peak showed approximately one day of delay under nonembryogenic conditions. Application of excess iron, or menadione at concentrations not decreasing cell viability ("mild oxidative stress") restored formation of the embryogenic cell type in the presence of low auxin (2,4-D) concentration (Pasternak et al. 2002). However, buffering of the medium pH by MES prevented embryogenic cell development in the presence of 10 μM 2,4-D. In both cases, the appearance of the peak of the endogenous IAA level strongly correlated with the fate of the cells (Pasternak et al. 2002). A similar peak of endogenous IAA level was observed in immature zygotic embryos of sunflower induced to form somatic embryos (Charriére et al. 1999). An auxin surge has been reported to take place after fertilization in carrot zygotic embryos as well (Ribnicky et al. 2001). All these observations emphasize the significance of temporal changes in endogenous auxin levels in the expression of cellular totipotency. The experiments that served to manipulate endogenous abscisic acid (ABA) levels in tobacco mesophyll protoplasts further support the link between stress responses, endogenous hormone levels, and cell fate (Senger et al. 2001). In these experiments, different approaches were used to reduce cellular ABA levels in *Nicotiana plumbaginifolia* plants: a homozygous transgenic line was produced which overexpressed an antiabscisic acid single-chain variable fragment (ScFv) antibody, wild type cultures were treated with the ABA synthesis inhibitor fluoridone, and abscisic acid synthesis mutants (*aba1* and *aba2*) also used. In all cases, ABA deficiency disturbed the morphogenic response of protoplasts and inhibited the formation of preglobular embryoids, which could be reverted by exogenous ABA application. In the alfalfa leaf protoplast system, the small, cytoplasm-rich embryogenic competent cells were shown to have a higher vacuolar pH of 1 or 2 values than the elongated, differentiating cells (Pasternak et al. 2002). Plant cells have different types of vacuoles which exhibit different functions from the large lytic type of vacuoles characteristic of differentiated cells to small storage-type vacuoles more characteristic of meristematic cells (for reviews Wink 1993; Marty 1999; Fig. 3.8). Supposedly the large difference in the vacuolar pH of the two different cell types in the alfalfa leaf protoplast cultures (Fig. 3.8), is related to the differences in the vacuolar functions linked to the fate of these cells: elongated, differentiated cells have the large central lytic vacuoles with more acidic pH while the small dedifferentiated cells have several small storage-type vacuoles characterized by low transparency under light microscopy and strong staining with toulidene blue indicative of high protein content, likely as storage proteins (Pasternak et al. 2002; see also Figs. 3.7, 3.8). It is also important to note that the vacuoles, in addition to storing amino acids, sugars, etc., sequestering toxic ions and xenobiotics,

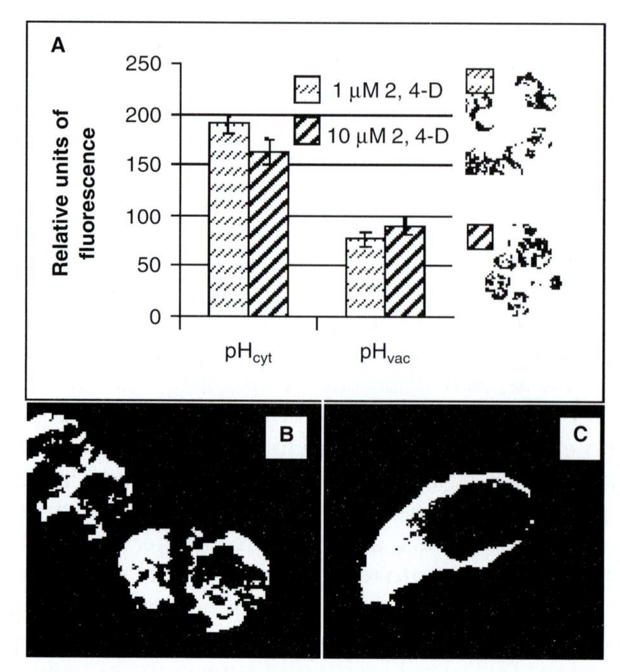

Fig. 3.8 Vacuole types of leaf protoplast-derived alfalfa cells. Alfalfa leaf protoplasts were cultured for three days in the presence of 1 and 10 μM 2,4-D and developed into elongated nonembryogenic and small, round embryogenic cell types as demonstrated in Fig. 3.6. In the embryogenic-type cells lower cytoplasmic and higher vacuolar pH could be detected after staining with the pH-dependent fluorescent dyes fluorescein diacetate and BCECF-AM respectively (A; for details see Pasternak et al. 2002). Indirect immunofluorescence staining using an antibody against the gamma-tonoplast intrinsic protein (Jauh et al. 1999) decorated many small vacuoles in embryogenic (B) and a large central vacuole in nonembryogenic (C) cells.

control the cell volume and regulate cytoplasmic ions and pH (for reviews, see Wink 1993; Marty 1999; Ratajczak 2000). It is assumed that vacuolar H^+ ATPase-driven osmotic uptake of water into the central vacuole facilitates cell expansion (Wink 1993). Gogarten et al. (1992) provided *in-vivo* proof of a direct link between vacuolar H^+ transport and morphogenesis: antisense mRNA mediated inhibition of the tonoplast ATPase in carrot resulted in reduced cell expansion. The *det3* mutant of *Arabidopsis* has organ-specific defects in cell elongation and growth due to a failure arresting the apical meristem (Schumacher et al. 1999). The *det3* gene has been identified to encode the C-subunit of the vacuolar H+-ATPase (Schumacher et al. 1999). It has also been shown recently that the δ-subunit of the vacuolar ATPase is also essential for embryonic development (Miura et al. 2003). These data support the direct involvement of V-ATPases in the control of cell elongation and meristematic activity (morphogenesis). Therefore it may be suggested that vacuolar functions are important with respect to the development of leaf protoplast-derived cells as well.

CONCLUDING REMARKS: NUDE CELLS = GOOD CELLS

To understand how plants develop and function, a wide range of experimental approaches must be employed. Some questions may be understood only at the level of the whole organism and others at the level of organs, tissues or single cells. For example, *in-vitro* cultures of dedifferentiated plant cells proved very useful experimental systems in understanding basic cellular functions such as the regulation of cell division (e.g. Magyar et al. 1993) but gave very limited scope for revealing the link between regulation of cell division and plant development. In contrast, the use of mutants or transgenic plants may highlight the developmental context of the activity of a certain gene, but cellular and biochemical studies are required to understand its exact function. Cultures of protoplasts isolated from differentiated plant cells may represent suitable experimental systems to link cellular and developmental levels in certain cases. In protoplast cultures, cells can preserve some of their differentiated functions but lose cellular connections and can be studied as independent entities. Single cells can be easily isolated and investigated and their fate can be individually followed. Being cultured in liquid medium, protoplasts can be readily treated with experimental agents and a more or less homogeneous response provoked. Due to this more or less synchronous response, protoplasts can also be handled as a homogeneous population of cells and used in biochemical and molecular studies. Protoplasts efficiently take up naked DNA that allows the transient expression of foreign genes and their functional analysis in single cells. Moreover, some specific questions such as dedifferentiation, cell wall synthesis, cell cycle reactivation, and cellular responses to stress can probably be best studied in these systems.

ACKNOWLEDGMENTS

Experiments on leaf protoplast-derived embryogenic alfalfa cells were supported by grants IC15-CT96-0906, OTKA T034818 and BIO-00062/2000. AF is also thankful for the "János Bólyai" Research Fellowship. The anti-γ Tonoplast intrinsic protein antibody was kindly provided by Prof. J.C. Rogers (Institute of Biological Chemistry, Washington State University, Pullman, WA, USA).

REFERENCES

Adler V, Yin ZM, Tew KD, Ronai Z (1999) Role of redox potential and reactive oxygen species in stress signaling. Oncogene 18:6104-6111.

Anand S, Prasad R (1989) Rise in intracellular pH is concurrent with 'start' progression of *Saccharomyces cerevisiae*. J. Gen. Microbiol 135:2173-9.

Bartel B (1997) Auxin biosynthesis. Ann. Rev. Plant Physiol Plant Molec Biol. 48:51-66.

Barthou H, Petitprez M, Briere C, Souvre A, Alibert G (1999) RGD-mediated membrane-matrix adhesion triggers agarose-induced embryoid formation in sunflower protoplasts. Protoplasma 206:143-151.

Belanger KD, Quatrano RS (2000) Polarity:the role of localized secretion. Curr. Opin. Plant Biol. 3:67-72.

Bergounioux C, Perennes C, Brown SC, Gadal P (1988) Cytometric analysis of growth-regulator-dependent transcription and cell cycle progression in *Petunia* protoplast cultures. Planta 175:500-505.

Berrios EF, Gentzbittel L, Mokrani L, Alibert G, Sarrafi A (2000) Genetic control of early events in protoplast division and regeneration pathways in sunflower. Theor. Appl. Genet. 101:606-612.

Bethke PC, Lonsdale JE, Fath A, Jones RL (1999) Hormonally regulated programmed cell death in barley aleurone cells. Plant Cell 11:1033-1045.

Blackman S, Yeung EC, Staves M. (2000). Effects of the disintegrin fragment, RGD, on somatic embryogenesis in *Daucus carota* L. Amer. Soci. Plant Physiologists Ann. Meeting, July, 2000. http://abstracts.aspb.org/pb2002/public/P44/0857.html.

Blackman SA, Miedema M, Yeung EC, Staves MP (2001) Effect of the tetrapeptide RGDS on somatic embryogenesis in *Daucus carota*. Physiol Plant 112:567-571.

Blaschek W, Koehler H, Semler U, Franz G (1982) Molecular weight distribution of cellulose in primary cell walls. Investigations with regenerating protoplasts, suspension cultured cells and mesophyll of tobacco. Planta 154:550-555.

Blervacq AS, Dubois T, Dubois J, Vasseur J (1995) First divisions of somatic embryogenic cells in Cichorium hybrid "474". Protoplasma 186:163-168.

Bögre L, Stefanov I, Ábrahám M, Somogyi I, Dudits D (1990) Differences in the responses to 2,4-dichlorophenoxyacetic acid (2,4-D) treatment between embryogenic and non-embryogenic lines of alfalfa. *In:* (HJJ Nijkamp, LHW, van der Plaas, and J. van Aartrijk J. (eds.). Progress in Plant Cellular and Molecular Biology. Kluwer Acad. Publ., Dordrecht, Netherlands, pp. 427–436.

Braam J (1999) If walls could talk. Curr. Opin. Plant Biol 2:521-524.

Brownlee C, Berger F, Bouget FY (1998) Signals involved in control of polarity, cell fate and developmental pattern in plants. Symp Soc Exp Biol 51:33-41.

Brudern A, Thiel G (1999) Effect of cell-wall-digesting enzymes on physiological state and competence of maize coleoptile cells. Protoplasma 209:246-255.

Butowt R, Niklas A, Rodriguez-Garcia MI, Majewska-Sawka A (1999) Involvement of JIM13-and JIM8-responsive carbohydrate epitopes in early stages of cell wall formation. J. Plant Res. 112:107-116.

Carle SA, Bates GW, Shannon TA (1998) Hormonal control of gene expression during reactivation of the cell cycle in tobacco mesophyll protoplasts. Plant Growth Regul. 17:221-230.

Caumont C, Petitprez M, Woynaroski S, Barthou H, et al. (1997) Agarose embedding affects cell wall regeneration and microtubule organization in sunflower hypocotyl protoplasts. Physiologia Plantarum 99:129-134.

Charriére F, Sotta B, Miginiac É, Hahne G (1999) Induction of adventitious or somatic embryos on *in vitro* cultured zygotic embryos of *Helianthus annuus*: Variation of endogenous hormone levels. Plant Physiol. Biochem. 37:751-757.

Cleary AL (2001) Plasma membrane-cell wall connections: roles in mitosis and cytokinesis revealed by plasmolysis of Tradescantia virginiana leaf epidermal cells. Protoplasma 215:21-34.

Cook R, Meyer Y (1981) Hormonal control of tobacco protoplast nucleic acid metabolism during *in vitro* culture. Planta 152:1-7.

Cosgrove DJ (1996) Plant cell enlargement and the action of expansins. Bioessays 18:533-540.

Cove DJ (2000) The generation and modification of cell polarity. J. Exp. Bot. 51:831-838.

Criqui MC, Parmentier Y, Derevier A, Shen WH, Dong AW, Genschik P (2000) Cell cycle-dependent proteolysis and ectopic over expression of cyclin B1 in tobacco BY2 cells. Plant J. 24:763-773.

Criqui MC, Plesse B, Durr A, Marbach J, (1992) Characterization of genes expressed in mesophyll protoplasts of *Nicotiana sylvestris* before the re-initiation of the DNA replicational activity. Mech Dev 38:121-132.

De Marco A, Roubelakis-Angelakis KA (1996a) The complexity of enzymic control of hydrogen peroxide concentration may affect the regeneration potential of plant protoplasts. Plant Physiol. 110:137-145.

De Marco A, RoubelakisAngelakis KA (1996b) Hydrogen peroxide plays a bivalent role in the regeneration of protoplasts. J. Plant Physiol. 149:109-114.

del Campillo E (1999) Multiple endo-1,4-beta-D-glucanase (cellulase) genes in Arabidopsis. Curr. Top. Dev. Biol. 46:39-61.

den Boer BG, Murray JA (2000) Triggering the cell cycle in plants. Trends Cell. Biol. 10:245-250.

Dijak M, Simmonds DH (1988) Microtubule organization during early direct embryogenesis from mesophyll protoplasts of *Medicago sativa L.* Plant Sci. 58:183-191.

Dudits D, Bögre L, Györgyey J (1991) Molecular and cellular approaches to the analysis of plant embryo development from somatic cells *in vitro*. J. Cell Sci. 99:475-484.

Fehér A, Dudits D (1994) Plant protoplasts for cell fusion and direct DNA uptake: culture and regeneration systems. *In:* IK, Vasil and TA, Thorpe (eds.), Plant Cell and Tissue Culture. Kluwer Acad. Publ. Dordrecht, Netherland pp. 71-118.

Fehér A, Pasternak T, Dudits D. (2003) Transition of somatic plant cells to an embryogenic state. Plant Cell Tissue and Organ Cult. 174: 201-228.

Fehér A, Pasternak T, Ötvös K, Miskolczi P, Dudits D (2002) Induction of embryogenic competence in somatic plant cells: a review. Biologia (Bratislava) 57:5-12.

Fleck J, Durr A (1980) Comparison of proteins synthetised in vivo and *in vitro* by mRNA of isolated protoplasts. Planta 148:453-454.

Fowler JE, Quatrano RS (1997) Plant cell morphogenesis: Plasma membrane interactions with the cytoskeleton and cell wall. Ann. Rev. Cell Dev. Biol. 13:697-743.

Frelin C, Vigne P, Ladoux A, Lazdunski M (1988) The regulation of the intracellular pH in cells from vertebrates. Eur J Biochem 174:3-14.

Gamaley IA, Klyubin IV (1999) Roles of reactive oxygen species: Signaling and regulation of cellular functions. Inter. Rev. Cytol.: Survey Cell Biol. 188:203-255.

Gaudino RJ, Pikaard CS (1997) Cytokinin induction of RNA polymerase I transcription in *Arabidopsis thaliana*. J Biol Chem 272:6799-6804.

Genschik P, Criqui MC, Parmentier Y, Derevier A, Fleck J (1998) Cell cycle-dependent proteolysis in plants: Identification of the destruction box pathway and metaphase arrest produced by the proteasome inhibitor MG132. Plant Cell 10:2063-2075.

Gogarten JP, Fichmann J, Braun Y, Morgan L, et al. (1992) The use of antisense mRNA to inhibit the tonoplast H+ ATPase in carrot. Plant Cell 4:851-864.

Gossett DR, Millhollon EP, Lucas MC, Banks SW, Marney MM (1994) The effects of NaCl on antioxidant enzyme-activities in callus- tissue of salt-tolerant and salt-sensitive cotton cultivars (*Gossypium-Hirsutum* L). Plant Cell Rep. 13:498-503.

Gregory PD, Horz W (1998) Life with nucleosomes: chromatin remodelling in gene regulation. Curr Opin Cell Biol 10:339-345.

Grosset J, Marty I, Chartier Y, Meyer Y (1990) mRNAs newly synthetised by tobacco mesophyll protoplasts are wound-inducible. Plant Mol Biol 15:485-496.

Hasezawa S, Nazaki H (1999) Role of cortical microtubules in the orientation of cellulose microfibril deposition in higher-plant cells. Protoplasma 209:98-104.

Hemerly AS, Ferreira P, de Almeida E, Van Montagu M, Engler G, Inze D (1993) cdc2a expression in *Arabidopsis* is linked with competence for cell division. Plant Cell 5:1711-1723.

Hernandez-Verdun D (1991) The nucleolus today. J Cell Sci 99 (Pt 3):465-471.

Hirt H, Pay A, Gyorgyey J, Bako L, et al. (1991) Complementation of a yeast cell cycle mutant by an alfalfa cDNA encoding a protein kinase homologous to p34cdc2. Proc. Natl. Acad. Sci. USA 88:1636-1640.

Ingvardsen C, Veierskov B (2001) Ubiquitin- and proteasome-dependent proteolysis in plants. Physiologia Plantarum 112:451-459.

Jamet E, Durr A, Parmentier Y, Criqui MC, Fleck J (1990) Is ubiquitin involved in the dedifferentiation of higher plant cells? Cell. Differ. Dev. 29:37-46.

Jauh GY, Phillips TE, Rogers JC (1999) Tonoplast intrinsic protein isoforms as markers for vacuolar functions. Plant Cell 11:1867-1882.

Jouanneau JP, Tandeau de Marsac (1973) Stepwise effects of cytokinin activity and DNA synthesis upon mitotic cycle events in partially synchronised tobacco cells. Exp Cell Res 77:167-174.

Jurgens G, Grebe M, Steinmann T (1997) Establishment of cell polarity during early plant development. Curr Opin Cell Biol 9:849-52.

Kato N, Esaka M (2000) Expansion of transgenic tobacco protoplasts expressing pumpkin ascorbate oxidase is more rapid than that of wild-type protoplasts. Planta 210:1018-1022.

Klein AS, Montezinos D, Delmer DP (1981) Cellulose and 1,3-glucan synthesis during the early stages of wall regeneration in soybean protoplasts. Planta 152:105-114.

Knox JP (1997) The use of antibodies to study the architecture and developmental regulation of plant cell walls. Int. Rev. Cytol. 171:79-120.

Kohorn BD (2000) Plasma membrane-cell wall contacts. Plant Physiol. 124:31-38.

Kohorn BD (2001) WAKs; cell wall associated kinases. Curr. Opin. Cell. Biol. 13:529-533.

Laureys F, Dewitte W, Witters E, Van Montagu M, Inze D, Van Onckelen H (1998) Zeatin is indispensable for the G2-M transition in tobacco BY-2 cells. FEBS Lett. 426:29-32.

Levina NN, Heath IB, Lew RR (2000) Rapid wound responses of *Saprolegnia ferax* hyphae depend upon actin and Ca2+-involving deposition of callose plugs. Protoplasma 214:199-209.

Leyser O (2001) Auxin signalling: the beginning, the middle and the end. Curr. Opin. Plant Biol. 4:382-386.

Luo Y, Koop HU (1997) Somatic embryogenesis in cultured immature zygotic embryos and leaf protoplasts of *Arabidopsis thaliana* ecotypes. Planta 202:387-396.

Magyar Z, Bako L, Bogre L, Dedeoglu D, Kapros T, Dudits D (1993) Active cdc2 genes and cell cycle phase-specific cdc2-related kinase complexes in hormone stimulated alfalfa cells. Plant J 4:151-161.

Maheshwari SC, Gill R, Maheshwari N, Gharyal PK (1986) Isolation and regeneration of protoplasts from higher plants. *In:* J, Reinert and H, Binding (eds.), Differentiation of Protoplasts and of Transformed Plant Cells Springer-Verlag, Berlin:pp. 3-36.

Majewska-Sawka A, Fernandez MC, M'rani-Alaoui M, Munster A, Rodriguez-Garcia MI (2002) Cell wall reformation by pollen tube protoplasts of olive (*Olea europaea* L.): structural comparison with the pollen tube wall. Sex. Plant Reprod. 15:21-29.

Majewska-Sawka A, Nothnagel EA (2000) The multiple roles of arabinogalactan proteins in plant development. Plant Physiol. 122:3-9.

Malinowski R, Filipecki M (2002) The role of cell wall in plant embryogenesis. Cell. Molec. Biol. Lett. 7:1137-1151.

Marty F (1999) Plant vacuoles. Plant Cell 11:587-600.

Marty I, Brugidou C, Chartier Y, Meyer Y (1993) Growth-related gene expression in *Nicotiana tabacum* mesophyll protoplasts. Plant J. 4:265-278.

McCabe PF, Valentine TA, Forsberg LS, Pennell RI (1997) Soluble signals from cells identified at the cell wall establish a developmental pathway in carrot. Plant Cell 9:2225-2241.

Meszaros T, Miskolczi P, Ayaydin F, Pettko-Szandtner A, et al. (2000) Multiple cyclin-dependent kinase complexes and phosphatases control G(2)/M progression in alfalfa cells. Plant Molec. Biol. 43:595-605.

Meyer P (2001) Chromatin remodelling. Curr. Opin. Plant Biol. 4:457-462.

Miao GH, Hong Z, Verma DP (1993) Two functional soybean genes encoding p34cdc2 protein kinases are regulated by different plant developmental pathways. Proc. Natl. Acad. Sci USA 90:943-947.

Miura GI, Froelick GJ, Marsh DJ, Stark KL, Palmiter DP (2003) The d subunit of the vacuolar ATPase (*Atp6d*) is essential for embryonic development. Transgenic Res. 12:131-133.

Muramatsu M, Busch H (1965) Studies on the nuclear and nucleolar ribonucleic acid of regenerating rat liver. J. Biol. Chem. 240:3960-3966.

Nagata T, Ishida S, Hasezawa S, Takahashi Y (1994) Genes involved in the dedifferentiation of plant cells. Int. J. Dev. Biol. 38:321-327.

Nakano Y, Asada K (1987) Purification of ascorbate peroxidase in spinach chloroplasts: its inactivation in ascorbate-depleted medium and reactivation by monodehydroascorbate radical. Plant Cell Physiol. 28:131-140.

Negrutiu I, Shillito R, Potrykus, I, Biasini, G, Sala F (1987) Hybrid genes in the analysis of transformation conditions: I. Setting up a simple method for direct gene transfer to protoplasts. Plant Mol. Biol. 8:363-373.

Nomura K, Komamine A (1985) Identification and isolation of single cells that produce somatic embryos at a high frequency in a carrot cell suspension culture. Plant Physiology 79:988-991.

Nomura K, Komamine A (1995) Physiological and biological aspects of somatic embryogenesis TA Thorpe (ed.). *In: In vitro* embryogenesis in plants. Kluwer Acad. Publ. pp. 249-266.

Papadakis AK, Roubelakis-Angelakis KA (1999) The generation of active oxygen species differs in tobacco and grapevine mesophyll protoplasts. Plant Physiol. 121:197-206.

Papadakis AK, Siminis CI, Roubelakis-Angelakis KA (2001) Reduced activity of antioxidant machinery is correlated with suppression of totipotency in plant protoplasts. Plant Physiol 126:434-444.

Pasternak T, Miskolczi P, Ayaydin F, Mészáros T, Dudits D, Fehér A (2000) Exogenous auxin and cytokinin dependent activation of CDKs and cell division in leaf protoplast-derived cells of alfalfa. Plant Growth Reg. 32:129-141.

Pasternak TP, Prinsen E, Ayaydin F, Miskolczi P, et al. (2002) The role of auxin, pH and stress in the activation of embryogenic cell division in leaf protoplast-derived cells of alfalfa. Plant Physiol. 129:1807-1819.

Pennell RI, Cronk QC, Forsberg LS, Stohr C, et al. (1995) Cell-context signalling. Phil. Trans. R. Soc. Lond. B Biol. Sci. 350:87-93.

Perry SE, Lehti MD, Fernandez DE (1999) The MADS-domain protein AGAMOUS-like 15 accumulates in embryonic tissues with diverse origins. Plant Physiol. 120:121-30.

Petitprez M, Briere C, Borin C, Kallerhoff J, Souvre A, Alibert G (1995) Characterization of Protoplasts from Hypocotyls of *Helianthus annuus* in Relation to their tissue origin. Plant Cell Tiss. Org. Cult. 41:33-40.

Phillips R, Darrell NJ (1988) A simple technique for single-cell cloning of crown gall tumor tissue: *Petunia* protoplast regeneration without exogenous hormones. Plant Physiol 133:447-451.

Ratajczak R (2000). Structure, function and regulation of the plant vacuolar H(+)-translocating ATPase. Biochim. Biophys. Acta 1465:17-36.

Redig P, Shaul O, Inze D, Van Montagu M, Van Onckelen H (1996) Levels of endogenous cytokinins, indole-3-acetic acid and abscisic acid during the cell cycle of synchronized tobacco BY-2 cells. FEBS Lett 391:175-180.

Ribnicky DM, Cohen JD, Hu WS, Cooke TJ (2001) An auxin surge following fertilization in carrots: a mechanism for regulating plant totipotency. Planta. 214: 505-509.

Rock KL, Gramm C, Rothstein L, Clark K, et al. (1994) Inhibitors of the proteasome block the degradation of most cell proteins and the generation of peptides presented on MHC class I molecules. Cell 78:761-771.

Rogg LE, Bartel B (2001) Auxin signaling. derepression through regulated proteolysis. Dev Cell. 604 (1):595-604.

Russell SD (1993) The egg cell: Development and role in fertilization and early embryogenesis. Plant Cell 5:1349-1359.

Scheres B, Benfey PN (1999) Asymmetric cell division in plants. Ann. Rev. Plant Physiol. and Plant Molec. Biol. 50:505-537.

Schumacher K, Vafeados D, McCarthy M, Sze H, Wilkins T, Chory J (1999) The Arabidopsis det3 mutant reveals a central role for the vacuolar H(+)-ATPase in plant growth and development. Genes Dev 13:3259-70.

Senger S, Mock HP, Conrad U, Manteuffel R (2001) Immunomodulation of ABA function affects early events in somatic embryo development. Plant Cell Rep. 20:112-120.

Sheen J (2001) Signal transduction in maize and Arabidopsis mesophyll protoplasts. Plant Physiol. 127:1466-1475.

Showalter AM (2001) Arabinogalactan-proteins: structure, expression and function. Cell. Molec. Life Sci. 58:1399-1417.

Sim GE, Loh CS, Goh CJ (1988) Direct somatic embryogenesis from protoplasts of *Citrus mitis* Blanco. Plant Cell Rep 7:418-420.

Siminis CI, Kanellis AK, Roubelakis-Angelakis KA (1993) Differences in protein synthesis and peroxidase isoenzymes between recalcitrant and regenerating protoplasts. Physiologia Plantarum 87:263-270.

Song J, Sorensen EL, Liang GH (1990) Direct embryogenesis from single mesophyll protoplasts in alfalfa (*Medicago sativa* L.). Plant Cell Rep. 9:21-25.

Sotiriou P, Spyropoulos CG (2002) Cell wall regeneration and galactomannan biosynthesis in protoplasts from carob. Plant Cell Tiss. Org. Cult. 71:15-22.

Souter M, Lindsey K. (2000) Polarity and signalling in plant embryogenesis. J. Exp. Bot. 51:971-983.

Stals H, Inze D (2001) When plant cells decide to divide. Trends Plant Sci 6:359-364.

Steffens B, Feckler C, Palme K, Christian M, Bottger M, Luthen H (2001) The auxin signal for protoplast swelling is perceived by extracellular ABP1. Plant J. 27:591-599.

Swann K, Whitaker MJ (1990) Second messengers at fertilization in sea-urchin eggs. J. Reprod. Fertil. Suppl. 42:141-53.

Tao W, Verbelen JP (1999) Switching on and off cell division and cell expansion in cultured mesophyll protoplasts of tobacco. Plant Sci. 116: 107-115.

Taylor JE, Abram B, Boorse G, Tallman G (1998) Approaches to evaluating the extent to which guard cell protoplasts of *Nicotiana glauca* (tree tobacco) retain their characteristics when cultured under conditions that affect their survival, growth, and differentiation. J. Exp. Bot. 49:377-386.

Tepfer DA, Fosket DE (1978) Hormone-mediated translational control of protein synthesis in cultured cells of *Glycine max*. Dev. Biol. 62:486-497.

Thomas C, Bronner R, Molinier J, Prinsen E, Van Onckelen H, Hahne G (2002) Immuno-cytochemical localization of indole-3-acetic acid during induction of somatic embryogenesis in cultured sunflower embryos. Planta 215:577-583.

Toonen MAJ, Hendriks T, Schmidt EDL, Verhoeven HA, van Kammen A, de Vries SC (1994) Description of somatic-embryo forming single cells in carrot suspension cultures employing video cell tracking. Planta 194:565-572.

Trehin C, Planchais S, Glab N, Perennes C, Tregear J, Bergounioux C (1998) Cell cycle regulation by plant growth regulators: involvement of auxin and cytokinin in the re-entry of *Petunia* protoplasts into the cell cycle. Planta 206:215-224.

Tsukiyama T, Wu C (1997) Chromatin remodeling and transcription. Curr. Opin. Genet. Deve. 7:182-191.

Udvardy A (1996) The role of controlled proteolysis in cell-cycle regulation. Eur. J. Biochem. 240:307-313.

Valente P, Tao W, Verbelen JP (1998) Auxins and cytokinins control DNA endoreduplication and deduplication in single cells of tobacco. Plant Sci. 134:207-215.

Vallee N, Briere C, Petitprez M, Barthou H, Souvre A, Alibert G (1997) Studies on ion channel antagonist-binding sites in sunflower protoplasts. FEBS Lett. 411:115-118.

Vallee N, Briere C, Petitprez M, Barthou H, Souvre A, Alibert G (1999) Cytolocalization of ion-channel antagonist binding sites in sunflower protoplasts during the early steps of culture. Protoplasma 210:36-44.

Van Breusegem F, Vranova E, Dat JF, Inze D (2001) The role of active oxygen species in plant signal transduction. Plant Sci. 161:405-414.

vanAmstel TNM, Kengen HMP (1996) Callose deposition in the primary wall of suspension cells and regenerating protoplasts, and its relationship to patterned cellulose synthesis. Revue Canadienne de Botanique 74:1040-1049.

Vardi A, Spiegel-Roy P, Galun E (1987) Plant regeneration from citrus protoplasts: variability in methodological requirements among cultivars and species. Theor Appl Genet 62:171-176.

Varga-Weisz PD, Becker PB (1998) Chromatin-remodeling factors: Machines that regulate? Curr. Opin. Cell Biol. 10:346-353.

Vernet T, Fleck J, Durr A, Fritsch C, Pinck M, Hirth L (1982) Expression of the gene coding for the small subunit for ribulose biphosphate carboxylase during differentiation of tobacco plant protoplasts. Eur. J. Biochem. 126:489-494.

Vissenberg K, Quelo AH, Van Gestel K, Olyslaegers G, Verbelen JP (2000) From hormone signal, via the cytoskeleton, to cell growth in single cells of tobacco. Cell Bio. Inter. 24:343-349.

Vranova E, Inze D, Van Breusegem F (2002) Signal transduction during oxidative stress. J. Exp. Bot. 53:1227-1236.

Whitaker MJ (1990) Cell cycle control proteins are second messenger targets at fertilization in sea-urchin eggs. J. Reprod. Fertil. Suppl. 42:199-204:199-204.

Wink M (1993) The plant vacuole: a multifunctional compartment. J. Exp. Bot. 44:231-246.

Wojtaszek P (2001) Organismal view of a plant and a plant cell. Acta Biochim Pol 48:443-451.

Xu XH, Briere C, Vallee N, Borin C, et al. (1999) *In-vivo* labeling of sunflower embryonic tissues by fluorescently labeled phenylalkylamine. Protoplasma 210:52-58.

Yu HJ, Moon MS, Lee HS, Mun JH, Kwon YM, Kim SG (1999) Analysis of cDNAs expressed during first cell division of petunia petal protoplast cultures using expressed sequence tags. Molec. Cells. 9:258-264.

Zhang K, Letham DS, John PC (1996) Cytokinin controls the cell cycle at mitosis by stimulating the tyrosine dephosphorylation and activation of p34cdc2-like H1 histone kinase. Planta 200:2-12.

Zhao J, Morozova N, Williams L, Libs L, Avivi Y, Grafi G (2001) Two phases of chromatin decondensation during dedifferentiation of plant cells: Distinction between competence for cell fate switch and a commitment for S phase. J. Biol. Chem. 276: 22772-22778.

4

Protoplasts: Consequences and Opportunities of Cellular Nudity

K. Peter Pauls
Department of Plant Agriculture
Ontario Agricultural College
University of Guelph,
Guelph, Ontario, Canada N1G 2W1

INTRODUCTION

The discovery that naked plant cells, called protoplasts, could be produced in large quantities by simple treatments of somatic tissues with cell wall-degrading enzymes (Cocking 1960) opened an enormous range of uses for plant cells in basic and applied studies. In the 40 years since this seminal work, protoplasts have been extensively used to study many aspects of cell function and for genome manipulation by cell fusion and DNA uptake.

EARLY DAYS AND MILESTONES

Ironically, in the first 10 years after large-scale production of protoplasts was achieved, research focused on establishing culture systems for these fragile cells (Fig. 4.1) that would support the regrowth of their cell walls and allow them to reinitiate cell division (Fig. 4.2). The media and culture systems devised in these experiments (Murashige and Skoog 1962; Gamborg et al. 1968; Schenk and Hildebrandt 1972; Kao and Michayluk 1975) form the basis of most modern protoplast culture protocols (Nagata and Bajaj 2001).

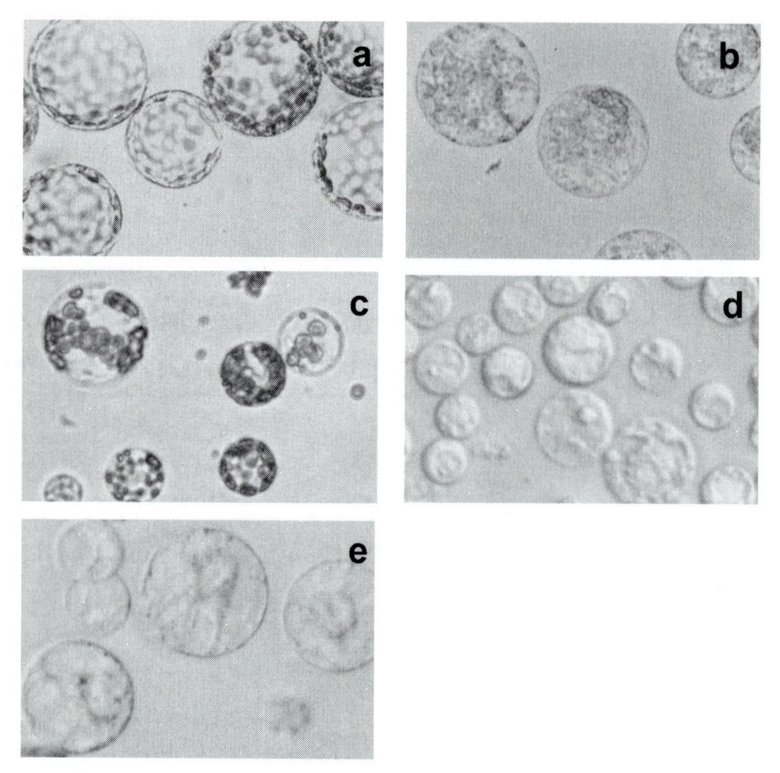

Fig. 4.1 Protoplasts from various tissues. (a) Diploid *B. napus* mesophyll protoplasts. (b) Diploid hypocotyl *B. napus* protoplasts. (c) Haploid *B. napus* stem peel protoplasts. (d) Haploid *B. napus* root protoplast. (e) Tetraploid alfalfa (*Medicago sativa*) suspension culture protoplasts.

Tissue culture research culminated in the regeneration of entire plants from tobacco protoplasts (Takebe et al. 1971). In time, procedures were developed to allow large numbers of plants to be recovered from protoplast cultures of a wide range of plant species. However, for many years low frequency or no regeneration was observed for most monocot protoplasts (Potrykus 1990). This limitation was overcome incrementally by identifying appropriate explants and genotypes for protoplast culture and by optimizing media and culture conditions. Currently, plants can be regenerated from more than 300 species, including genotypes of the most important cereal grains (Roest and Gilisen 1989, 1993).

Reports of somaclonal variants occurring at frequencies of several percent in populations of plants regenerated from protoplasts were disturbing to early researchers who were interested in using protoplast culture to produce large numbers of identical plants (Secor and Shepard

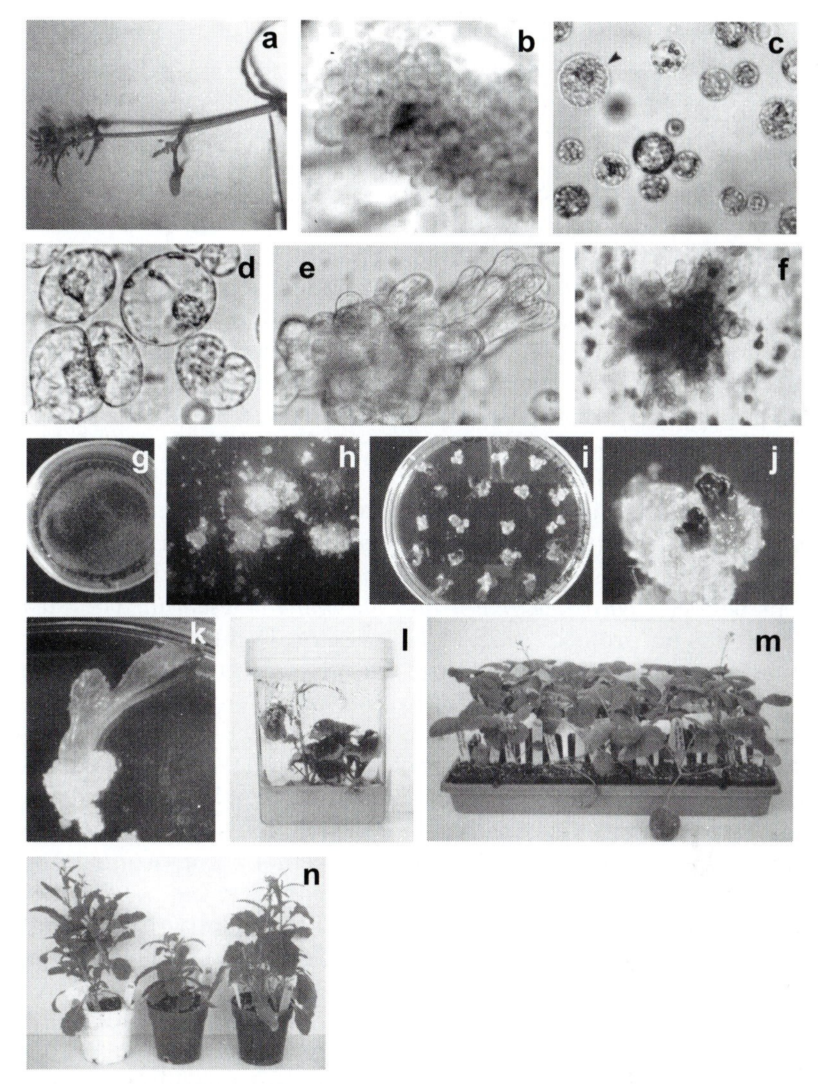

Fig. 4.2 Regeneration from *B. napus* haploid stem protoplasts. (a) Stem explants from haploid plants used to isolate stem peel protoplasts. (b) Stem peel explants 24 h after treatment with 1.5 % cellulase and 3% macerase, 10 % mannitol and CPW salts showing release of protoplasts. (c) Haploid stem peel protoplasts. (d) First cell divisions induced in protoplasts 3-5 days after culture in VN medium (Chuong et al. 1987). (e-f) Stages in microcallus production between 5 and 20 days in culture. (g) microcalli derived from haploid stem protoplasts cultured in a liquid film on agarose (Chuong et al. 1987) (h) Microcalli developing in liquid VN medium plated on ½ strength VN plus 0.3% agarose. (i-l) Shoot regeneration from protoplast derived callus cultured on 2N medium. (m) Rooting plantlets regenerated from protoplasts. (n) Haploid plants regenerated from stem peel protoplast.

1981). However, others suggested that the variation arising from protoplast may be useful for plant improvement (Larkin and Scowcroft 1981) and some variants with resistance to diseases or abiotic stress were recovered from protoplast cultures (Evans and Sharp 1983; Miller et al. 1985; Shahin and Spivey 1986; Barden et al. 1986; Smith and Murakishi 1987).

Studies were conducted to define the process leading to somaclonal variation and to reduce the incidence of mutants arising from the cultural procedure. An experiment with tobacco protoplasts provided important information on the source of variation observed in tissue culture-derived plants (Lorz and Scowcroft 1983). This experiment confirmed that most of the variation was induced by the cultural procedure. Several mechanisms for somaclonal variation have been proposed including activation of transposon activity in tissue cultures (Walbot and Cullis 1985). Hirochika (1993) observed an increase in transcription from a tobacco retrotransposon during protoplast formation.

The potential for using protoplasts to modify plant genomes was the major practical justification for work on protoplasts in the 1960s and 1970s. In 1972 Carlson et al. produced the first somatic hybrid plant from fusions between *Nicotiana glauca* and *Nicotiana langsdorffii*. They used a high salt fusion protocol (0.25 M $NaNO_3$ for 30 min) and selection on media that supported the growth of amphiploid cells but not cells from the individual donor species. Thirty-three calli were produced from 2×10^7 cells and shoots regenerated from the calli were grafted onto *N. glauca* stems. The biochemical and morphological characteristics of the shoots from the fusion experiments were identical to sexually produced amphiploids between these two species. In addition, the somatic chromosome number (42) was the summation of the parental chromosome numbers (24 + 18), which was distinct from whole ploidy changes in cells of either parent. In a note added during proofing of the manuscript the authors reported that the somatic hybrid produced flowers and set fertile seed. In addition to being the first report of the production of a somatic hybrid, this publication addressed all the steps common to protoplast fusion studies including: protoplast production, fusion induction, hybrid selection, protoplast culture, plant regeneration and hybridity verification (Li et al. 1996).

Significant improvements in the frequency of protoplast fusion were achieved by utilizing polyethylene glycol as a fusigenic agent (Kao and Michayluk 1974; Wallin et al. 1974; Fig. 4.3). Melchers et al. (1978) were the first to report the recovery of intergenic somatic hybrid plants from protoplast fusions induced by polyethylene glycol between potato and tomato protoplasts. There were early reports that this procedure was successful in transferring chilling tolerance into tomato from potato (Smillie et al. 1979). Also in this period, the first asymmetric somatic

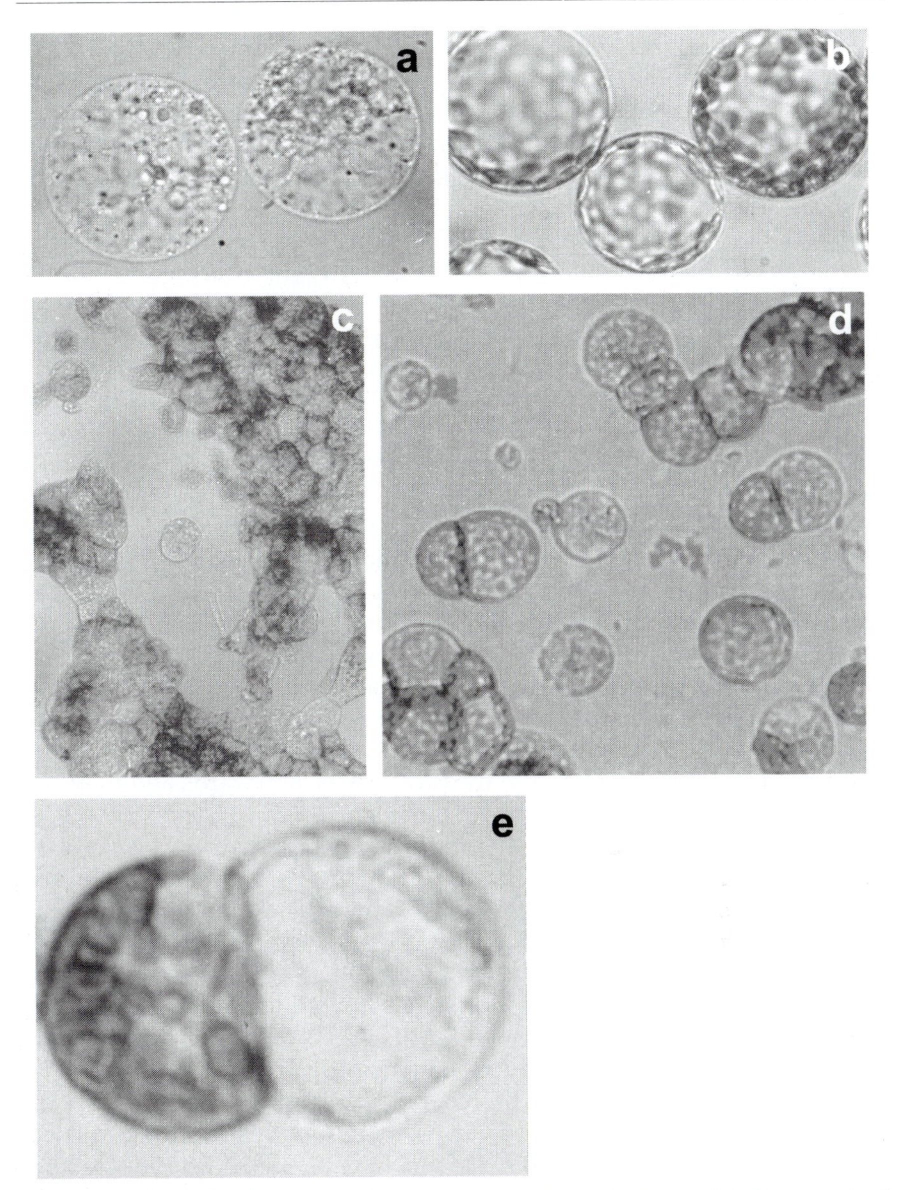

Fig. 4.3 Protoplast fusion. (a) *B. napus* hypocotyl protoplasts. (b) *B. napus* leaf protoplasts (c) Aggregation of leaf mesophyll protoplasts and hypocotyl protoplasts induced by 30% polyethylene glycol (PEG) 8000 in 9% mannitol, 30 mM $CaCl_2$. (d) Fusion mixture of *B. napus* leaf protoplasts and hypocotyl protoplasts after dilution showing cells in various stages of aggregation and fusion. (e) Early stage cell fusion between a *B. napus* leaf protoplast and an alfalfa cell culture protoplast.

hybrid/cybrid plants were produced to transfer cytoplasmic genomes or create new combinations of chloroplasts and mitochondria in plant cytoplasms (Zelcer et al. 1978; Aviv et al. 1980). For these experiments one of the fusion partners was treated with ionizing radiation to reduce its contribution to the nuclear makeup of the hybrid and increase the chances of obtaining an asymmetric fusion product.

Interkingdom (plant/animal) cell fusions were reported soon after development of the PEG fusion system between tobacco and human (HeLa) cells (Dudits et al. 1976) and *Drosophila* cells and soybean protoplasts (Hadlaczky et al. 1980). The hybrids produced in these experiments were maintained as cultures but no plants were regenerated. More recently, tobacco plants were regenerated from PEG mediated fusions between mouse spleen cells and tobacco mesophyll cells that produced mouse immunoglobulin gamma-3 heavy and lambda light chains (Makonkawkeyoon et al. 1995).

Other approaches to protoplast fusion have been developed, including electrofusion (Senda et al. 1979; Zimmerman and Scheurich 1981). In an electrofusion apparatus an oscillating field causes the cells to migrate to the two electrodes by diaelectrophoresis and positions them along field lines in pairs or "pearl chains". A strong direct current pulse breaks down the membranes between adjacent protoplasts and allows them to fuse, whereupon the membranes spontaneously reform. The developers of the electric fusion systems suggested that it was superior to chemical systems because there is no need to wash the protoplasts afterwards and there are no residual chemical effects.

Because success in a somatic hybridization experiment always involves striking a balance between a high fusion frequency and maintaining cell viability the percentage of heterokaryons in a fusion mixture is usually much less than 10%. This has necessitated the development of methods for enriching cultures for fusion products, either by *in vitro* selection or physical sorting. The selection approach is limited to plant materials that have different resistances determined by their genetic makeup (Donaldson et al. 1995) or deficiencies induced by chemical treatments that can be rescued by the fusion partner (Bottcher et al. 1989). In some cases hybrid vigor has been observed in the fusion products and this was used to select heterokaryons (Grosser et al. 1988; Taguchi et al. 1993). Physical sorting protocols utilize micromanipulators (Mendis et al. 1991), flow cytometers (Galbraith and Galbraith 1979; Harkins and Galbraith 1987; Pauls and Chuong 1987) or computer-driven image recognition and pipetting systems (Schweiger et al. 1987). The potential advantage of the sorting methods is that they can be applied to any plant material, usually after staining the cells with fluorescent dyes or by choosing cell types that autofluoresce (Pauls and Chuong 1987).

Direct DNA uptake by protoplasts (Ohyama 1983) has been developed for the production of transgenic plants and to study the behavior of genes and promoters during transient expression. The green fluorescent protein (GFP) from jellyfish has been an important tool for transient expression studies (Galbraith et al. 1995; Sheen et al. 1995; Chiu et al. 1996) because it can be used with live cells (Fig. 4.4).

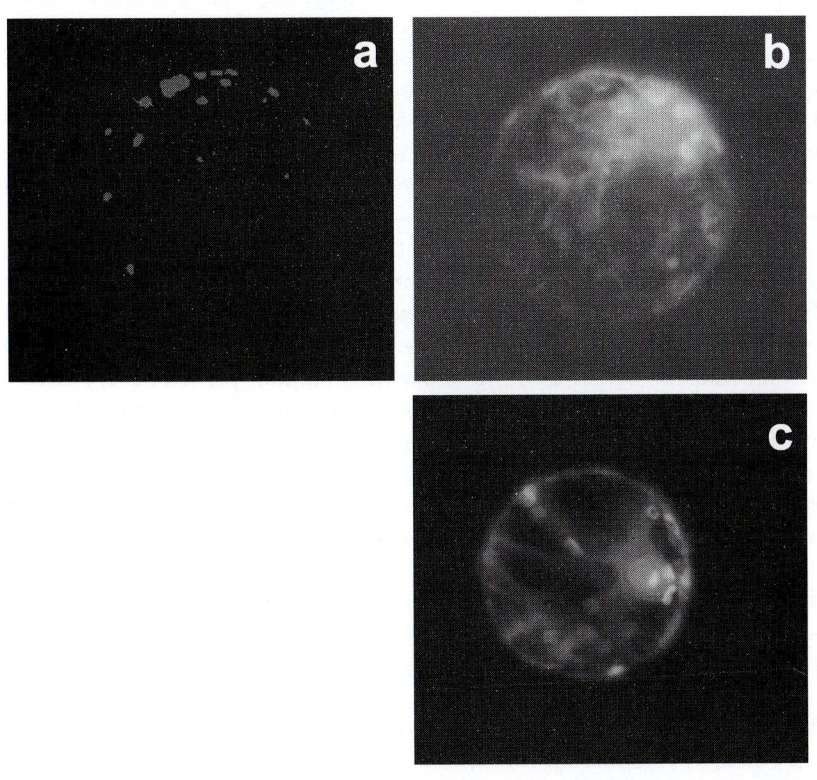

Fig. 4.4 Tobacco mesophyll protoplasts with and without green fluorescent protein(GFP). (a) Nontransformed protoplast showing red fluorescence from protoplasts. (b,c) Protoplasts from a plant stably transformed with 35S GFP showing intense green fluorescence in cytoplasmic strands and around the nucleus.

PROTOPLAST ISOLATION AND CULTURE PROTOCOLS

Many species-specific and variety-specific protocols for protoplast isolation and culture have been developed but some general procedures are useful starting points when developing new protocols (Banks and Evans 1976; Glimelius et al. 1986; Menczel et al. 1981; Barsby et al. 1986). Greenhouse-grown plants or *in vitro* cultured material are typical explants

(Fig. 4.5). For leaf material pretreatment for 24 to 48 h in darkness often improves recovery. Greenhouse-grown material must be surface sterilized, usually with a low percentage of hypochlorite followed by extensive washing with sterile water. In some cases the epidermis can be abraded by rubbing with fine carborundum powder or with a fine nylon brush. Explants are cut into segments or scored into strips (0.5 to 2 mm) with a scalpel and incubated for 1 h in plasmolysing buffer containing 0.3M sorbitol and 0.05 M $CaCl_2$. Cell walls are digested by treatment with 1% cellulase and 0.1% mazerocyme in K3 medium (Nagy and Maliga 1976) containing 0.4 M sucrose and incubation for 16 to 18 h at 20 – 22 °C in darkness on a rocking tray at low speed or for shorter periods (4–5 h) without agitation for more sensitive material. In the latter case agitation to release the protoplast just before purification may be useful. The enzyme-treated material is filtered through a nylon mesh (50 μm pore size) to separate nondegraded material and the filtrate containing the protoplasts is mixed with an equal volume of a salt solution with 16% sucrose (CPW16; Banks and Evans 1976). The mixed solutions are centrifuged in a swingout rotor at 100 g for 7 min. Intact protoplasts will float to the surface and can be gently removed with a Pasteur pipette. The protoplast suspension is diluted about 10 times with wash solution containing 0.5 M sorbitol (Menczel et al. 1981) and centrifuged at 100 g for 3 min. Flotation and washing steps may be repeated and the protoplasts finally pelleted by centrifugation in culture medium (KM8p, Glimelius et al. 1986) and cultured at a density of $2.5 - 5.0 \times 10^4$ cells/ml^{-1}.

A variety of protoplast culture formats have been developed including: simple liquid culture systems, droplet cultures, hanging droplets, and agarose embedding procedures (Shillito et al. 1983; Horita et al. 2002) that may include feeder or nurse cultures (Kyozuka et al. 1989). Protoplast embedding techniques with low melting agarose allow media to be easily changed and avoid the labor involved in colony picking that individual cell culture systems require. Most often the culture protocols are divided into cell proliferation and regeneration phases. Proliferation media typically contain higher concentrations of auxins than cytokinins, whereas the reverse is generally true for regeneration media.

Fig. 4.5 General protoplast isolation and culture protocol. Explants can be made from greenhouse-grown or *in vitro* grown material. Segments are scored into 0.5 to 2.0 mm strips and incubated with 1% cellulase and 0.1% mazerocyme in medium containing 0.4 M sucrose for 5 to 18 h at 20 – 22 °C in darkness with or without gentle agitation. The enzyme-treated material is filtered through a nylon mesh (50 :m pore size) to separate undegraded material and mixed with an equal volume of the salt solution containing 16% sucrose. The mixed solutions are centrifuged in a swingout rotor at 100 × g for 7 min and intact protoplasts are

Contd.

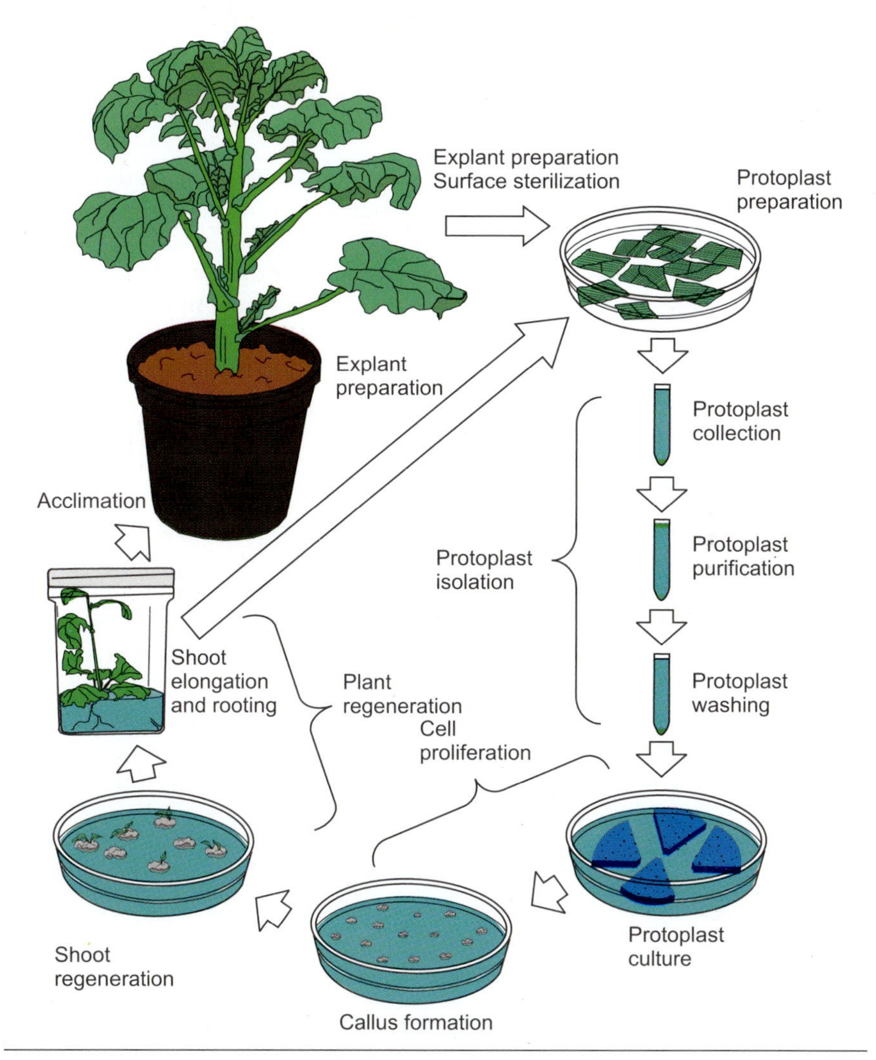

Fig. 4.5 Contd.

collected from the top with a Pasteur pipette. The protoplast suspension is diluted tenfold with a salt solution containing 0.5 M sorbitol and the protoplasts are pelleted by centrifugation at 100 g for 3 min. The protoplasts are finally pelleted by centrifugation in culture medium and cultured at a density of $2.5 - 5.0 \times 10^4$ cells ml^{-1} in culture medium. Protoplasts can be embedded in agarose by mixing with a low temperature agarose solution. The embedded cultures are cut into quarters, transferred to a larger petri plate and covered with liquid culture media containing a mixture of auxins and cytokinins to induce cell division and callus proliferation. Regeneration is induced by transfer of calli to media with higher cytokinin/auxin ratios. Shoot elongation and rooting reestablishes the entire plant.

APPLICATIONS

1. Increased Understanding: Cell Physiology Studies with Plant Protoplasts

Plant protoplast preparations are excellent systems for studying cell function because they retain their physiological activities and sensitivities, including: the ability to photosynthesize and respire and to respond to ions, hormones, elicitors and light, and to maintain their membrane potentials. In these experiments protoplast functions have been monitored by patch clamping (Moran et al. 1984; Schroeder et al. 1984; Maathuis and Sanders 1994; Cho and Spalding 1996; Bauer et al. 2000; Downey et al. 2000; Hamilton et al. 2000; Pandey 2002), microscopy (Asai et al. 2002) and flow cytometry (Harkins et al. 1990; Galbraith et al. 1999; Hagenbeek 2001). Alternatively, the properties of protoplasts from mutants or stable transgenics have been compared (Assmann 2001; Murata et al. 2001; Allen et al. 2002; Coursol et al. 2003; Stoelzle et al. 2003).

The endogenous α-amylase gene in barley aleurone protoplasts has been shown to be regulated by ABA and GA in a manner analogous to that observed in the intact seed (Jacobsen and Beach 1985) and blue light stimulates ion movements in broad bean guard cell protoplasts (Assman et al. 1985) and *Arabidopsis* mesophyll cells (Stoelzle et al. 2003). Patch-clamp studies of protoplasts have led to new understandings of the functions of ion channels in plants and the regulation of ion movements by light, stress, and hormones (Moran et al. 1984; Schroeder et al. 1984, 2001; Maathuis and Sanders 1994; Cho and Spalding 1996; Bauer et al. 2000; Downey et al. 2000; Hamilton et al. 2000). Protoplasts have also been used to study calcium signaling in plants (Gilroy and Jones 1992; Trewavas 1999; Pei et al. 2000; Pandey et al. 2002).

2. Transient Protoplast Expression Systems

Because large numbers of homogeneous and physiologically active protoplasts can be transfected with nucleic acids by simple chemical or physical methods they have been extensively used to examine gene expression, protein targeting, and protein-protein interactions. More than 95% of freshly isolated, intact *Arabidopsis* or maize mesophyll protoplasts retain their viability for more than 48 h in simple mannitol solutions (Sheen 2001), which is two to four times longer than required to complete most transient expression assay experiments. The information obtained from transient protoplast expression systems has been shown to be highly relevant to understanding tissue and whole plant function (Sheen 2001).

The first report of the introduction of nucleic acids into protoplasts was a demonstration that tobacco mesophyll protoplasts could be infected by tobacco mosaic virus RNA (Aoki and Tabeke 1969). Large-scale use of protoplast transient expression systems occurred after the development of efficient methods for introducing plasmid DNA into protoplasts using PEG (Krens et al. 1982; Potrykus et al. 1985; Koop et al. 1996), electroporation (Fromm et al. 1985; Hauptmann et al. 1987; Jones 1995), microinjection (Crossway et al. 1986; Hillmer et al. 1992), or a laser microbeam (Weber et al. 1989). Transfection efficiencies as high as 90% have been reported for tobacco mesophyll protoplasts (Sheen 2001). Development of good reporter genes including β-glucoronidase (Jefferson et al. 1987), firefly luciferase (Luehrsen et al. 1992) and green fluorescent protein (Sheen et al. 1995; Chiu et al. 1996) was also important to incremental use of protoplasts for gene expression studies. Derivatives of GFP have been developed that fluoresce at other wavelengths (Stewart 2001) which may allow simultaneous measurements of several parameters with protoplasts in future.

Early work in this area used fusions of plant promoters and reporter genes to examine gene regulation in a variety of plants. For example, the promoter from the wheat *Em* gene was shown to be responsive to ABA when introduced into rice protoplasts (Marcotte Jr et al. 1988). Parsley protoplasts were used to identify cis-acting elements responsive to UV in a chalchone synthase gene from snapdragon (Lipphardt et al. 1988) and cis-acting elements for an elicitor in a pathogenesis-related gene (van de Locht et al. 1990). Gibberellic acid regulation of the α-amylase promoter was demonstrated in oat (Huttley and Baulcombe 1989) and barley (Gopalakrishnan et al. 1991; Jacobsen and Close 1991) aleurone protoplasts. Phenypropanoid pathway intermediates regulated transient expression of a chalcone synthase gene promoter in alfalfa protoplasts (Loake et al. 1991). Regulation of several photosynthetic gene promoters by light and metabolism was studied with maize mesophyll protoplasts (Sheen 1990).

These early transient expression studies showed that the protoplast systems retained the transacting factors and signal transduction systems required for specific gene expression. Moreover, cell specificity was observed in this regulation since expression in maize of an rbcS promoter in dicot protoplasts was restricted (Schaeffner and Sheen 1991) as were promoters from photosynthesis-associated genes different tobacco cell types (Harkins et al. 1990). Also, loss of promoter activity in parsley protoplasts and transgenic plants occurred with the deletion of a specific region in the promoter of the 4CL-1 gene (Hauffe et al. 1991).

More recent studies have used protoplast transient expression systems to examine a large range of gene expression processes. These studies have led to better understandings of RNA transcription, splicing, transport and translation (Callis et al. 1987; Gallie et al. 1987, 1989; Goodall and Filipowicz 1989; Waibel and Filipowicz 1990; Carneiro et al. 1993; Gallie and Bailey-Serres 1997; Qu and Morris 2000; Neeleman et al. 2001); the roles of cis-acting elements and transacting factors in gene expression (McCartey et al. 1991; Hattori et al. 1992; Kao et al. 1996; Urao et al. 1996; Gubler et al. 1999; Ballas et al. 1993; Abel and Theologis 1994, 1996; Ulmasov et al. 1997a b, 1999; Guilfoyle et al. 1998; Sheen 1990, 1993; Frohnmeyer et al. 1994; Graham et al., 1994, Sadka et al. 1994; Liu et al. 1994; Ni et al., 1996; Yanagisawa and Sheen 1998; Ito et al. 2001, Rushton et al. 2002; Sprenger-Haussels and Weisshaar 2000; Heise et al. 2002), and the regulation of gene expression by hormones and elicitors (Eulgern et al., 1999; Sheen 1996, 1998; Kovtan et al., 1998, 2000; Hwang and Sheen 2001; Tena et al. 2001). Protoplast transient expression systems have also been used to study protein dynamics in cells including targeting and trafficking (Chang et al. 1999; Kleiner et al. 1999; Kubitsheck et al. 2000; Nimchuk et al., 2000; Jin et al., 2001; Ueda et al., 2001; Sohn et al. 2003); turnover (Worley et al. 2000; Ramos et al. 2001; Tiwari 2001); and protein-protein interactions (Subramanian et al. 2001; Dhonukshe and Madella 2003). Other processes that have been studied with transient protoplast expression systems include retrotransposon expression (Pouteau et al. 1991; Pauls et al. 1994); regulation of cell death (Asai et al. 2000; Fath et al. 2002); cell cycle regulation (Pasternak et al. 2002); movement of avirulent proteins and their interaction with resistance gene proteins (Leister and Katagiri 2000; Wu et al. 2003); virus movement protein cellular distribution patterns (Heinlein et al. 1995; McLean et al. 1995; Huang et al. 2000; Cohen et al. 2000); heat shock protein aggregation (Czarnecka-Verner et al. 2000; Kirschner et al. 2000); and auxin-induced cell enlargement (Steffens 2001).

Although the production of transgenic plants is efficient for species such as *Arabidopsis*, the process introduces considerable variation into expression studies because of transgene position effects. Transient protoplast expression systems are ideal for high throughput candidate gene screening (Chory and Wu 2001; Hwang and Sheen 2001; Kovtun et al. 2000; Cheng et al. 2001; Tena et al. 2001) and promoter screening (Rushton et al. 2002). These systems are also suited to analyzing genes that cause lethality when overexpressed or deleted in plants. Testing of arrays of constitutively active or dominant negative mutants (Kovtun et al. 2000; Tena et al. 2001) in transient systems can narrow the options for more extensive studies with transgenic plants. In a number of cases the discovery of signaling pathways or signaling components in protoplasts have been confirmed by studies with mutants or transgenic plants. For example, studies with maize protoplasts suggested hexokinase is a sugar sensor which has been validated by transgenic plant studies (Jang et al. 1997; Dai et al. 1999; Rolland et al. 2002). ABA signaling components identified with maize protoplasts (Sheen 1998) have been confirmed by the isolation of *Arabidopsis* mutants (Gosti et al. 1999; Hoth et al. 2002). Roles of auxin response factors and Aux/IAA proteins first defined and characterized in protoplast transient expression assays (Abel and Theologis 1996; Ulmasov et al. 1997ab; Guilfoyle et al. 1998; Ulmasov et al. 1999; Tiwari, 2001) have been confirmed by the identification of auxin signaling mutants (Gray et al. 2002). An ABA-regulated gene in a protoplast transient expression assay has been shown to be upregulated in an ABA hypersensitive mutant and repressed in an ABA insensitive mutant (Uno et al. 2000). The roles of components of the cytokinin signaling pathway that were identified with a protoplast assay (D'Agostino et al. 2000; Hwang and Sheen 2001) are consistent with results from *Arabidopsis* cytokinin mutants (Kakimoto 1996; Hwang and Sheen 2001; Inoue et al. 2001; Hwang et al. 2002) and patterns of cytokinin-induced gene regulation in transgenic plants (D'Agostino et al. 2000; Che et al. 2002). The functions of members of an MAP kinase signaling cascade induced by bacterial elicitors were identified with an *A. thaliana* protoplast system and were confirmed in tests with transgenics constitutively expressing members of the cascade (Asai et al. 2002). Plant Rac-like GTPases have been shown to mediate the auxin signal to downstream responsive genes in protoplasts and transgenic tobacco (Tao et al. 2002).

3. Novel Germplasm from Protoplast Fusion

PEG treatments, electroporation, and particle bombardment treatments have been used to produce stable transgenic cell lines and plants in a variety of species. Direct transfer of DNA into protoplasts was the first method leading to the production of transgenic monocot plants (Toriyama et al. 1988). PEG treatments of protoplasts has also been shown to be an effective method for producing plastid transformants (Koop et al. 1996; Kofer et al. 1998) and has been used for targeted disruption of plastid genes (De Santis-Maciossek et al. 1999).

Fusion outcomes

Fusions between two complete protoplasts can result in different cells with many novel combinations of genomes. Generally, the calli and plants obtained from fusions consist of mixtures of nuclear DNA that can cover the complete range of possibilities, from complete copies of both fusion partners, to fragments from both partners in varying amounts and complete elimination of one of the donor genomes (Li et al. 1996). Irradiation of one of the parents to produce asymmetric hybrids has been shown to influence the direction of the DNA elimination (Wijbrandi et al. 1990a,b; Melzer and O'Connell 1992). But some results indicate that irradiation does not affect the extent of DNA elimination, rather that the relative amounts of DNA from two sources determine the outcome.

The cytoplasmic genomes in cells and plants obtained from protoplast fusions each appear to sort into single mitochondrial and choroplastic cytotypes (Scowcroft and Larkin 1981; Rose et al. 1990; Walters and Earle 1993). However, recombination among mitochondria can occur in heterokaryons to produce novel cytotypes, so that the fusion product can contain one of the donor genomes or a recombination product (Chetrit et al. 1985; Rose et al. 1990; Earle et al 1991; Landgren and Glimelius 1994; Fig. 4.6).

Applications of protoplast fusion

The rich array of possibilities has attracted researchers to use protoplast fusion to understand genome interactions in plants and to obtain germplasm that may be useful for plant improvement. Specific reasons for using protoplast fusion for plant improvement include: rapid production of novel nuclear/cytoplasmic combinations by fusing protoplasts treated with an inhibitor to inactivate cytoplasmic genomes and protoplasts treated with ionizing irradiation to inactivate their nuclear genomes (Zelcer et al. 1978; Menzel et al. 1981; Tanno-Suenaga et al. 1988); transfer of single gene traits (Lelivelt et al. 1993; Liu et al. 1995); transfer of polygenic traits for which the molecular bases are not known (Mattheij and Puite 1992); improvement of plants such as banana, cassava, potato, sweet potato, sugarcane or yam in which sexual reproduction is difficult or absent; and improvement of plants sold for novel foliage or flower characteristics, e.g. ornamentals.

In a review of the use of somatic hybridization for plant improvement Li et al. (1996) tabulated almost 140 studies that resulted in the production of somatic hybrid/cybrid plants for plant breeding purposes The table excluded studies that did not produce plants or were produced for basic research purposes or were conducted to create genetic diversity without specific breeding objectives. The studies included 33 species in 22 genera of 12 families. The families (and species) most commonly used included: Solanaceae (*Solanum*, *Nicotiana*, *Lycopersicon*), Cruciferae (*Brassica*), Rutaceae (*Citrus*) and Leguminosae (*Medicago*). The list also included hybrid/cybrid plants produced in the families Carophyllaceae, Compositeae, Gramineae, Curcurbiaceae, Labiatae, Liliaceae, Passiflloraceae and Umbelliferae but no intergeneric somatic hybrids for breeding in the last five families. Fusion was induced almost exclusively by PEG or electrofusion techniques with an increase in the use of the latter technique in the 1990s.

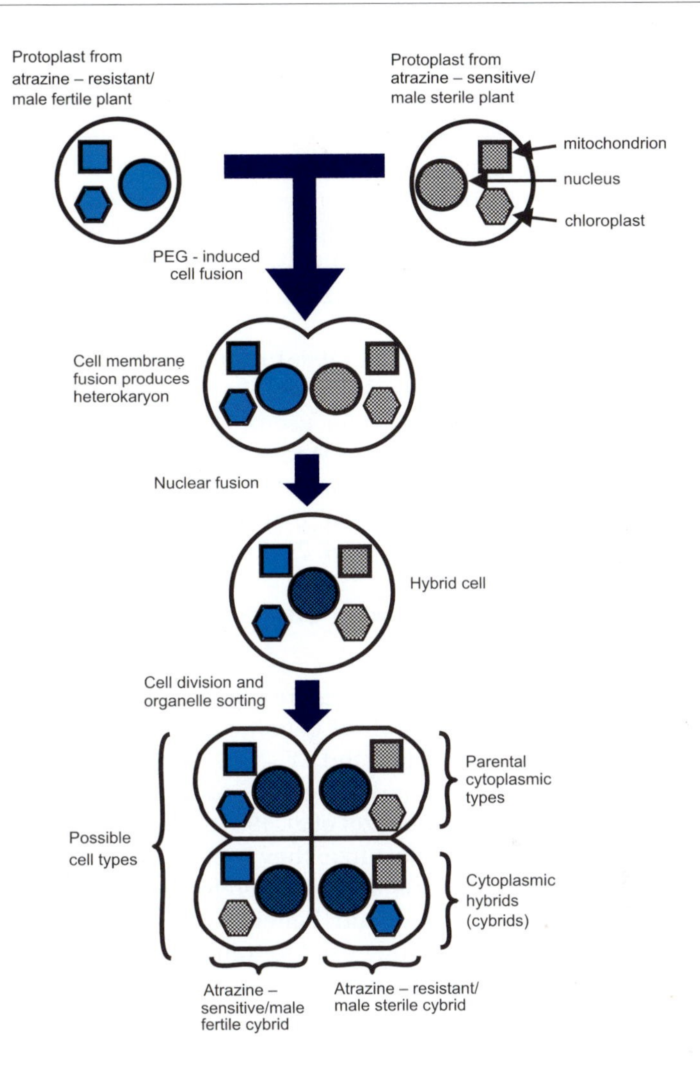

Fig. 4.6 Heterokaryon production and possible combinations of nuclear and cytoplasmic genomes from fusions between protoplasts from atrazine resistant/ male fertile plants and protoplasts from atrazine sensitive/cytoplasmic male sterile plants. Atrazine resistance is carried by the chloroplasts of one parent and male sterility is determined by the mitochondria of the other parent. In the callus that results from the division of the heterokaryon, one type of cell ultimately predominates and gives rise to a plant which has a hybrid nuclear condition and a parental cytoplasmic combination or a new combination of chloroplasts and mitochondria (cybrid). Note that often the chloroplastic and mitochondrial genomes are not altered.

Differential inactivation of the parent protoplasts with IOA or (γ-rays) or differential regenerability (hybrid vigor) were the most common hybrid selection methods, but flow cytometry was used to isolate Cruciferae hybrids (Glimelius et al. 1986). For hybrid/cybrid verification various tests were used, including: morphological comparisons, isozyme analysis, chromosome counting, and various molecular marker methods (such as RFLP, RAPD). The most frequent plant improvement objective in the fusion experiments was to increase disease resistance, followed by CMS and pest resistance. CMS was the trait transferred with the greatest success by protoplast fusion.

The literature since the Li et al. (1996) publication indicates that the general trends and conclusions described in that review are still true. Some specific changes include: wider use of flow cytometry to confirm hybridity and to sort heterokaryons (Waara et al. 1998; Dushenkov et al. 2002); successes in protoplast fusion in additional plant families and identification of universal molecular markers for identifying hybrids (Bastia et al. 2001; Przetakiewicz et al. 2002).

4. Manipulating Cytoplasmic Male Sterility (CMS) by Protoplast Fusion

Cytoplasmic male sterility is widespread in plants and is due to molecular rearrangements in the mitochondrial genome that result in the production of chimeric genes (Schnable and Wise 1998) and incompatibilities between nuclear and mitochondrial gene products that result in the failure of pollen to develop (Menczel et al. 1983; Newton 1988). Although CMS plants are unable to self-fertilize, they will set seed when fertilized with pollen from normal plants and are therefore valuable for the production of hybrid seed (Budar and Pelletier 2001). Protoplast fusion experiments have been successfully used to transfer CMS between species (Bannerot et al. 1974) and create new CMS sources (Liu et al. 1996).

CMS *in Brassica*

Hybrid variety development is a goal in plant breeding for a number of crops including canola because plants grown from hybrid seed (produced by crossing two inbreds) are uniform, more vigorous and higher yielding than their parents (due to heterosis); further, the interests of the plant breeders are protected because the next generation grown from seed produced on the F_1 plants will not be uniform in characteristics.

Low-temperature chlorosis in CMS cytoplasm from the Japanese radish Ogura (*Raphanus sativus*), which resulted from dysfunction between Ogura chloroplasts and Brassica sp. genome, resulted in low-temperature chlorosis (Dickson 1985) that limited the usefulness of this CMS system for hybrid production in the *Brassicas*. Protoplast fusion, followed by mitochondrial recombination was used to study the cause of male sterility (Pelletier et al. 1983; Budar and Pelletier 2001). And protoplast fusion followed by organelle assortment solved the problem. Researchers were able to identify fusion products in which the dysfunctional Ogura chloroplasts had been replaced with *Brassica* chloroplasts but had the desired mitochondria with mtDNA encoding the CMS trait (Sigareva and Earle 1996).

In many plants chloroplasts and mitochondria are maternally inherited (Fig. 4.7). For these plants the replacement of cytoplasmic organelles can sometimes be accomplished by wide hybridization followed by extensive backcrossing, but this is a very time-consuming process. By contrast, cytoplasmic replacement can be achieved in one step by a donor-recipient protoplast fusion procedure, wherein donor protoplasts are X or (γ-irradiated) and fused with iodoacetamide (IOA)-treated protoplasts (Zelcer et al. 1978; Menzel et al. 1987; Tanno-Suenaga et al. 1988). For example, the transfer of CMS from spring-type to winter-type oilseed rape (*Brassica napus*) required approximately 9 months (Barsby et al. 1987). Sigareva and Earle (1996) transferred Ogura CMS from broccoli to cabbage by protoplast fusion and produced plants large enough to plant in soil 8 months after somatic hybridization.

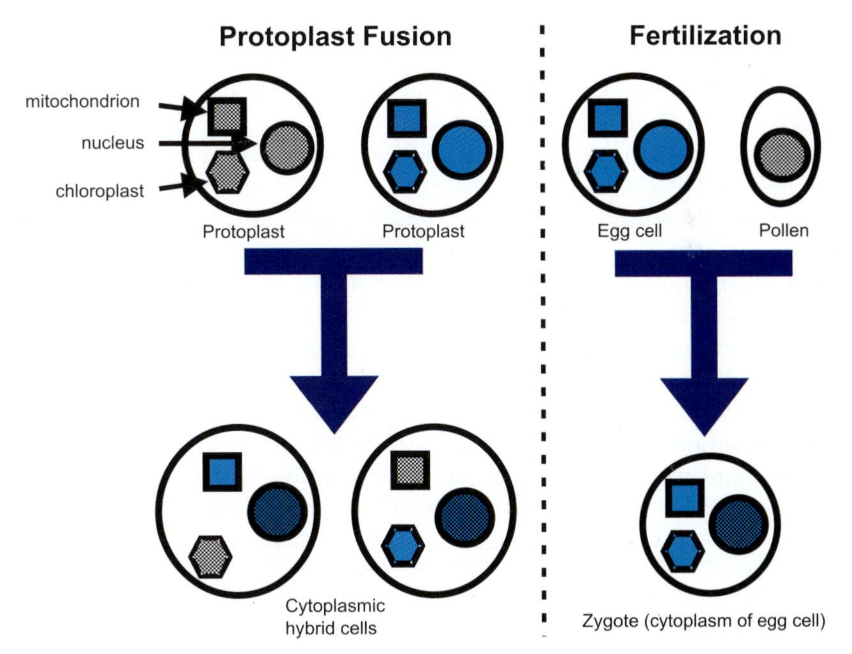

Fig. 4.7 Comparison of the transmission of organelles during protoplast fusion and fertilization.

CMS/ATR *plants*

Creation of new combinations of chloroplasts and mitochondria is difficult to achieve by conventional breeding because the organelles are most often maternally inherited (Fig. 4.7). Thus the zygote always resembles the egg cell with respect to its cytoplasmically coded traits. In contrast, the heterokaryon derived from protoplast fusion carries a mixture of chloroplasts and mitochondria derived from both cells. In subsequent cell divisions these organelles sort out randomly and can give rise to a plant with chloroplasts from one parent and mitochondria from the other.

Protoplast fusion was used to combine two cytoplasmic traits, namely, atrazine resistance and cytoplasmic male sterility in one plant (Fig. 4.8). Such a plant is useful for the production of hybrid canola seed by a patented scheme proposed by Beversdorf et al. (1985, Fig. 4.9). In this scheme the cytoplasmic male sterile/atrazine resistant (CMS/ATR) plants are pollinated by male fertile/atrazine sensitive plant carrying a restorer gene for fertility. After the pollinators are removed by atrazine applications, hybrid seed can be harvested from the CMS/ATR plants.

Cells from haploid atrazine resistant (Fig. 4.8a) and haploid Polima CMS plants (Chuong et al. 1997, 1988a; Fig. 4.8b) were used for the canola protoplast fusion experiments. Polima CMS (Liu et al. 1996) was used because fertility restorer lines exist (Fan et al. 1986), most varieties serve as maintainers, and it did not show low-temperature chlorosis. Haploid cells were used in order to increase the chances that the plants regenerated from heterokaryons would be diploid.

Cell fusion was induced by treatment with polyethylene glycol (PEG) and the protoplast mixtures were sequentially cultured on protoplast culture, callus growth and shoot induction media (Chuong et al. 1988a; Fig. 4.2). Of the 261 plants regenerated (Fig. 4.8c), 43% were diploid and one was atrazine resistant and CMS (Fig. 4.8d). The CMS/ATR plant was diploid and had reduced stamens (Fig. 4.8g). Although it produced very little pollen, and was effectively male sterile, it readily set seed when fertilized with pollen from another plant. The progeny were resistant to the herbicide atrazine (Fig. 4.8h). Furthermore, in accordance with the maternal inheritance of cytoplasmic traits, all the progeny were atrazine resistant and male sterile when crossed with any other canola plant not carrying restorer genes. Confirmation of the hybrid nature of the cytoplasm

Fig. 4.8 Somatic transfer of cytoplasmic traits in *Brassica napus* by haploid protoplast fusion between haploid atrazine sensitive/male sterile and haploid atrazine resistant/male fertile protoplasts. (a) Flower of the atrazine resistant parent. (b) Flower of Polima CMS parent (c) Plantlets regenerated from unselected protoplast fusion experiments between haploid atrazine sensitive/male sterile and haploid atrazine resistant/male fertile protoplasts. (d) A diploid ATR/CMS plant

Contd.

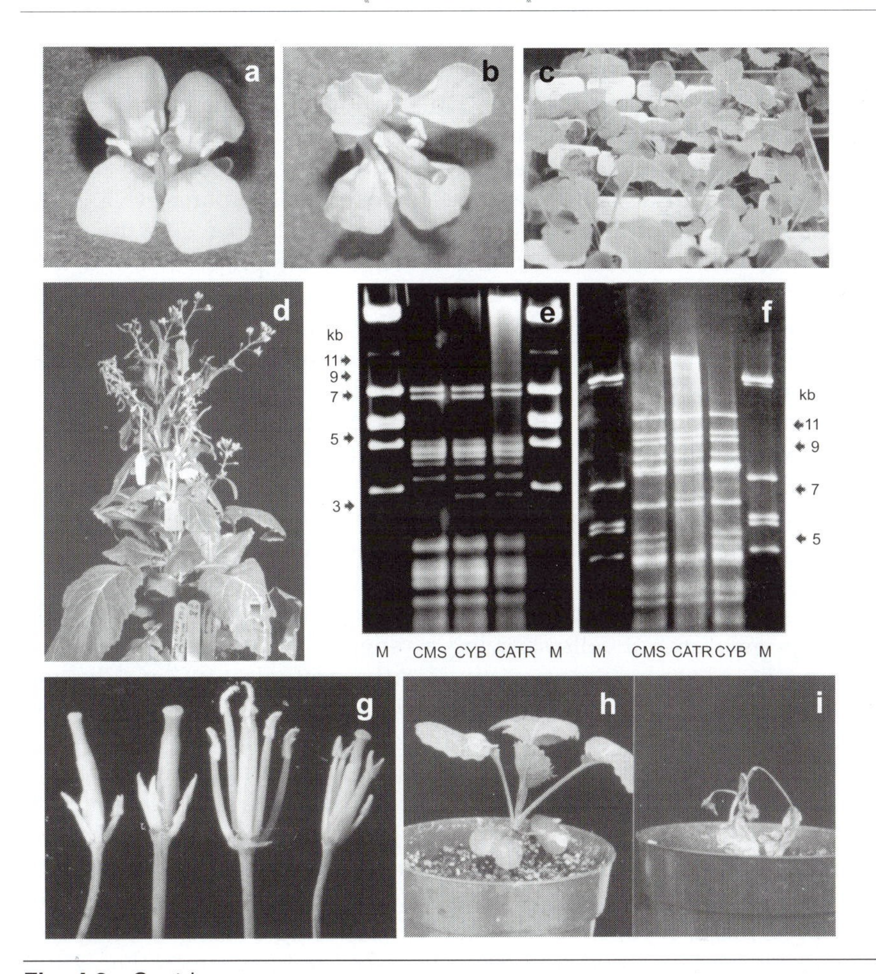

Fig. 4.8 Contd.
regenerated from fusion experiments between haploid atrazine sensitive/male sterile and haploid atrazine resistant/male fertile protoplasts. The CMS/ATR plant was diploid and had reduced stamens. Although it produced very little pollen and was effectively male sterile, it readily set seed when fertilized with pollen from another plant. (e) The *Eco*RI restriction fragment pattern of the chloroplast DNA from the protoplast fusion product (CYB) showing an identical pattern to the one obtained from the atrazine resistant parent (CATR). (f) Restriction fragment patterns of mitochondrial DNA showing that the pattern of the fusion product (CYB) was the same as the CMS parent. (g) Morphology of flower parts from the atrazine sensitive/male sterile parent (left) the CMS/ATR resistant protoplast fusion product (centre/left), a male fertile atrazine sensitive fusion product (centre/right) and the male fertile atrazine resistant parent (right). (h) Plant grown from seed produced on the ATR resistant/CMS plant pollinated with pollen from normal canola after treatment with atrazine. (i) Plant grown from seed produced on normal canola after treatment with atrazine.

was obtained from studies of the mitochondrial and chloroplastic DNA of the CMS/ATR plant. The *Eco*RI restriction fragment pattern of the chloroplast DNA from the protoplast fusion product was identical to the one obtained from the ATR parent (Fig. 4.8e) and the fragment pattern of the mitochondrial DNA was the same as the CMS parent (Fig. 4.8f). Although the use of atrazine resistance, has been superceded by other sources of herbicide resistance, the success of the project showed that protoplast fusion could be used to obtain plants with attributes useful for plant breeding, especially hybrid material (Fig. 4.9).

The low frequency of cybrid production in the unselected protoplast fusion study prompted investigation into flow cytometric methods for enriching the cultures for fusion products. Methods were developed to detect fused cells in mixtures of nonpigmented protoplasts stained with fluorescein isothiocyanate (FITC), which fluoresced green (Fig. 4.10a), and chlorophyll-containing cells which fluoresced red (Fig. 4.10d). After treatment with PEG the fused cells fluoresced red and green (Fig. 4.10h). Two- and three-dimensional flow cytometric profiles of unfused FITC-stained cells (Fig. 4.10b,c) or chlorophyll-containing cells (Fig. 4.10e,f) had single populations of cells showing a range of fluorescence intensities on the red and green axes respectively. After protoplast fusion a distinct population of cells, that represented approximately 1-5% of all the cells, was visible in the flow cytometric profiles (Fig. 4.10i,j).

For fusions between haploid cells, nonpigmented cells were obtained from roots developed on stem cuttings and these were stained with FITC and fused with stem peel protoplasts. Protoplasts that fluoresced red and green after PEG treatment were sorted with a flow cytometer into microwell plates containing *B. napus* microspores (Chuong et al. 1988b). The protoplasts recovered form the flow sorting process were intact (Fig. 4.10k) and included cells that fluoresced red and green (Fig. 4.10k inset). Cell division was observed in the sorted cells (Fig. 4.10 l,m). The microspores in the wells into which the protoplasts were sorted served a nurse culture function since protoplast division and microcalli development only occurred in the wells containing microspores (Fig. 4.10n). Twenty-six plants from 7 calli were regenerated (Fig. 4.10o). All the plants were male fertile and atrazine sensitive, suggesting that they were cybrids and all appeared to be diploid based on their seed set after selfing and chromosome counts of squashes prepared from their roots.

CMS in Rice

Methods for transferring CMS to rice cultivars have been developed to a high degree in rice (*Oryza sativa*). The motivation for this work is a 15% yield advantage that hybrid rice varieties have over open-pollinated varieties (Yuan 1994) and the fact that it takes approximately 3 years to

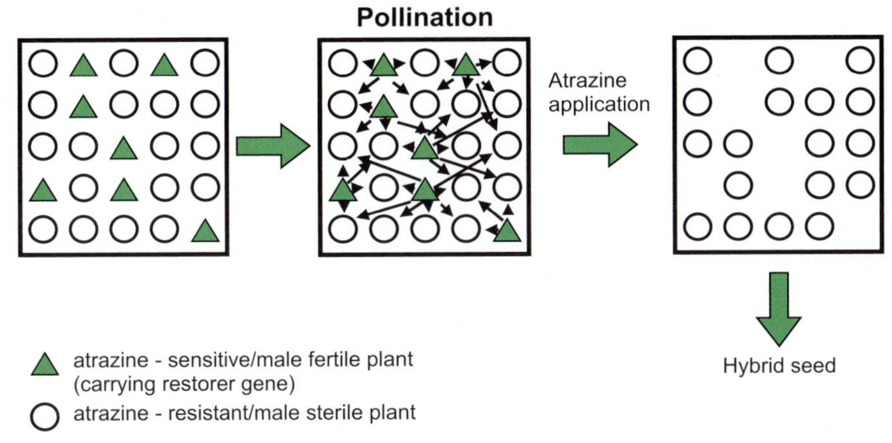

Fig. 4.9 Scheme for producing hybrid canola seed using cytoplasmic atrazine resistant/male sterile plants as female parents and plants carrying fertility restorer genes (that are atrazine sensitive/ male fertile) as pollinators (Beversdorf et al. 1985). Hybrid seed would be harvested from the field after the pollinators were eliminated by treatment with atrazine.

convert a conventional line to a CMS line by backcrossing (Akagi 2001). Cybrid rice plants were first created by donor-recipient protoplast fusions between CMS lines and fertile japonica rice cultivars (Akagi et al. 1989; Kyozuka et al. 1989; Yang et al. 1989). Gupta et al. (1996) obtained putative cybrid plants by asymmetric protoplast fusion of Indica rice.

An important source of CMS in rice is called ms-boro BT-type found in the cultivar Chinsurah Boro II (Indica rice) (Shinjyo 1969), which is restored by a Rf-1 gene in the nucleus (Shinjyo 1975). CMS in a fertile Japanese cultivar was obtained by asymmetric protoplast fusion between X-ray-treated Chinsurah Boro II protoplasts and IOA-treated Nipponbare protoplasts (Kyozuka et al. 1989). A plant regenerated from these experiments was completely male sterile and this trait was maternally transmitted when it was fertilized with Nippobare pollen. Furthermore, male fertility was restored when progeny were crossed with a plant containing *Rf-1* and DNA analysis confirmed that mitochondria from the Chinsurah Boro II parent was incorporated into the male sterile plant obtained by protoplast fusion.

Recently Akagi (2001) transferred the cytoplasm of Chinsurah Boro II to 40 Japanese cultivars by asymmetric protoplast fusion. F1 seeds were produced with these CMS lines and several hybrids and were tested in different locations in Japan. Two combinations were identified that had yields 30-40% higher than leading varieties for the areas in which they were grown and also good food qualities.

SOMATIC HYBRIDIZATION APPLICATIONS IN CITRUS BREEDING

Production of Triploids

One of the main objectives in citrus breeding is the production of seedless varieties. Seedlessness occurs at a high frequency in triploid citrus germplasm, which can be produced by crossing a tetraploid (female) with a diploid (male) (Soost and Cameron 1985). Somatic hybridization can therefore be very useful for broadening the tetraploid germplasm pool that can be used for creating triploids. Ohgawara and Kobayashi and coworkers developed a somatic hybridization system using Trovita (*Citrus sinensis*) nucellar cells and trifoliate (*Poncirus trifoliata*) mesophyll cells (Ohgawara et al. 1985) and produced somatic hybrids between a number of citrus species (Ohgwara el al 1997). The somatic hybrids were male and female fertile and produced triploid plants when crossed with a diploid cultivar such as Clementine mandarin (*C. clementia*) (Oiyama et al. 1991). In 1995 four somatic hybrids were registered in Japan as Citrus parental line No.1, No.2, No.3 and No.4 (Ohgwara el al 1997).

Root Stock Improvemnt

Tetraploids that are produced spontaneously or by symmetric somatic hybridization between dipolid cells are not directly useful for scion improvement because they have stout stems, thick leaves, and fruit with thick rough rinds. These traits are, however, desirable for rootstock. Also, progress in breeding rootstock is slow because the heterozygous nature of the parental material that necessitates evaluation of large populations of sexual crosses and because several years are required to allow material

---

Fig. 4.10 Flow cytometric analysis and sorting of *Brassica* protoplasts. (a) FITC-stained hypocotyl protoplasts fluorescing green. (b,c) Flow cytometric profiles of *Brassica* FITC-stained hypocotyl protoplasts showing a range of green fluorescence values; the origin in the three-dimensional plot (c) is in the back right-hand corner. Ten thousand protoplasts were analyzed (Pauls and Chuong, 1988) (d) Leaf protoplasts autofluorescing red. (e,f) Flow cytometric profiles of *Brassica* leaf protoplasts showing a range of red fluorescence values; the origin in the three-dimensional plot (f) is in the back right-hand corner. (g) Fusion product between a leaf protoplast and a hypocotyl protoplast visualized by light microscopy. (h) Fusion product between a leaf protoplast and an FITC-stained hypocotyl protoplast visualized by fluorescence microscopy showing green fluorescence on the left and red fluorescence on the right half of the cell.(i,j) Flow cytometric profiles of a protoplast fusion mixture between *Brassica* FITC-stained hypocotyl protoplasts and leaf protoplasts showing three populations, including: unfused leaf protoplasts with a range of red fluorescence values, unfused FITC-stained hypocotyl protoplasts showing a range of green fluorescence values, and a

Contd.

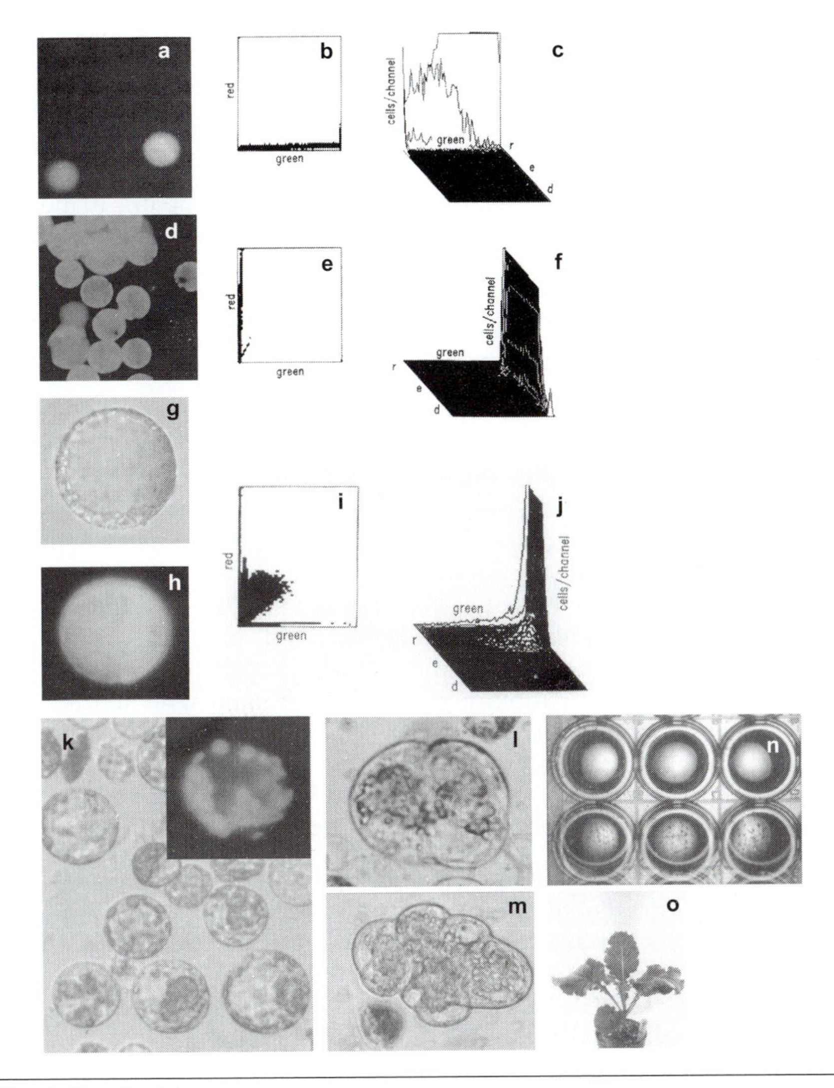

Fig. 4.10 Contd.

population of fused cells showing red and green flurescence; the origin in the three-dimensional plot (f) is in the back right-hand corner. (k) Flow-sorted cells from fusions between haploid *B. napus* stem peel protoplasts and haploid *B. napus* FITC-stained root protoplasts. Inset shows the red and green fluorescence of a sorted cell. (l,m) Divisions in sorted cells. (n) Microcalli production from flow sorted protoplasts in microwell plates (bottom row). The wells in the bottom row contained micropsore cells that acted as nurse cells for the sorted protoplasts. (o) Plant regenerated from a sorted protoplast culture. The plant was atrazine sensitive and male fertile, indicating that it was a cytoplasmic hybrid (Chuong et al. 1988a).

to mature before it can be evaluated. Somatic hybridization is thought to be a promising technique for directly combining dominant traits into material for rootstock development. Grosser and coworkers have produced a number of intergeneric hybrids between sexually incompatible plants by protoplast fusion , including between Hamlin orange (*C. sinensis*) and *Siverinia disticha* (Grosser et al. 1988) and between Cleopatra mandarin (*C. reticulata*) and *Citropsis gilletiana* (Grosser et al. 1990), that may be useful for rootstock development because they are salt tolerant and disease resistant.

Somatic Hybridization to Improve Potato

Potato (*Solanum tuberosum*) is the fourth most important food in the world but improvement through traditional breeding is complicated by tetrasomic inheritance, high heterozygosity, self-incompatibility, and sterility in cultivated genotypes. Also, because potatoes are propagated vegetatively their gene pool is limited. Although wild *Solanum* species have many traits of value for potato improvement, e.g. disease resistance and stress tolerance, their use has been limited because of differences in endosperm balance numbers (Hanneman and Bamberg 1986) that prevent sexual hybridization. Somatic hybridization has been used to overcome sexual incompatibility between potato and wild species, e.g. *Solanum commersonii* or *Solanum bulbocastanum*, to introduce a variety of desirable traits into potato.

Potato/*Solanum commersonii* Fusion Products

A number of fusions between potato and *Solanum commersonii* protoplasts were made to transfer resistance to biotic and abiotic stress to potato (Cardi et al. 1993a, b; Kim et al. 1993; Carotenuto et al. 1996, 1997; Carputo et al. 1997a, b; Nyman and Waara 1997; Waara et al. 1998; Chen 1999a, b, c; Laferriere et al. 1999; Parrella and Cardi 1999; Bastia et al. 2000; Seppanen 2000). *In vitro* grown plantlets were used for protoplast isolation and electrofusion was used for most of the experiments. Hybrid vigor *in vitro* was used to select the fusion products in most studies and one study used flow sorting (Waara et al. 1998). The somatic hybrids were most often female fertile but male sterility was a common problem. Progeny obtained from selfing the somatic hybrids, crossing the somatic hybrids or crossing the somatic hybrids with potato showed that the procedure was useful for capturing genes for a variety of traits, including frost tolerance, tolerance to oxidative stress, resistance to blackleg, tuber soft rot, *Verticillium* wilt, and common scab (Cardi 2001). Most often multiple trait transfers were achieved.

Potato/*Solanum bulbocastanum* Fusion Products

Phytophthora infestans, the cause of late blight in potatoes, is a major problem for potato production. A Mexican species *S. bulbocastanum* is highly resistant to late blight and is a potential source of durable resistance genes for potato breeding but is sexually incompatible with potato. Helgeson et al. (1998) used PEG- mediated fusion of leaf cells of *S. bulbocastanum* ($2n = 2x = 24$) and *S. tuberosum* ($2n = 4x = 48$) to produce hexaploid ($2n = 6x = 72$) somatic hybrids that were resistant to *P. infestans*. Crosses of the hybrids with potato cultivars yielded BC_1s that segregated for resistance to *P. infestans*. Resistant BC_1s were crossed with susceptible cultivars to produce BC_2 populations that also segregated for resistance to the fungus. Further crossing and field tests indicated that the resistance could be passed on undiminished and appeared to be controlled by a single dominant gene or a tightly linked cluster of genes. A molecular marker study mapped the resistance to the long arm of chromosome 8 and identified its linkage to the RFLP marker CT88 and RAPD marker OPG02-625 (Naess et al. 2000). Recently, Song et al. (2003) showed that the gene for late blight resistance in *S. bulbocastanum* exists in a cluster of resistance gene analogues and is of the CC-NB-LRR class of disease resistance genes.

INTERKINGDOM PLANT/ANIMAL FUSIONS

Interkingdom plant/animal cell fusions were made soon after the protoplast fusion method was developed, between HeLA cells and carrot protoplasts (Dudits et al. 1976), and *Drosophila* cells and soybean protoplasts (Hadlaczky et al. 1980). Given the more recent interest in expressing animal proteins, especially antibodies and biopharmaceuticals in plants (Daniell et al., 2001), an experiment that resulted in a tobacco plant that produces mouse immunoglobulin G is intriguing. Makonkawkeyoon et al. (1995) fused mouse sperm cells with tobacco mesophyll protoplasts with PEG. Calli and plants obtained from the fused cells produced gamma-3 heavy and lambda light chains as detected by ELISA, Western blotting, and Southern blotting. However, no indication of the efficiency of the procedure was given and no evidence for transmission of the ability to the next generation was reported. Nevertheless, plant animal cell fusions with highly regenerable cells, such as tobacco protoplast, might be an attractive system for producing a range of animal proteins.

UNIVERSAL PCR PRIMERS TO VERIFY FUSION

To simplify determinations of the cytoplasmic genomes of plants from protoplast fusion experiments, a set of universal PCR primers were designed to amplify polymorphic intergenic and intronic regions of mitochondrial and chloroplastic genomes (Bastia et al. 2001). These were tested with 13 genotypes of *Nicotiana tobacum, Solanum commersonii, Solanum tuberosum, Solanum etubersosum, Solanum phereja, Brassica oleracea, Brassica rapa,* 'Anand' CMS *B. rapa,* 'Chiang' CMS *B. oleracea* and 'Ogura' CMS *B. oleracea* and interspecific somatic hybrids and cybrids. The authors concluded that the method was simpler and less expensive than direct analysis of organellar DNA and did not require specific cpDNA or mt DNA isolation. Thus, it might be applied to early stage screening of potential hybrids from fusion experiments.

In addition, an amplification-based procedure with arbitary (10-mers) generating RAPDs or semirandom primer for intron-exon sequences was used to verify potato somatic hybrids (Przetakiewicz et al. 2002).

FUTURE OF PROTOPLAST RESEARCH

Forty years after large-scale production of protoplasts became possible the range of work with these isolated plant cells is varied and innovative. Some of the original enthusiasm surrounding the use of somatic hybridization to create fantastic new species for agriculture is somewhat tempered but there has been steady progress in the recovery of fertile plants from narrower hybridizations that have novel characteristics of real value to potato, rice, brassica and citrus breeding. Methods for identifying fusion products, including the use of plant cells expressing GFP (Olivares-Fuster et al. 2002) are also steadily improving. In future it is likely that these developments will lead directly to the introduction of varieties from protoplast fusions with improved disease resistance and hybrid vigor.

Ironically, the antipathy of some parts of the world to the use of genetically modified organisms (GMOs) may increase the use of protoplast fusion to circumvent species boundaries by "natural" hybridization methods. For those jurisdictions in which no distinction is made regarding the method used for introducing genetic variability the ability to transform the plastome by PEG transformation of protoplast (Koop et al. 1996; De Santis-Maciossek et al. 1999) might be increasingly important for introducing genes into plants because it reduces the probability of gene flow. In general, the protoplast could be considered to be a better target for transformation than multicellular systems, if regeneration protocols are equally efficient, because their use eliminates the concern about the selection of chimeras.

The use of protoplasts for exploring basic processes in plant biology appears to be the fastest growing area in protoplast research. Most of the molecular and cellular tools available for animal cell biology have now been developed for plant cells, including a recent demonstration of gene targeting in plants (Terada 2002). These procedures will allow results from transient expression analyses with plant protoplasts to be rapidly confirmed with transgenic plant studies. It may be that for future studies of basic processes in biology or for determining the functions of the thousands of genes that have been sequenced in the various genomics projects, plant protoplasts will be the preferred material to use because of their totipotent nature.

REFERENCES

Abel S, Theologis A (1994) Transient transformation of *Arabidopsis* leaf protoplasts: a versatile experimental system to study gene expression. Plant J 5:421-427.

Abel S, Theologis A (1996) Early genes and auxin action. Plant Physiol 111:9-17.

Akagi H (2001) Cybridization in *Oryza sativa* L. (Rice), *In:* T Nagata and YPS Bajaj, (eds.), Biotechnology in Agriculture and Forestry, Somatic Hybridization in Crop Improvement II. Springer-Verlag, Berlin, Vol 49, pp. 17-34.

Akagi H, Sakamoto M, Negishi T, Fujimura T (1989) Construction of rice cybrid plants. Molec. Gen. Genet. 215:501-506.

Allen GJ, Murata Y, Chu SP, Nafisi M, Schroeder JI (2002) Hypersensitivity of abscisic acid-induced cytosolic calcium increases in the *Arabidopsis* farnesyl-transferase mutant era1–2. Plant Cell 14:1649-1662.

Aoki S, Tabeke I (1969) Infection of tobacco mesophyll protoplasts by tobacco mosaic virus ribonucleic acid. Virology 39:439-448.

Asai T, Stone JM, Head JE, Kovtun Y, et al. (2000) Fumonisin B1-induced cell death in *Arabidopsis* protoplasts requires jasmonate-, ethylene-, and salicylate-dependent signaling pathways. Plant Cell 12:1823-1835.

Asai, T, Tena G, Plotnikova J, Willmann MR, et al. (2002) MAP kinase signaling cascade in *Arabidopsis* innate immunity. Nature 415:977-983.

Assmann SM (2001) From proton pump to proteome. Twenty-five years of research on ion transport in higher plants. Plant Phys 125:139-141.

Assmann SM, Simoncini L, Schroeder JI (1985) Blue light activates electrogenic ion pumping in guard cell protoplasts of *Vicia faba*. Nature 318:285-287.

Aviv D, Fluhr R, Edelman M, Galun E (1980) Progeny analysis of the interspecific somatic hybrids: *Nicotiana tabacum* (CMS) + *Nicotiana sylvestris* with respect to nuclear and chloroplast markers. Theor. Appl. Genet. 56:145-150.

Ballas N, Wong LM, Theologis A (1993) Identification of the auxin-responsive element, AuxRE, in the primary indoleacetic acid-inducible gene, PS-IAA4/5, of pea (*Pisum sativum*). J Molec. Biol. 233:580-596.

Banks MS, Evans PK (1976) A comparison of the isolation and culture of mesophyll protoplasts from several *Nicotiana* species and their hybrids. Plant. Sci. Lett. 7:409-416.

Bannerot H, Boulidard L, Cauderon Y, Tempe J, (1974) Transfer of cytoplasmic male sterility from *Raphanus sativus* to *Brassica oleracea*. Proc. EUCARPIA, Crop Section *Cruciferae* 25:52-54.

Barden KA, Smith SLS, Murakishi HH (1986) Regeneration and screening of tomato somaclones for resistance to tobacco mosaic virus. Plant Sci. 45:209-213.

Barsby TL, Yarrow SA, Shepard JF (1986) A rapid and efficient alternative procedure for the regeneration of plants from hypocotyl protoplasts of *Brassica napus*. Plant Cell Rep. 5:101-103.

Barsby TL, Yarrow SA, Kemble RJ, Grant I (1987) The transfer of cytoplasmic male sterility to winter-type oilseed rape (*Brassica napus* L.) by protoplast fusion. Plant Sci. 53:243-248.

Bastia T, Scotti N, Cardi T (2001) Organelle DNA analysis of *Solanum* and *Brassica* somatic hybrids by PCR with 'Universal Primers'. Theor. Appl. Genet. 102:1265-1272.

Bastia T, Carotenuto N, Basile B, Zoina A, Cardi T (2000) Induction of novel organelle DNA variation and transfer of resistance to frost and *Verticillium* wilt in *Solanum tuberosum* through somatic hybridization with 1EBN *S. commersonii.* Euphytica 116:1-10.

Bauer CS, Hoth S, Haga K, Philippar K, Aoki N, Hedrich R (2000) Differential expression and regulation of K^+ channels in the maize coleoptile: molecular and biophysical analysis of cells isolated from cortex and vasculature. Plant J. 24:139-145.

Bethke PC, Jones RL (2001) Cell death of barley aleurone protoplasts is mediated by reactive oxygen species. Plant J. 25:19-29.

Beversdorf WD, Erickson LR, Grant I (1985) Hybridization process utilizing a combination of cytoplasmic male sterility and herbicide tolerance. US Patent and Trademark Office No 43:511.

Bonnett HT and Glimelius K (1983) Somatic hybridization in *Nicotiana*: behavior of organelles after fusion of protoplasts from male-fertile and male-sterile cultivars. Theor. Appl. Genet. 65:213-217.

Bottcher UF, Aviv D, Gahn E (1989) Complementation between protoplasts treated with either of two metabolic inhibitors results in somatic hybrid plants. Plant Sci. 63:67-77.

Budar F, Pelletier G (2001) Male sterility in plants: occurrence, determinism, significance and use. C.R. Acad.Sci. Series III, Sciences de la Vie 324:543-550.

Callis J, Fromm M, Walbot V (1987) Expression of mRNA electroporated into plant and animal cells. Nucl. Acid. Res. 15:5823-5831.

Cardi T (2001) Somatic hybridization between *Solanum commersonii* Dun. and *S. tuberosum* L. (Potato). *In:* T. Nagata and YPS Bajaj. (eds.), Biotechnology in Agriculture and Forestry: Somatic Hybridization in Crop Improvement II. Springer-Verlag, Berlin, Vol 49, pp. 245-263.

Cardi T, D'Ambrosio F, Consoli D, Puite K.J. Ramulu KS (1993a) Production of somatic hybrids between frost tolerant *Solanum commersonii* and *S. tuberosum*: characterization of hybrid plants. Theor. Appl. Genet. 87:193-200.

Cardi T, Puite KJ, Ramulu KS, D'Ambrosio F, Frusciante L (1993b) Production of somatic hybrids between frost-tolerant *Solanum commersonii* and *S. tuberosum*: protoplast fusion, regeneration and isozyme analysis. Amer. Potato J 70:753-564.

Carlson PS, Smith Hit, Dearing RD (1972) Parasexual interspecific plant hybridization. Proc. Natl. Acad. Sci. USA 69: 2292-2244.

Carneiro VT, Pelletier G, Small I (1993) Transfer RNA-mediated suppression of stop codons in protoplasts and transgenic plants. Plant Molec. Biolec. 22:681-90.

Carotenuto N (1997). Valutazione agronomicae caratterizzazione di cloni derivanti da ibridzione somatica tra *Solanum tuberosumi* e *S. tuberosum.* PhD Thesis, Univ. Naples, Federico II, Portici, Italy.

Carotenuto N, Cardi T (1996) Evaluation of frost tolerance of somatic hybrids *Solanum commersonii* (+) *Solanum tuberosum* XL. Ann. Meet. Italian Soc. Agric. Gene., Perugia, Italy, pp. 218-219.

Carotenuto N, Martin R, Mok MC, Mok DWS (1996) SSCP analysis for chromosomal characterization of somatic hybrids in the *Solanum* genus. Xl Ann. Meet. Italian Soc. Agric. Genet., Perugia, Italy, 58 pp.

Carputo D, Barone A, Cardi T, Sebastiona A, Frusciante L, Peloquin SJ (1997a) Endosperm balance number manipulation for direct *in vivo* germplasm introgression to potato from a sexually isolated relative (*Solanum commersonii* Dun). Proc. Natl. Acad. Sci USA 94:12031-12017.

Carputo D, Cardi T, Speggiorin M, Zoina ALF (1997b) Resistance to blackleg and tuber soft rot in sexual and somatic interspecific hybrids with different genetic background. Amer. Potato J. 74:161-172.

Chang C-C, Sheen J, Bligny M, Niwa Y, Lerbs-Mache S, Stern DB (1999) Functional analysis of two maize cDNAs encoding T7-like RNA polymerases. Plant Cell 11:911-926.

Che P, Gingerich DJ, Lall S, Howell SH (2002) Global and hormone-induced gene expression changes during shoot development in Arabidopsis. Plant Cell 14:2771-2785.

Chen, Y-K-H, Bamberg, JB, Palta JP (1999a) Expression of freezing tolerance in the inerspecific F1 and somatic hybrids of potatoes. Theor Appl. Genet. 98:995-1004.

Chen, Y-K-H, Palta JP, Bamberg, JB (1996b) Freezing tolerance and tuber production in selfed and backcross progenies derived from somatic hybrids between *Solanum tuberosum* L. and *S. commersonii* Dun. Theor. Appl. Genet. 99:100-107.

Chen, Y-K-H, Palta, JP, Bamberg JB, Kim H, Haberlach GT, Helgeson JP (1999c) Expression of nonacclimated freezing tolerance and cold acclimation capacity in somatic hybrids between hardy wild *Solanum* species and cultivated potatoes. Euphytica 107:1-8.

Cheng S-H, Sheen J, Gerrish C, Bolwell GP (2001) Molecular identification of phenylalanine ammonia-lyase as a substrate of a specific constitutively active *Arabidopsis* CDPK expressed in maize protoplasts. FEBS Lett 503:185-188.

Chetrit P, Mathieu C, Vedel F, Pelletier G, Primard C (1985) Mitochnodrial DNA polymophism induced by protoplast fusion in *Cruciferae.* Theor. Appl. Genet. 69:361-366.

Chiu W-L, Niwa Y, Zeng W, Hirano T, Kobayashi H, Sheen J (1996) Engineered GFP as a vital reporter in plants. Curr Biol 6:325-330.

Cho MH, Spalding EP (1996) An anion channel in *Arabidopsis* hypocotyls activated by blue light. Proc Natl Acad Sci USA 93:8134-8138

Chory J, Wu D (2001) Weaving the complex web of signal transduction. Plant Phys. 125:77-80.

Chuong PV, Beversdorf WD, Pauls KP (1987) Plant regeneration from haploid stem peel protoplasts of *Brassica napus* L. J Plant Phys. 130:57-65.
Chuong PV, Beversdorf WD, Powell AD, Pauls KP (1988a) Somatic transfer of cytoplasmic traits in *Brassica napus* L. by haploid protoplast fusion. Mol. Gen. Genet. 211:197- 201.
Chuong PV, Beversdorf WD, Powell AD, Pauls KP (1988b) The use of haploid protoplast fusion to combine cytoplasmic atrazine resistance and cytoplasmic male sterility in *Brassica napus*. Plant Cell Tissue Organ Culture 12:181-184.
Cocking EC (1960) A method for the isolation of plant protoplasts and vacuoles. Nature 187:927-929.
CohenY, Qu F, Gisel A, Morris TJ, Zambryski PC (2000) Nuclear localization of turnip crinkle virus movement protein p8. Virology 273:276-285.
Coursol S, Fan LM, Le Stunff H, Spiegel S, Gilroy S, Assmann SM. (2003) Sphingolipid signaling in *Arabidopsis* guard cells involves heterotrimeric G proteins. Nature 423:651-654.
Crossway A, Oakes JV, Irvine JM, Ward B, Knauf VC, Shewmaker CK (1986) Integration of foreign DNA following microinjection of tobacco mesophyll protoplasts. Molec. Gen. Genet. 202:179-185.
Czarnecka-Verner E, Yuan CX, Scharf KD, Englich G, Gurley WB (2000) Plants contain a novel multi-member class of heat shock factors without transcriptional activator potential. Plant Molec. Biol. 43:459-471.
D'Agostino IB, Deruere J, Kieber JJ (2000) Characterization of the response of the Arabidopsis regulator gene family to cytokinin. Plant Phys. 124:1706-1717.
Dai N, Schaffer A, Petreikov J, Shahak Y, et al. (1999) Overexpression of *Arabidopsis* hexokinase in tomato plants inhibits growth, reduces photosynthesis, and induces rapid senescence. Plant Cell 11:177-189.
Daniell H, Streatfield SJ, Wycoff K (2001) Medical molecular farming: production of antibodies, biopharmaceuticals and edible vaccines in plants. Trends Plant Sci 6:219-225.
De Santis-Maciossek G, Kofer W, Bock A, Schoch S, et al. (1999) Targeted disruption of the plastid RNA polymerase genes *rpoA, B* and *C1*: molecular biology, biochemistry and ultrastructure. Plant J. 18:477-489.
Dhonukshe P, Gadella Jr TWJ (2003) Alteration of microtubule dynamic instability during preprophase band formation revealed by yellow fluorescent protein–clip170 microtubule plus-end labeling. Plant Cell 15:597-611.
Dickson MH (1985) Male sterile persistent white curd cauliflower NY 7642 A and its maintainer NY 7642B. HortSci 20:957.
Doelling JH, Pikaard CS (1993) Transient expression in *Arabidopsis thaliana* protoplasts derived from rapidly established cell suspension cultures. Plant Cell Rep 12:241-244.
Donaldson, PA, Bevis E, Pandeya R, Gleddie S (1995) Rare symmetric and asymmetric *Nicotiana tabacum* (plus) *N. megalosiphon* somatic hybrids recovered by selection for nuclear encoded resistance genes and in the absence of genome inactivation. Theor. Appl. Genet. 91:747-755.
Downey P, Szabo I, Ivashikina N, Negro A, et al. (2000) KDC1, a novel carrot root hair K^+ channel. Cloning, characterization, and expression in mammalian cells. J. Biol. Chem. 275:39420-39426.

Dudits D, Rasko I, Hadlaczky G, Lima-de-Faria A. (1976) Fusion of human cells with carrot protoplasts induced by polyethylene glycol. Hereditas 82:121-3.

Dushenkov S, Skarzhinskaya M, Glimelius K, Gleba D, Raskin I. (2002) Bioengineering of a phytoremediation plant by means of somatic hybridization. Int. J. Phytoremed. 2002:117-26.

Dyer TA (1985) The chloroplast genome and its products. Oxford Surveys. Plant Molec. Cell. Biol. 2:147-177.

Earle E.D, Temple M., Walters TW (1991) Organelle assortment and mitochondrial DNA rearrangements in *Brassica* somatic hybrids and cybrids. Phys. Planta 85:325-333.

Eulgesn T, Rushton PJ, Schmelzer E, Hahlbrock K, Somssich IE (1999) Early nuclear events in plant defense signaling:rapid gene activation by WRKY transcription factors. EMBO J. 18:4689-4699.

Evans DA, Sharp. WR, 1983. Single gene mutations in tomato plants regenerated from tissue culture. Science 22:949-951.

Fan Z, Stefansoon BR, Sernyk JL (1986) Maintainers and restorers for three male sterility inducting cytoplasms in Rape, *Brassica napus*. Can. J. Plant Sci. 66:229-234.

Fath A, Bethke P, Beligni V, Jones R (2002) Active oxygen and cell death in cereal aleurone cells J. Exp. Bot. 53:1273-1282.

Frohnmeyer H, Hahlbrock K, Schafer E (1994) A light-responsive in vitro transcription system from evacuolated parsley protoplasts. Plant J. 5:437-449.

Fromm M, Taylor LP, Walbot V (1985) Expression of genes transferred into monocot and dicot plant cells by electroporation. Proc. Natl. Acad. Sci. USA 82:5824-5828.

Galbraith DW, Galbraith JEC (1979) A method for identification of fusion of plant protoplats derived from tissue cultures. Z Pflanzenphys 93:149-158.

Galbraith DW, Hazenberg LA, Anderson MT (1999) Flow cytometric analysis of transgene expression in higher plants: green fluorescent protein. Meth. Enzym. 302:296-315.

Galbraith DW, Lambert GM, Grebenok RJ, Sheen J. (1995) Flow cytometric analysis of transgene expression in higher plants: green-fluorescent protein. Meth. Cell. Biol. 50:3-14.

Gallie DR, Bailey-Serres J (1997) Eyes off transcription! The wonderful world of post-transcriptional regulation. Plant Cell 9:667-673.

Gallie DR, Lucas WJ, Walbot V (1989) Visualizing mRNA expression in plant protoplasts: factors influencing efficient mRNA uptake and translation. Plant Cell 1:301-311.

Gallie DR, Sleat DE, Watts JW, Turner PC, Wilson TM (1987) A comparison of eukaryotic viral 5′-leader sequences as enhancers of mRNA expression in vivo. Nucl. Acid. Res. 15:8693-8711.

Galun E (1982) Somatic cell fusion for inducing cytoplasmic exchange: a new biological system for cytoplasmic genetics in higher plants. *In:* I Vasil, WR Scowcroft, KJ Frey, (eds.), Plant Improvement and Somatic Cell Genetics. Acad. Press Inc., pp. 205-219.

Gamborg OL, Miller RA, Ojima K (1968) Nutrient requirements of suspension cultures of soybean root cells. Exp. Cell Res. 50:151-158.

Gilroy S, Jones RL (1992) Gibberellic acid and abscisic acid coordinately regulate cytoplasmic calcium and secretory activity in barley aleurone protoplasts. Proc. Natl. Acad. Sci. USA 89:3591-3995.

Glimelius K, Djupsjobacka M, Fellner-Felddegg H (1986) Selection and enrichment of plant protoplast heterokaryons of *Brasicaceae* by flow sorting Plant Sci. 45:133-141.

Goodall GJ, Filipowicz W (1989) The AU-rich sequences present in the introns of plant nuclear pre-mRNAs are required for splicing. Cell 58:473-483.

Gopalakrishnan B, Sonthayanon B, Ramatullah R, Muthukrishnan S (1991) Barley aleurone layer cell protoplasts as a transient expression system. Plant Molec. Biol. 16:463-467.

Gosti F, Beaudoin N, Serizet C, Webb AAR, Vartanian N, Giraudat J (1999) ABI1 protein phosphotase 2C is a negative regulator of ABA signaling. Plant. Cell. 11:1897-1909.

Graham IA, Baker CJ, Leaver CJ (1994) Analysis of the cucumber malate synthase gene promoter by transient expression and gel retardation assays. Plant J. 6:893-902.

Gray WM, Hellmann H, Dharmasiri S, Estelle M (2002) Role of the *Arabidopsis* RING-H2 protein RBX1 in RUB modification and SCF function. Plant Cell 14:2137-2144.

Grosser JW, Gmitter Jr FG, Chandler JL (1988) Intergeneric somatic hybrid plants of *Citrus sinensis* cv Hamlin and *Poncirus trifoliata* cv Flying Dragan. Plant Cell Rep. 7:5-8.

Grosser, JW, Gmitter Jr FG, Chandler JL (1988) Intergeneric somatic hybrid plants from sexually incompatible woody species: *Citrus sinensis* and *Severinia disticha.* Theor. Appl. Genet. 75:397-401.

Grosser, JW, Gmitter Jr FG, Tusa, N, Chandler JL (1990) Somatic hybrid plants from sexually incompatible woody species: *Citrus reticulata* and *Citropis gilletiara.* Plant Cell Rep. 8:656-659.

Gubler F, Raventos D, Keys M, Watts R, Mundy J, Jacobsen JV (1999) Target genes and regulatory domains of the GAMYB transcriptional activator in cereal aleurone. Plant J 17:1-9.

Guilfoyle T, Hagen G, Ulmasov T, Murfett J (1998) How does auxin turn on genes? Plant Physiol. 118:341-347.

Gupta HS, Bhattacharjee B, Pattanayak A (1996) Transfer of cytoplasmic male sterility in Indica rice through protoplast fusion. IRRN 21:33-34.

Hadlaczky G, Burg K, Maroy P, Dudits D (1980) DNA synthesis and division in interkingdom heterokaryons. *In Vitro* 16:647-50.

Hagenbeek D, Rock CD (2001) Quantitative analysis by flow cytometry of abscisic acid-inducible gene expression in transiently transformed rice protoplasts. Cytometry 45:170-179.

Hamilton DW, Hills A, Kohler B, Blatt MR (2000) Ca^{2+} channels at the plasma membrane of stomatal guard cells are activated by hyperpolarization and abscisic acid. Proc. Natl. Acad. Sci. USA 97:4967-4972.

Hanneman RE, Bamberg JB (1986) Inventory of tuber bearing *Solanum* species. Bull. 533, Univ. Wisconsin, Madison, WI.

Harkins KR, Galbraith DW. (1987) Factors governing the flow cytometric analysis and sorting of large biological particles. Cytometry 8:60-70.

Harkins KR, Jefferson RA, Kavanagh TA, Bevan MW, Galbraith DW (1990) Expression of photosynthesis-related gene fusions is restricted by cell type in transgenic plants and in transfected protoplasts. Proc. Natl. Acad. Sci. USA 87:816-820.

Hattori T, Vasil V, Rosenkrans L, Hannah LC, McCarty DR, Vasil IK (1992) The Viviparous-1 gene and abscisic acid activate the C1 regulatory gene for anthocyanin biosynthesis during seed maturation in maize. Genes. Dev. 6:609-618.

Hauffe KD, Paszkowski U, Schulze-Lefert P, Hahlbrock K, Dangl JL, Douglas CJ (1991) A parsley 4CL-1 promoter fragment specifies complex expression patterns in transgenic tobacco. Plant Cell 3:435-443.

Hauptmann RM, Ozias-Akins P, Vasil V, Tabaeizadeh Z, et al. (1987) Transient expression of electroporated DNA in monocotyledonous and dicotyledonous species. Plant Cell Rep. 6:265-270.

Heinlein M, Epel BL, Padgett HS, Beachy RN (1995) Interaction of tobamovirus movement proteins with the plant cytoskeleton. Science 270:1983-1985.

Heise A, Lippok B, Kirsch C, Hahlbrock K (2002) Two immediate early pathogen responsive members of the AtCMPG gene family in *Arabidopsis thaliana* and the W-box-containing elicitor-response element of AtCMPG1. Proc. Natl. Acad. Sci. USA 99:9049-9054.

Hillmer S, Gilroy S, Jones RL (1992) Visualizing enzyme secretion from individual barley *(Hordeum vulgare)* aleurone protoplasts. Plant Physiol. 102:279-286.

Hirochika, H (1993) Activation of retrotransposons during tissue culture. EMBO J. 12:2521-2528.

Horita M, Morohashi H, Komai F (2002) Regeneration of flowering plants from difficult lily protoplasts by means of a nurse culture. Planta 215:880-884.

Hoth S, Morgante M, Sanchez J-P, Hanafey MK, Tingey SV, Chua N-H (2002) Genome-wide gene expression profiling in *Arabidopsis thaliana* reveals new targets of abscisic acid and largely impaired gene regulation in the abi1-1 mutant. J. Cell. Sci. 115:4891-4900.

Huang Z, Han Y, Howell SH (2000) Formation of surface tubules and fluorescent foci in *Arabidopsis thaliana* protoplasts expressing a fusion between the green fluorescent protein and the cauliflower mosaic virus movement protein. Virology 271:58-64.

Huttley AK, Baulcombe DC (1989) A wheat ∀-Amy-2 promoter is regulated by gibberellin in transformed oat aleurone protoplasts. EMBO J. 8:1907-1913.

Hwang I, Sheen J (2001) Two-component circuitry in *Arabidopsis* cytokinin signal transduction. Nature 413:383-389.

Hwang I, Chen H-C, Sheen J (2002) Two-component signal transduction pathways in *Arabidopsis*. Plant Phys 129:500-515.

Inoue T, Higuchi M, Hashimoto Y, Seki M, et al. (2001) Identification of CRE1 as a cytokinin receptor from *Arabidopsis*. Nature 409:1060-1063.

Ito M, Araki S, Matsunaga S, Itoh T, et al. (2001) G2/M-phase-specific transcription during the cell cycle is mediated by c-Myb-like transcription factors. Plant Cell 13:1891-1905.

Jacobsen JV, Beach LR (1985) Control of transcription of ∀-amylase and rRNA genes in barley aleurone protoplasts by gibberellin and abscisic acid. Nature 316:275-277.

Jacobsen JV, Close TJ (1991) Control of transient expression of chimeric genes by gibberellic acid and abscisic acid in protoplasts prepared from mature barley aleurone layers. Plant Mol. Biol. 16:713-724.

Jang JC, Leon P, Zhou L, Sheen J (1997) Hexokinase as a sugar sensor in higher plants. Plant Cell 9:5-19.

Jefferson RA, Kavanagh TA, Bevan MW (1967) GUS Fusion: b-glucoronidase as a sensitive and versatile gene fusion marker in higher plants. EmBOJ. 6: 3901-3907.

Jin JB, Kim YA, Kim SJ, Lee SH, et al. (2001) A new dynamin-like protein, ADL6, is involved in trafficking from the trans-Golgi network to the central vacuole in *Arabidopsis*. Plant Cell 13:1511-1526.

Jones H(1995) Gene transfer into plant protoplasts by electroporation. Methods Molec. Biol. 49:107-12.

Jourdan PS, Earle ED, Mutschler MA (1988) Synthesis of male sterile, triazine-resistant *Brassica napus* by somatic hybridization between cytoplasmic male sterile *B. oleracea* and atrazine-resistant B. campestris. Theor. Appl. Genet. 78:445-455.

Kakimoto T (1996) CKI1, a histidine kinase homolog implicated in cytokinin signal transduction. Science 174:982-985.

Kao CY, Cocciolone SM, Vasil IK, McCarty DR (1996) Localization and interaction of the cis-acting elements for abscisic acid, VIVIPAROUS1, and light activation of the C1 gene of maize. Plant Cell 8:1171-1179.

Kao HM, Keller WA, Gleddie S, Brown GG (1991) Synthesis of *Brassica oleracea/ Brassica napus* somatic hybrid plants with novel organelle DNA compositions. Theor. Appl. Genet. 83:313-320.

Kao KN, Michayluk MR (1974) A method for high-frequency intergeneric fusion of plant protoplasts. Planta 115:355-367.

Kao KN, Michayluk MR (1975) Nutritional requirements for growth of *Vicia hajastanai* cells and protoplasts at a very low population density in liquid media. Planta 126:105-110.

Kemble RJ, Yarrow SA, Wu S-C, Barsby TL (1987). Abscence of mitochondrial and chloroplast DNA recombinations in *Brassica napus* plants regenerated from protoplasts, protoplast fusions and anther culture. Theor. Appl. Genet. 75:875-881.

Kim H, Choi Su, Chae MD, Wielgus SM, Helgeson JP (1993) Identification of somatic hybrids produced by protoplast fusion between *Solanum commersonii* and *S. tuberosum* haploid. Korean J Plant Tiss. Cult. 20:337-344.

Kirschner M, Winkelhaus S, Thierfelder JM, Nover L (2000) Transient expression and heat-stress-induced co-aggregation of endogenous and heterologous small heat-stress proteins in tobacco protoplasts. Plant J. 24:397-411.

Kleiner O, Kircher S, Harter K, Batschauer A (1999) Nuclear localization of the *Arabidopsis* blue light receptor cryptochrome. Plant J. 19:289-296.

Kofer W, Glimelius K, Bonnett H. (1991). Fusion of male-sterile tobacco causes modifications of mtDNA leading to changes in floral morphology and restoration of fertility in cybrid plants. Phys. Planta 85:334-338.

Kofer W, Eibl C, Steinmuller K, Koop H-U (1998) PEG-mediated plastid transformation in higher plants. In Vitro Cell Dev. Biol. Plant 34:303-309.

Koop HU, Steinmuller K, Wagner H, Rossler C, Eibl C, Sacher L (1996) Integration of foreign sequences into the tobacco plastome via polyethylene glycol-mediated protoplast transformation. Planta 199:193-201.

Kovtun Y, Chiu W-L, Zeng W, Sheen J (1998) Suppression of auxin signal transduction by MAPK cascade in higher plants. Nature 395:716-720.

Kovtun Y, Chiu W-L, Tena G, Sheen J (2000) Functional analysis of oxidative stress-activated MAPK cascade in plants. Proc. Natl. Acad. Sci. USA 97:2940-2945.

Krens FA, Molendijk L, Wullems GJ, Schilperoot RA (1982) *In vitro* transformation of plant protoplasts with Ti-plasmid DNA. Nature 296:72-74.

Kubitscheck U, Homann U, Thiel G (2000) Osmotically evoked shrinking of guard-cell protoplasts causes vesicular retrieval of plasma membrane into the cytoplasm Planta 210:423-431.

Kyozuka J, Hayashi Y, Shimamoto S (1987) High frequency plant regeneration from rice protoplasts by novel nurse culture methods Molec. Gen. Genet. 206:408-413.

Kyozuka J, Kaneda T, Shimamoto K (1989) Production of cytoplasmic male sterile rice (*Oryza sativa*) by cell fusion. Bio./Tech. 7:1171-1174.

Laferriere LT, Helgeson, JP, Allen C (1999) Fertile *Solanum tuberosum* + *S. commersonii* somatic hybrids as sources of resistance to bacterial wilt caused by *Ralstonia solanacearum*. Theor. Appl. Genet. 98:1272-1278.

Landgren M, Glimelius K. (1994). A high frequency of intergenomic mitochondrial recombination and an overall biased segregation of *B. campestris* or recombined *B. campestris* mitochondria were found in somatic hybrids made within *Brassicaceae.* Theor. Appl. Genet. 87:854-862.

Larkin PJ, Scowcroft WR (1981) Somaclonal variation—a novel source of variability from cell cultures for plant improvement. Theor. Appl. Genet. 60:197-214.

Leister RT, Katagiri F (2000) A resistance gene product of the nucleotide binding site leucine rich repeats class can form a complex with bacterial avirulence proteins *in vivo.* Plant J. 22:345-354.

Lelivelt CLC, Leunissen EHM, Frederiks HJ, Helsper JPFG, Krens FA (1993) Transfer of resistance to the beet cyst nematode (*Heterodera schachtii* Schm.) from *Sinapis alba* L. (white mustard) to the *Brassica napus* L. gene pool by means of sexual and somatic hybridization. Theor. Appl. Genet. 85:688-696.

Li Y-G, Stoutjestijk PA, Larkin PJ (1999) Somatic hybridization for plants. *In:* W-Y Soh, SS Bhojwani, (eds.) Improvement In Morphogenesis in Plant Tissue Cultures. Kluwer Acad. Publ. Boston, MA, pp. 363–418.

Li Y-G, Tanner G J, Delves AC, Larkin J (1993). Asymmetric somatic hybrid plants between *Medicago sativa* L. (alfalfa, lucerne) and *Onobrychis viciifolia* Scop. (sainfoin). Theor. Appl. Genet. 87:455-463.

Li Y-G, Tanner K, Stoutjesdijk P, and Larkin P (1996). A highly sensitive method for screening of condensed tannins and its application in plants. *In:* J Vercauteren et al., (eds.) Polyphenols Communications 96, Vol. 1: Groupe Polyphenols, Universe Bordeaux 2, Bordeaux Cedex, France, pp.195-196.

Lipphardt S, Brettschneider R, Kreuzaler F, Schell J, Dangl JL (1988) UV-inducible transient expression in parsley protoplasts identifies regulatory cis-elements of a chimeric *Antirrhinum majus* chalcone synthase gene. EMBO J. 7:4027-4033.

Liu J-H, Dixeluis C, Eriksson I, Glimelius K. (1995) *Brassica napus* (+) *B. tournefortii*, a somatic hybrid containing traits of agronomic importance for rapeseed breeding. Plant Sci. 109:75-86.

Liu J-H, Landgren M, Glimelius K (1996) Transfer of the *Brassica tournefortii* cytoplasm to *B. napus* for the production of cytoplasmic male sterile *B. napus*. Phys. Plant 96:123-129.

Liu ZB, Ulmsov T, Shi X, Hagen G, Guilfoyle TJ (1994) Soybean GH3 promoter contains multiple auxin-inducible elements. Plant Cell 6:645-657.

Loake GJ, Choudhary AD, Harrison MJ, Mavandad M, Lamb CJ, Dixon RA (1991) Phenylpropanoid pathway intermediates regulate transient expression of a chalcone synthase gene promoter. Plant Cell 3:829-840.

Lorz H, Scowcroft WR (1983) Variability among plants and their progeny regenerated from protoplasts of Su/su heterozygotes of *Nicotiana tabacum*. Theor. Appl. Genet. 66:67-75.

Luehrsen KR, de Wet JR, Walbot V (1992) Transient expression analysis in plants using firefly luciferase reporter gene. Meth Enz 216:397-414.

Maathuis FJ, Sanders D (1994) Mechanism of high-affinity potassium uptake in roots of *Arabidopsis thaliana*. Proc. Natl. Acad. Sci. USA 91:9272-9276.

Makonkawkeyoon S, Smitamana P, Hirunpetharat C, Maneekarn N (1995) Production of mouse immunoglobulin G by a hybrid plant derived from tobacco-mouse cell fusions. Experientia 51:19-25.

Marcotte Jr WR, Bayley CC, Quatrano RS (1988) Regulation of a wheat promoter by abscisic acid in rice protoplasts. Nature 335:454-457.

Mattheij WM, Puite KJ (1992) Tetraploid potato hybrids through protoplast fusions and analysis of their performance in the field. Theor. Appl. Genet. 83:807-812.

McCarty DR, Hattori T, Carson CB, Vasil V, Lazar M, Vasil IK (1991) The viviparous-1 developmental gene of maize encodes a novel transcriptional activator. Cell 66:895-905.

McLean BG, Zupan J, Zambryski PC (1995) Tobacco mosaic virus movement protein associates with the cytoskeleton in tobacco cells. Plant Cell 7:2101-2114.

Medgyesy P, Menczel L, Maliga P (1980) The use of cytoplasmic streptomycin resistance: chloroplast transfer from *Nicotiana tabacum* into *Nicotiana sylvestris*, and isolation of their somatic hybrids. Molec. Gen. Genet. 179:693-698.

Melchers G, Sacristan MD and Holder AA (1978) Somatic hybrid plants of potato and tomato regenerated from fused protoplasts. Carlsberg Res. Comm. 43:203-218.

Melzer JM, O'Connell MA (1992) Effect of radiation dose on the production of and the extent of asymmetry in tomato asymmetric somatic hybrids. Theor. Appl. Genet. 83:337-344.

Menczel L, Nagy F, Kiss ZsR, Maliga P (1981) Streptomycin resistant and sensitive somatic hybrids of *Nicotiana tabacum* + *Nicotiana knightiana*: correlation of resistance to *N. tabacum* plastids. Theor. Appl. Genet. 59:191-195.

Menczel L, Nagy F, Lazar G, Maliga P (1983) Transfer of cytoplasmic male sterility by selection for streptomycin resistance after protoplast fusion in *Nicotiana*. Molec. Gen. Genet. 189:365-369.

Mendis MH, Power JB, Davey MR (1991) Somatic hybrids of the forage legumes *Medicago sativa* L. and *M. falcata*. J. Exp. Bot 42:1565-1573.

Miller SA, Williams GR, Medina-Filho H, Evans DA (1985) A somaclonal variant of tomato resistant to race 2 of *Fusarium oxysporum* f. sp.*lycopersici.* Phytopath. 75:1354.

Moran N, Ehrensten G, Iwasa K, Bare C, Mischke C (1984) Ion channels in plasmalemma of wheat protoplasts. Science 226:835-838.

Murashige T, Skoog F (1962). A revised medium for rapid growth and bioassays with tobacco tissue cultures. Physiol. Plant 15:473-497.

Murata Y, Pei Z-M, Mori IC, Schroeder J (2001) Abscisic acid activation of plasma membrane Ca^{2+} channels in guard cells requires cytosolic NAD(P)H and is differentially disrupted upstream and downstream of reactive oxygen species production in abil-1 and abi2-1 protein phosphatase 2C mutants. Plant Cell13:2513-2523.

Naess, SK, Bradeen JM, Wielgus SM, Haberlach GT, McGrath JM, Helgeson JP (2000) Resistance to late blight in *Solanum bulbocastanum* is mapped to chromosome 8. Theor. Appl. Genet. 101:697-704.

Naess, SK, Bradeen JM, Wielgus SM, Haberlach GT, McGrath JM, Helgeson JP (2001) Analysis of introgression of *Solanum bulbocastanum* DNA into potato breeding lines. Molec. Genet. Genom 265:694-704.

Nagata T, Bajaj YPS (eds.) (2001) Somatic Hybridization in Crop Improvement: Biotechnology in Agriculture and Forestry 49. Springer-Verlag Berlin.

Nagy JI and Maliga Z (1976) Callus induction and plant regeneration from mesophyll protoplasts of *Nicotiana sylvestris.* Z. Pflanzenphysiol. 78:453.

Neeleman L, Olsthoorn RCL, Linthorst HJM, Bol JF (2001) Translation of a nonpolyadenylated viral RNA is enhanced by binding of viral coat protein or polyadenylation of the RNA. Proc. Natl. Acad. Sci. USA. 98:14286-14291.

Newton KJ (1988) Plant mitochondrial genomes: organization, expression and variation. Ann. Rev. Plant Physiol. Plant Molec. Biol. 39:503-532.

Ni M, Dehesh K, Tepperman JM, Quail PH (1996) GT-2: *in vivo* transcriptional activation activity and definition of novel twin DNA binding domains with reciprocal target sequence selectivity. Plant Cell 8:1041-1059.

Nimchuk Z, Marois E, Kjemtrup S, Leister RT, Katagiri F, Dangl JL (2000) Eukaryotic fatty acylation drives plasma membrane targeting and enhances function of several type III effector proteins from *Pseudomonas syringae.* Cell 101:353-363.

Nyman M, Waara S (1997) Characterization of somatic hybrids between *Solanum tuberosum* and its frost-tolerant relative *Solanum commersonii.* Theor. Appl. Genet. 95:1127-1132.

Oghawara T, Saito W, Kobayashi S (1977) Production of Hybrids and cybrids in the *Ruaceae* family and application to citrus breeding. Plant Biotech. 14:141-144.

Ohgawara T, Kobayashi S, Ohgawara E, Uchimiya H, Ishii S (1985) Somatic hybrid plants obtained by protoplast fusion between *Citrus sinensis* and *Poncirus trifoliata.* Theor. Appl. Gent. 78:609-612.

Ohyama K, (1983) Genetic transformation in plants. *In:* DA Evans, WR Sharp, PV Amminrato, Y Yamada, (eds.) Handbook of Plant Cell Culture. MacMillan Publi. Co., New York, NY, Vol 1, pp. 501-509.

Oiyama I, Kobayashi S, Yoshinga K, Ohgarawa T, Ishii S (1991) Use of pollen from somatic hybrid between *Citrus* and *Poncirus* in the production of triploids. HortSci 26:1082.

Olivares-Fuster O, Pena L, Duran-Vila, Navarro L (2002) Green fluorescent protein as a visual marker for somatic hybridization. Ann. Bot. 89:491-497.

Pandey S, Wang X, Coursol SA, Assmann SM (2002) Preparation and applications of *Arabidopsis thaliana* guard cell protoplasts. New Phytol. 153:517-526.

Parrella G, Cardi T (1999) Transfer of a new PVX resistance gene from *Solanum commersonii* to *S. tuberosum* through somatic hybridization. J. Genet. Breed 53:359-362.

Pasternak TP, Prinsen E, Ayaydin F, Miskolczi P, et al. (2002) The role of auxin, pH, and stress in the activation of embryogenic cell division in leaf protoplast-derived cells of alfalfa. Plant Physiol 129:1807-19.

Pauls, KP, Chuong PV (1987) Flow cytometric identification of fusion products formed after PEG induced fusion of *Brassica* protoplasts. Can. J. Bot. 65:834-838.

Pauls PK, Kunert K, Hattner E, Grandbastein MA (1994) Expression of Tnt1 retrotransposon promoter in heterologous species. Plant Molec. Biol. 16:393-902.

Pei ZM, Murata Y, Benning G, Thomine S, et al. (2000) Calcium channels activated by hydrogen peroxide mediate abscisic acid signaling in guard cells. Nature 406:731-734.

Pelletier G, Primard C, Vedel F, Chetrit P, et al. (1983). Intergeneric cytoplasmic hybridization in *Cruciferae* by protoplast fusion. Molec. Gen. Genet. 191:244-250.

Potrykus I (1990) Gene transfer to cereals: an assessment. Bio/Tech. 8:535-542.

Potrykus I, Shillito RD, Saul M, Paszkowski J (1985) Direct gene transfer: state of the art and future perspectives. Plant Molec. Biol. Rep. 3:117-128.

Pouteau S, Huttner E, Grandbastien MA, Caboche M (1991) Specific expression of the tobacco Tnt1 retrotransposon in protoplasts. EMBO J. 10:1911-1918.

Przetakiewicz J, Nadolska-Orczyk A, Orczyk W (2002) The use of RAPD and semi-random markers to verify somatic hybrids between diploid lines of *Solanum tuberosum* L. Cell. Molec. Biol. Lett. 7:671-676.

Qu F, Morris TJ (2000) Cap-independent translational enhancement of turnip crinkle virus genomic and subgenomic RNAs. J. Virol. 74:1085-1093.

Ramos JA, Zenser N, Leyserc O, Callis J (2001) Rapid degradation of auxin/ indoleacetic acid proteins requires conserved amino acids of domain II and is proteasome dependent. Plant Cell 13:2349-2360.

Reisch B (1983) Genetic variability in regenerated plants. *In:* DA Evans, WR Sharp, PV Ammirato. Y Yamada, (eds.) Handbook of Plant Cell Culture. Macmillan Publ. Co., New York, NY, Vol. 1, pp. 748-769.

Roest R and Gilissen JW (1989) Regeneration from protoplasts: a literature review. Acta. Bot. Neerl. 38:1-23.

Roest R, Gilissen JW (1993) Regeneration from protoplasts: a supplementary literature review. Acta. Bot. Neerl. 42:1-23.

Rolland F, Moore B, Sheen J (2002) Sugar sensing and signaling in plants. Plant Cell 14:S185-S205.

Rose RJ, Thomas MR, Fitter JT (1990) The transfer of cytoplasmic and nuclear genomes by somatic hybridization. Aust. J. Plant. Phys. 17:303-321.

Rushton PJ, Reinstädler A, Lipka V, Lippok B, Somssich IE (2002) Synthetic plant promoters containing defined regulatory elements provide novel insights into pathogen-and wound-induced signaling. Plant Cell 14:749-762.

Sadka A, DeWald DB, May GD, Park WD, Mullet JE (1994) Phosphate modulates transcription of soybean VspB and other sugar-inducible genes. Plant Cell 6:737-749.

Schaeffner AR, Sheen J (1991) Maize rbcS promoter activity depends on sequence elements not found in dicot rbcS promoters. Plant Cell 3:997-1012.

Schenk RU, Hildebrandt AC (1972) Medium and techniques for induction and growth of monocotyledonous and dicotyledonous plant cell cultures. Can. J. Bot. 50:199-204.

Schieder O. 1982. Somatic hybridisation: a new method for plant improvement. *In:* IK Vasil, WR Scowcroft, KJ Frey, (eds.) Plant Improvement and Somatic Cell Genetics. Acad. Press Inc., London, UK, pp. 239-253.

Schnable PS, Wise RP (1998) The molecular basis of cytoplasmic male sterility and fertility restoration Trends. Plant Sci. 3:175-180.

Schroeder JI, Hedrich R, Fernandez JM (1984) Potassium-selective single channels in guard cell protoplasts of *Vicia faba*. Nature 312:361-362.

Schwieger HG, Dirk J, Koop HU, Kranz E, Neuhaus G, Spangenkerg G, Wolf D (1987) Individual selection, culture and manipulation of higher plant cells. Theor. Appl. Genet. 73:769-783.

Scowcroft WR, Larkin PJ (1981) Chloroplast DNA assorts randomly in interspecific somatic hybrids of *Nicotiana debneyi*. Theor. Appl. Genet. 60:179-184.

Secor GA, Shepard JF (1981) Variability of protoplast-derived potato clones. Crop Sci. 21:102-105.

Senda M, Takeda J, Abe S, Nakamura T (1979) Induction of cell fusion of plant protoplasts by electrical stimulation. Plant Phys. 20:1441-1443.

Seppänen MM, Cardi T, Borg Hyökki M, Pehu E (2000) Characterization and expression of cold-induced glutathione S-transferase in freezing-tolerant *Solanum commersonii*, sensitive *S. tuberosum* and their interspecific somatic hybrids. Plant Sci 153:125-133.

Shahin EA, Spivey R (1986) A single dominant gene for Fusarium wilt resistance in protoplast-derived tomato plants. Theor. Appl. Genet. 73:164-169.

Sheen J (1990) Metabolic repression of transcription in higher plants. Plant Cell 2:1027-1038.

Sheen J (1993) Protein phosphatase activity is required for light-inducible gene expression in maize. EMBO J. 12:3497-3505.

Sheen J (1996) Specific Ca^{2+}-dependent protein kinase in stress signal transduction. Science 274:1900-1902.

Sheen J (1998) Mutational analysis of protein phosphatase 2C involved in abscisic acid signal transduction in higher plants. Proc. Natl. Acad. Sci. USA 98:975-980.

Sheen J (2001) Signal transduction in maize and Arabidopsis mesophyll protoplasts. Plant Physiol. 127:1466-1475.

Sheen J, Hwang S, Niwa Y, Kobayashi H, Galbraith DW (1995) Green-fluorescent protein as a new vital marker in plant cells. Plant J. 8:777-784.

Shillito RD, Paszkowski J, Potrykus I (1983) Agarose plating and a bead type culture technique enable and stimulate development of protoplast-derived colonies in a number of plant species. Plant Cell Rep. 2:244-247.

Shinjyo C (1969) Cytoplasmic-genetic male sterility in cultivated rice, *Oryza sativa* L. II. The inheritance of male sterility. Jpn. J. Genet. 44: 149-156.

Shinjyo C (1975) Genetical studies of cytoplasmic male sterility and fertility restoration in rice, *Oryza sativa* L. Sci. Bull. Coll. Agric. Univ. Ryukus. 22:1-51.

Sigareva MA, Earle ED (1996) Direct transfer of a cold-tolerant Ogura male-sterile cytoplasm into cabbage (*Brassica oleracea* ssp. *capitata*) via protoplast fusion. Theor. Appl. Genet. 94:213-220.

Smillie RM, Melchers G, von Wettstein D (1979) Chilling resistance of somatic hybrids of tomato and potato. Carlsberg Res. Commun. 44: 127-132.

Smith SLS, Murakishi HH (1987) Inheritance of resistance to tomato mosaic virus (ToMV-0) in tomato somaclones. TGC Rep. 37:65-66.

Sohn EJ, Kim ES, Zhao M, Kim SJ, et al. (2003) Rha1, an *Arabidopsis* Rab5 homolog, plays a critical role in the vacuolar trafficking of soluble cargo proteins. Plant Cell 15:1057-1070.

Song J, Bradeen JM, Naess SK, Raasch JA, Wielgus SM, Haloerlach GT, Liu J, Kuang H, Austin-Phillips S, Buell CR, Helgeson JP, Jiang J (2003) Gene RB cloned from *Solanum bulbocastanum* confersbroad spectrum resistance to potato late blight Proc. Natl. Acad. Sci. USA 100:9128-9133.

Soost RK, Cameron JM (1985) 'Melogold'a triploid Pummelo-grapefuit hybrid. HortSci 20:1134-1135.

Sprenger-Haussels M, Weisshaar B (2000) Transactivation properties of parsley proline-rich bZIP transcription factors. Plant J. 22:1-8.

Stoelzle S, Kagawa T, Wada M, Hedrich R DietricH P (2003) Blue light activates calcium-permeable channels in *Arabidopsis* mesophyll cells via the phototropin signaling pathway. Proc. Natl. Acad. Sci. USA100:1456-461.

Steffens B, Feckler C, Palme K, Christian M, Bottger M, Luthen H. (2001) The auxin signal for protoplast swelling is perceived by extracellular ABP1. Plant J 27:591-599.

Stewart DN Jr. (2001) The utility of green fluorescent protein in transgenic plants. Plant Cell Rep. 20:376-382.

Stoelzle S, Kagawa T, Wada M, Hedrich R DietricH P (2003) Blue light activates calcium permeable channels in *Arabidopsis* mesophyll cells via the photoropin signaling pathway. Proc. Natl. Acad. Sci. USA 100: 1456-461.

Subramanian R, Desveaux D, Spickler C, Michnick SW, Brisson N (2001) Direct visualization of protein interactions in plant cells. Nature Biotech. 19:769-772.

Taguchi T, Sakamoto K, Terada M (1993) Fertile somatic hybrids between *Petunia hybrida* and a wild species, *Petunia variabilis*. Theor. Appl. Genet. 87:75-80.

Takebe I, Labib G, Melchers G (1971) Regeneration of whole plants from isolated mesophyll protoplasts of tobacco. Naturwissen 58:318-320.

Tanno-Suenaga L, Ichikawa H, Imamura J (1988) Transfer of CMS trait in *Daucus carota* L. by donor recipient protoplast fusion. Theor. Appl. Genet. 76:855-860.

Tao L-Z, Cheung AY, Wua H (2002) Plant Rac-Like GTPases are activated by auxin and mediate auxin-responsive gene expression. Plant Cell 14:2745-2760.

Tena G, Asai T, Chiu W-L, Sheen J (2001) Plant MAP kinase signaling cascades. Curr. Opin. Plant Biol. 4:392-400.

Terada R, Urawa H, Inagaki Y, Tsugane K, Iida S (2002) Efficient targeting by homologous recombination in rice. Nature Biotech. 20:1030-1034.

Thomzik JE, Hain R (1988) Transfer and segregation of triazine tolerant chloroplasts in *Brassica napus* L. Theor. Appl. Genet. 76:165-171.

Tiwari SB, Wang X-J, Hagen G, Guilfoyle TJ (2001) AUX/IAA proteins are active repressors, and their stability and activity are modulated by auxin. Plant Cell 13: 2809-2822.

Toriyama K, Arimoto Y, Uchimiya H, Hinata K (1998) Trangenic rice plants after direct gene transfer into protoplasts. Bio/Tech. 6:1072-1074.

Trewavas A (1999) Le Calcium, C'est la vie: Calcium makes waves. Plant Physiol. 120:1-6.

Ueda T, Yamaguchi M, Uchimiya H, Nakano A (2001) Ara6, a plant-unique novel type Rab GTPase, functions in the endocytic pathway of *Arabidopsis thaliana*. EMBO J. 20:4730-4741.

Ulmasov T, Hagen G, Guilfoyle TJ (1997a) ARF1, a transcription factor that binds to auxin response elements. Science 276:1865-1868.

Ulmasov T, Murfett J, Hagen G, Guilfoyle TJ (1997b) Aux/IAa proteins repress expression of reporter genes containing natural and highly active synthetic auxin response elements. Plant Cell 9: 1963-1971.

Ulmasov T, Hagen G, Guilfoyle TJ (1999) Activation and repression of transcription by auxin-response factors. Proc. Natl. Acad. Sci. USA 96: 5844-5849.

Uno Y, Furihata T, Abe H, Yoshida R, et al. (2000) *Arabidopsis* basic leucine zipper transcription factors involved in an abscisic acid-dependent signal transduction pathway under drought and high-salinity conditions. Proc. Natl. Acad. Sci. USA 97:11632-11637.

Urao T, Noji M-A, Yamaguchi-Shinozaki K, Shinozaki K (1996) A transcriptional activation domain of ATMYB2, a drought-inducible *Arabidopsis* Myb-related protein. Plant Cell 10:1145-1148.

van de Locht U, Meier I, Hahlbrock K, Somssich IE (1990) A 125 bp promoter fragment is sufficient for strong elicitor-mediated gene activation in parsley. EMBO J 9: 2945-2950.

Waara S, Nyman M, Johannison A (1998) Efficient selection of potato heterokaryons by flow cytometric sorting and regeneration of hybrid plants. Euphytica 101:293-299.

Waibel F, Filipowicz W (1990) RNA-polymerase specificity of transcription of *Arabidopsis* U snRNA genes determined by promoter element spacing. Nature 346:199-202.

Walbot V, Cullis CA (1985) Rapid genomic change in higher plants. Ann. Rev. Plant. Phys. 36:367-396.

Wallin A, Glimelius K, Eriksson T (1974) The induction of aggregation and fusion of *Daucus carota* protoplasts by polyethylene glycol. Z. Pflanzenphysiol 74:64-80.

Walters TW, Earle ED (1993) Organellar segregation, rearrangement and recombination in protoplast fusion-derived *Brassica oleracea* calli. Theor. Appl. Genet. 85:761-769.

Weber G, Monajembashi S, Greulich KO, Wolfrum J (1989) Uptake of DNA in chloroplasts of *Brassica napus* (L.) facilitated by a UV-laser microbeam. Eur. J. Cell Biol. 49:73-79.

Wijbrandi J, Wolters AMA, Koornneef M (1990a) Asymmetric somatic hybrids between *Lycopersicon esculentum* and irradiated *Lycopersicon peruvianum*. 2. Analysis with marker genes. Theor. Appl. Genet. 80:665-672.

Wijbrandi J, Zabel P, Koornneef M (1990b) Restriction fragment length polymorphism analysis of somatic hybrids between *Lycopersicon esculentum* and irradiated *L. peruvianum;* evidence for limited donor genome elimination and extensive chromosome rearrangements. Molec. Gen. Genet. 222:270-277.

Worley CK, Zenser N, Ramos J, Rouse D, et al. (2000) Degradation of Aux/IAA proteins is essential for normal auxin signaling. Plant J 21:553-562.

Wu Y, Wood MD, TaoY, Katagiri F (2003) Direct delivery of bacterial avirulence proteins into resistant Arabidopsis protoplasts leads to hypersensitive cell death. Plant J. 33:131.

Yanagisawa S, Sheen J (1998) Involvement of maize Dof zinc finger proteins in tissue-specific and light-regulated gene expression. Plant Cell 10:75-89.

Yang Z-Q, Shikanai T, Mori K, Yamada Y (1989) Plant regeneration from cytoplasmic hybrids of rice (*Oryza sativa* L.). Theor. Appl. Genet. 77:305-310.

Yuan LP (1994) Increasing the yield potential in rice by exploitation of heterosis. *In:* SS Virmanii (ed.) Hybrid Rice Technology. New Developments and Future Prospects. IRRI, Manila, Philippines, pp. 1-6.

Zelcer A, Aviv D, Galun E (1978) Interspecific transfer of cytoplasmic male sterility by fusion between protoplasts on normal *Nicotiana sylvestris* and X-ray-irradiated protoplasts of male sterile *N. tabacum*. Z Pflanzenphys 90:397- 407.

Zimmermann U, Scheurich P (1981) Fusion of *Avena sativa* mesophyll cell protoplasts by electrical breakdown. Biochim. Biophys. Acta. 641:160-165.

5

Plant Coated Vesicles Exposed: Ultrastructural and Biochemical Studies of Cultured Cells and Protoplasts

L.C. Fowke
Department of Biology,
University of Saskatchewan,
112 Science Place, Saskatchewan SK S7N 5E2,
CANADA

This story highlights the work of the dedicated young scientists who worked in my laboratory between 1978 and 1988 as well as important collaborations forged with other research laboratories here in North America and abroad. This story illustrates the importance of applying ultrastructural and bio-chemical approaches to tissue culture systems (suspension cultures and protoplasts) in order to study plant cell organelles. The research I am about to describe resulted in key discoveries which helped to unravel the mystery of plant coated vesicles. By the late 1970s it was clear that animal-coated vesicles were involved in receptor mediated endocytosis, a specific mechanism used by animal cells to internalize proteins (e.g. Goldstein, Anderson, Brown 1979). We now know that in animal cells a variety of proteins bind to specific receptors localized in clathrin-coated plasma membrane invaginations called coated pits. These pits invaginate to form coated vesicles which uncoat and deliver the proteins to different locations in the cells including the lysosome for destruction (e.g. insulin), transport back to the surface following release of cargo (e.g. transferrin), and transport to the opposite cell surface (e.g. IgA) (see Low and Chandra, 1994). Following the intial report of coated vesicles in plants by Bonnett and Newcomb in 1966, these tiny clathrin coated membrane vesicles were observed in a variety of plant tissues (Newcomb 1980) but their function was not known. The general concensus at the time was that coated vesicles were likely involved in exocytosis of wall materials since they were most abundant in rapidly growing cells. The possibility that they might be involved in endocytosis was also considered (Ryser 1979).

STRUCTURE AND DISTRIBUTION OF PLANT COATED VESICLES

The first serious effort in my laboratory to understand plant coated vesicles began in 1978 with the arrival of Pieter Van der Valk, a postdoctoral fellow from the Netherlands. Pieter had a strong background in fungal cell wall structure and was keen to explore the structure and distribution of coated vesicles in cultured plant protoplasts which were actively synthesizing new cell walls in culture. It quickly became clear from his transmission electron micrographs of fixed tobacco protoplasts that coated pits (clathrin coated indentations of the plasma membrane) and coated vesicles were numerous (Plate VI, 1). Pieter produced high resolution images of coated vesicles near the Golgi as well as coated pits and vesicles at the cell surface (Plate VI, 2-5) but from these static images it was not possible to determine whether pits formed vesicles or vice versa. At the time we favored the latter interpretation and envisaged the coated vesicles moving wall matrix polysaccharides from the Golgi to the cell surface. We were to learn later that we were wrong.

Pieter also examined large fragments of protoplast plasma membrane, often referred to as protoplast ghosts or footprints, to examine the distribution of coated pits/vesicles. This method was devised by Marchant and Hines (1979) for examining the inner surface of protoplast plasma membranes of green algae. The method involves fastening protoplasts to plastic-coated electron microscope grids and then osmotically bursting the protoplasts to leave large fragments of plasma membrane attached to the grid. The membrane fragments are rinsed thoroughly to remove cell contents and then stained to reveal organelles associated with the inner surface of the plasma membrane (Plate VI, 6 Van der Valk et al. 1980; Van der Valk and Fowke 1981). Using this method Pieter showed coated structures were abundant on the plasma membranes of tobacco protoplasts from suspension culture cells which were actively synthesizing cell walls. He counted approximately 8 coated structures (pits and/or vesicles) per μm^2 membrane surface (Plate VI, 6; Van der Valk and Fowke 1981). Subsequent studies of the plasma membrane of tobacco leaf protoplasts which were slow to generate new walls revealed only approximately 1 coated structure (pit and/or vesicle) per μm^2 membrane surface (Plate VI, 7; Fowke et al. 1983), suggesting a strong correlation of coated pits and vesicles with wall formation. Pieter's micrographs of the inner surface of negatively stained plasma membrane also clearly illustrated the nature of the patterned clathrin coats consisting of a lattice of hexagons and pentagons (Plate VI, 8). Similar coated structures were reported on the cytoplasmic surface of onion guard cell plasma membrane (Doohan and Palevitz 1980).

ISOLATION AND CHARACTERIZATION OF COATED VESICLES

Professor Eldon Newcomb, University of Wisconsin, Madison, WI and I met at the International Cell Biology Congress in Berlin in 1980 and discussed the possibility of establishing a collaboration in order to isolate plant coated vesicles. In December of 1980 I was invited to Madison to talk about our recent research and investigate further the possibility of working together. Brent Mersey, a PhD student with Eldon, was keen to participate in the project and a collaboration was formalized. A small team was organized including Fred Constabel at the Prairie Regional Laboratory (now Plant Biotechnology Institute) of the National Research Council of Canada in Saskatoon, who agreed to supply cell suspension cultures of tobacco (*Nicotiana tobacum*) and periwinkle (*Catharanthus roseus*), and my research assistant Pat Clay (nee Rennie), with expertise in electron microscopy. In May of 1981 Brent traveled to Saskatoon for a 2-week period to initiate the isolation of coated vesicles. Using a method which involved two sequential sucrose density gradients, Brent was able to prepare the first plant fraction enriched in coated vesicles (Mersey et al. 1982). Negative stained preparations revealed the characteristic basket-like clathrin coats (Plate VII, 9). SDS polyacrylamide gel electrophoresis revealed a putative plant clathrin of 190,000 daltons, 10,000 daltons larger than published values for animal clathrin. Similar results were obtained when coated vesicles were isolated from suspension-cultured cells of periwinkle (unpubl.).

In the fall of 1981 Brent moved to Saskatoon to take up a postdoctoral position with Fred Constabel at the Prairie Regional Laboratory. Although his main focus in his new position was not plant coated vesicles, over the next few years he was able to devote considerable time and effort to the isolation and characterization of these tiny membranous organelles. With the arrival in my laboratory of a new postdoctoral fellow, Larry Griffing, in the summer of 1982, we gained additional strength in the area of plant biochemistry. Larry was well versed in the coated vesicle literature, had expertise in organelle isolation, and was keen to investigate both the structure and function of this fascinating organelle. The team was attracted to the possibility of using protoplasts as a source of coated vesicles since they had proven useful for isolation of other cell organelles (e.g. Fowke and Gamborg 1980). Protoplasts which lack cell walls require only gentle homogenization methods and the absence of a wall eliminates entrapment of cytoplasm by cell wall debris. Furthermore, ultrastructural studies of soybean protoplasts revealed an abundance of cytoplasmic coated vesicles, particularly near the plasma membrane (Van der Valk and Fowke 1981). Using a new method, the team set out to isolate coated vesicles from plant protoplasts derived from suspension cultured cells of soybean (*Glycine max* L., line SB-1). Initial enrichment was achieved by isopycnic centrifugation of a protoplast homogenate through a linear sucrose gradient in a vertical rotor. The coated vesicle fractions from this gradient were pooled and centrifuged through a second linear sucrose gradient in a rate zonal fashion to remove larger contaminating membrane vesicles. Fractions highly enriched in coated vesicles were obtained using this relatively rapid isolation method (Mersey et al. 1985) and SDS polyacrylamide gel electrophoresis confirmed a molecular mass of 190 kdalton for plant clathrin (Plates VII, 10-12). Relatively pure fractions of bovine brain coated vesicles were also obtained using this method (Plate VII, 13) and comparisons of soybean and bovine clathrin clearly showed a difference in molecular mass (Plate VII, 10).

Further experiments using the new isolation method coupled with marker-enzyme analysis revealed that coated vesicles from soybean protoplasts were enriched for glucan-synthase I, an enzyme which may be important in cell wall biosynthesis. However, pulse chase experiments indicated that coated vesicles do not play a major role in matrix polysaccharide secretion during cell wall regeneration by protoplasts (Griffing et al. 1986).

This period of intense activity in my laboratory produced many lasting memories. One I shall never forget related to the isolation and characterization of coated vesicles. I knew that Brent and Larry were making regular trips to the abattoir in Saskatoon to obtain bovine brain material for their experiments but I was unaware of the details of this exercise. I recall my horror when I unexpectedly entered our wood working shop in the basement of the Biology building to find Brent and Larry, rather crude tools in hand, removing the brain from an intact cow's head. I almost fainted. I was assured that this was an emergency situation and that normally the procedure was completed in a proper laboratory in the College of Veterinary Medicine.

ENDOCYTOSIS BY COATED VESICLES IN PLANTS

Mike Tanchak graduated with high honors from Carleton University in the spring of 1982 and joined my laboratory as a graduate student later that summer. He was excited about the possibility of becoming involved in the coated vesicle research program. Mike set about drafting his research proposal which included a section in which he outlined different approaches to determine the direction of movement of coated vesicles in plant cells. One approach we discussed was the use of electron dense markers (e.g. ferritin, colloidal gold) which could be applied to the surface of protoplasts in order to determine whether endocytosis was feasible. At that time there were laboratories in the USA and Germany actively pursuing similar approaches despite the fact that some scientists argued that this type of endocytosis was physically impossible (Cram 1980). Mike was not deterred. He opted for soybean protoplasts as his experimental system and cationized ferritin which he argued should bind electrostatically to the outer surface of freshly isolated protoplasts and should be easily visible by conventional transmission electron microscopy in lightly stained or unstained sections of protoplasts. Mike's schedule of courses limited his research time; however, he managed to initiate labeling experiments using cationized ferritin.

I can clearly recall sitting in the coffee room, early in 1983 I believe, when Mike sauntered in and stood quietly beside me waiting for a break in the conversation. During a brief lull, he leaned over to me and in a quiet voice suggested that he had something in the electron microscope that might interest me. Just a casual suggestion, not a hint of excitement in his voice or eyes. Off we went to the electron microscope in the basement. Mike turned on the beam and I sat down and had a look through the binoculars. And there it was! Cationized ferritin in a coated vesicle within the cytoplasm, some distance from the plasma membrane. Wow! He had done it. The first convincing evidence that plant protoplasts were capable of internalizing material via coated membranes. In short order, Pat, Brent and Larry were at the microscope to have a look. It was indeed an exciting day in the laboratory. For the next few months it was difficult to tear Mike away from the microtome and electron microscope. He cut numerous embedded protoplasts, examined hundreds of sections and produced masses of photographs. Before long it was clear that cationized ferritin was being internalized by coated pits to coated vesicles which uncoated to form smooth vesicles, and then delivered the cationized ferritin to various membranous compartments in the cell including Golgi, a system of partially coated reticulum, multivesicular bodies, and small vacuoles. In November of 1983, Mike traveled to San Antonio, Texas to present his exciting story at the 23rd annual meeting of the American Society for Cell Biology (Tanchak et al. 1983). A more complete report was published in 1984 (Tanchak et al. 1984). By 1984 similar results had been obtained by Robinson's group in Germany working with bean leaf protoplasts (Joachim and Robinson 1984). Many questions remained to be answered. What was the time course of endocytosis and in what order were various cell organelles labeled? Were coated vesicles discrete structures or were they simply invaginations of the plasma membrane? What was the labeled partially coated reticulum near the Golgi and was it attached to the Golgi? Mike and others in the group set about answering these questions.

In order to determine a time course, Mike undertook a series of labeling experiments in which he fixed protoplasts at different times after exposure to cationized ferritin, extending from a few seconds to 3 hours. The analysis was painstaking and involved careful analysis of hundreds of sections. Gradually a picture began to emerge. Cationized ferritin uptake by coated pits (Plate VIII, 14) into coated vesicles (Plate VIII, 15) occurred within seconds at room temperature and very quickly some coated vesicles uncoated. Label appeared in the partially coated reticulum (Plate VIII, 16) within 2 minutes. The next organelle to label was the Golgi (Plate VIII, 17) with the earliest detection of cationized ferritin by 4 minutes. Most of the label appeared at the periphery of one or two Golgi cisternae. The first appearance of cationized ferritin in multivesicular bodies (Plate VIII, 18) occurred after about 6 minutes and the label accumulated with time. Finally after an hour clusters of cationized ferritin were detected in the central vacuole. Figure 19 (Plate IX) presents a diagrammatic interpretation of the uptake data. More complete discussions of the endocytosis of cationized ferritin by soybean protoplasts had been published earlier (Fowke et al. 1985, 1991).

The partially coated reticulum, a system of interconnected tubular membranes bearing clathrin-coated regions, was first described in higher plants by Pesacreta and Lucas (1985). Based on our uptake data, we favored the idea that this membrane system functions as an endosome in plant cells. Mike carefully examined the relationship between the partially coated reticulum and Golgi by serial sectioning techniques (Tanchak et al. 1988). While located close to one another, connections between them were rare. Examination of the multivesicular bodies by various techniques suggested (Griffing and Fowke 1985; Tanchak and Fowke 1987), that it may be an early lysosomal compartment in plants which transfers molecules to the central lysosomal compartment, the plant vacuole. The question of coated vesicle integrity was answered by a careful serial section analysis of soybean protoplasts involved in uptake of cationized ferritin (Fowke et al. 1989). Serial sections clearly demonstrated that cationized-labeled coated vesicles were separate organelles within the cytoplasm.

EXPLORING THE FUNCTION OF COATED VESICLE-MEDIATED ENDOCYTOSIS

Taken together, the uptake and structural data suggested two distinct pathways for uptake in plant cells (Plate IX, 19). First, there was clear evidence for movement from the plasma membrane to the Golgi via the partially coated reticulum. Plants incorporate large amounts of new membrane into their surface during the secretion of matrix polysaccharides by the Golgi. Endocytosis by coated vesicles provided an excellent method to recycle plasma membrane into the cell. The second proposed pathway utilizing coated vesicles involves the movement of specific molecules from the plasma membrane to the vacuole for degradative processes. Internalization of molecules such as plant growth regulators, fungal toxins and elicitors might occur by a process of receptor-mediated endocytosis. Sorting could occur in the partially coated reticulum and molecules destined for degradation would then be directed to the multivesicular bodies and/or vacuole (see Fowke et al. 1991 for a more complete discussion).

In the summer of 1988, Moira Galway completed her PhD at the Australian National University in Canberra and came to Saskatoon as a postdoctoral fellow. She was keen to explore the possibility that receptor-mediated endocytosis via coated vesicles might occur in plant cells. A collaboration was established with Professor Jurgen Ebel and his group in Germany. They had isolated a 1,3-glucan elicitor from the cell walls of the fungus *Phytophthora megasperma* which induced soybean tissues to produce and accumulate phytoalexins. The production of phytoalexins by the host tissue was thought to prevent growth of the invading organism at the site of infection. High affinity binding sites for the elicitor had been described in soybean plasma membrane-enriched fractions (Schmidt and Ebel 1987) and the elicitor bound specifically to the plasma membrane of soybean protoplasts *in vivo* (Coscio et al. 1988). It appeared to be the ideal system to search for receptor-mediated endocytosis. Moira traveled to Germany and undertook a series of experiments designed to visualize elicitor binding by electron microsopy with the hope of observing internalization. She tried a range of techniques but unfortunately could not retain bound elicitor on soybean protoplasts for structural visualization and eventually, frustrated and very disappointed, she had to abandon the project.

Moira's second international collaboration, this time with Professor Andrew Staehelin Boulder Colorado, proved more fruitful. She decided to examine coated vesicle-mediated endocytosis in protoplasts of the conifer white spruce (*Picea glauca*) to determine whether an endocytosis pathway similar to that established in soybean was present in gymnosperms. In addition, she wanted to use the latest in high pressure freeze fixation techniques to examine the possibility that the pathway for uptake of cationized ferritin or the organelles involved were altered by artefacts due to conventional chemical fixation. Professor Staehelin offered the use of his prototype Balzers HPM 010 high-pressure freezing apparatus and kindly agreed to provide Moira with assistance. Using protoplasts from somatic embryos of spruce, Moira demonstrated that the pathway for endoctyosis in the spruce protoplasts was essentially the same as that observed in soybean protoplasts, and that the sequence of events and the structure of the organelles involved was basically the same with conventional fixation and high-pressure freeze fixation (Galway et al. 1993).

SUBSEQUENT ISOLATION AND CHARACTERIZATION OF COATED VESICLES

From the early 1980s when coated vesicle isolation was first reported (Mersey et al. 1982, 1985) until the early 1990s, considerable progress was made in the development of methods for isolating and characterizing plant coated vesicles. Particularly noteworthy were the contributions of the research groups of David Robinson in Germany, Chris Hawes in England, and Leonard Beevers in USA. A series of papers appeared describing the isolation of coated vesicles from suspension cultured cells (e.g. carrot, Depta and Robinson 1986; Coleman et al. 1987) and plant tissues (e.g. bean leaves, Depta et al. 1987; pea cotyledons, Demmer et al. 1993). Robinson's group introduced a number of improvements to the isolation procedure including the substitution of Ficoll/D_2O for sucrose to avoid dissociation of the clathrin coat, a ribonuclease treatment to remove contaminant ribosomes, and a "cocktail" of protease inhibitors to preclude proteolysis of coated vesicles. The highest yields of coated vesicles were obtained from developing pea cotyledons (Harley and Beevers 1989, Robinson et al. 1991). The first descriptions of triskelions (clathrin subunits) were provided by Coleman et al. (1987,1991) and Lin et al. (1992). The triskelions composed of clathrin heavy and light chains, appeared morphologically identical to those described for animal cells, but larger. Four polypeptides (50, 46, 40, 31 kDa) were identified as possible candidates for the clathrin light chains (Lin et al. 1992). Studies of assembly and disassembly of coated vesicles were undertaken (Wiedenhoft et al. 1988; Holstein et al. 1994; Beevers 1995) and provided evidence for the existence in plant coated vesicles of adaptor peptides, molecules thought to localize specific receptors into coated vesicles for internalization.

SEARCHING FOR A ROLE FOR COATED VESICLES IN PLANTS

It is clear from published data that endocytosis by coated vesicles provides plant cells with an excellent mechanism for retrieving excess plasma membrane. Even in expanding cells an excess of membrane is delivered from the Golgi to the plasma membrane during secretion of cell wall matrix polysaccharides. Estimates of membrane recycling certainly argue for this role for coated vesicles. Steer (1985), for example, has estimated that plant cells endocytose the equivalent of their entire plasma membrane within 10 minutes to 3 hours. On the basis of coated pit densities in different plant cells and measurements of pit diameters and rates of internalization, Emons and Traas (1986) estimated that the complete plasma membrane is internalized every 20-115 minutes. Endocytosis via coated vesicles would not only retrieve excess membrane, but could also play an important role in repair of the plasma membrane or perhaps replacement of vital membrane components such as transport proteins and ATPase proteins.

Experiments with protoplasts clearly indicate that plant cells are capable of internalizing molecules via coated pits/coated vesicles and delivering them to cytoplasmic organelles including the partially coated reticulum, Golgi, multivesicular bodies and vacuole. Conclusive evidence that such endocytosis is receptor mediated and that it operates in intact plant cells is still lacking. The experiments of Low's group suggest the presence of biotin receptors on the surface of soybean cells which mediate internalization of macromolecules conjugated to biotin (Horn et al. 1990, 1992). However, specific biotin receptors were not demonstrated and ultrastructural data were not presented to demonstrate internalization of the probes into endocytic compartments within soybean cells. As mentioned earlier, fungal elicitors such as the 1,3-glucan elicitor from *Phytophthora megasperma*, are likely candidates for receptor-mediated endocytosis in plants. Low's group also reported receptor mediated endocytosis of a proteinaceous elicitor from *Verticillium dahliae* and a polygalacturonic acid elicitor from citrus pectin into cultured soybean cells (Horn et al. 1990). However, the study lacked ultrastructural data showing coated vesicle mediated uptake and delivery of the elicitors. The Xu and Mendgen study (1994) is particularly interesting in that it demonstrated internalization of 1,3-glucan by coated pits/coated vesicles following the infection of broad bean plants with cowpea rust fungus and subsequent localization in the partially coated reticulum and multivesicular bodies. The 1,3-glucan might be delivered to the multivesicular body for degradation since it has been argued that multivesicular bodies are lysosomal compartments in plants (Griffing, 1991).

FUTURE PROSPECTS

There has recently been considerable interest in using the lipophilic styryl dye FM 4-64 as a vital stain for membranes in order to examine endocytosis in plant protoplasts (e.g. Ueda et al. 2001). Results suggest uptake of the plasma membrane and delivery to an endosomal compartment, then to the vacuole. This approach is interesting but does not address the question as to whether plant cells utilize receptor-mediated endocytosis. To answer this question specific receptors at the plasma membrane surface must be isolated and characterized. Then using a combination of light and electron microscopy, it is imperative that the movement of these receptors and/or bound ligands into plant cells via coated pits/vesicles be demonstrated. It may be possible to track the movement of receptors, ligands or components of coated vesicles using specific antibodies and GFP constructs with confocal microscopy and immunogold electron microscopy. Protoplasts provide useful tools for labeling experiments since they lack a cell wall; however, ultimately demonstration of uptake via coated pits/vesicles into intact plant cells would be necessary. Well characterized cell suspensions such as soybean SB-1 and tobacco BY-2 should facilitate this approach. It should also be possible to manipulate the endocytosis pathway using modern molecular techniques such as upregulation of the clathrin gene or suppression of the same gene using antisense technology. After a long hiatus, there appears to be a revival of interest in endocytosis by plant cells and within the next few years we should know whether plant cells use receptor-mediated methods to regulate uptake of specific molecules.

ACKNOWLEDGMENTS

I thank the team of dedicated young scientists who worked so hard to characterize plant coated vesicles. Thanks also to the technicians and support staff involved in the projects. Continuing financial support by the Natural Sciences and Engineering Research Council of Canada is gratefully acknowledged.

REFERENCES

Beevers L (1995) Clathrin-coated vesicles in plants. Int. Rev. Cytol. 167: 1-35.

Bonnett TH, Newcomb EH (1966) Coated vesicles and other cytoplasmic components of growing root hairs of radish. Protoplasma 62:59-75.

Coleman J, Evans DE, Horsley D, Hawes CR (1991) The molecular structure of plant clathrin coated vesicles. Semin. Ser. - Soc. Exp. Biol. 45:41-63.

Coleman J, Evans D, Hawes C, Horsley D, Cole L (1987). Structure and molecular organization of higher plant coated vesicles. J. Cell Sci. 88:35-45.

Coscio E, Popperyl H, Schmidt WE, Ebel. J. (1988) High affinity binding of fungal -glucan fragments to soybean (*Glycine max* L.) microsomal fractions and protoplasts. Eur. J. Biochem. 175:309-315.

Cram WJ (1980) Pinocytosis in plants. New Phytol. 84:1-17.
Demmer A, Holstein SEH, Hinz G, Schauermann G, Robinson DG (1993). Improved coated vesicle isolation allows better characterization of clathrin polypeptides. J. Exp. Bot. 44:23-33.
Depta H, Robinson DG (1986) The isolation and enrichment of coated vesicles from suspension-cultured carrot cells. Protoplasma 130:162-170.
Depta H, Freundt H, Hartmann D, Robinson DG (1987) Preparation of a homogeneous coated vesicle fraction from bean leaves. Protoplasma 136: 154-160.
Doohan ME, Palevitz BA (1980) Microtubules and coated vesicles in guard-cell protoplasts of *Allium cepa* L. Planta 149:389-401.
Emons AMC, Traas JA (1986) Coated pits and coated vesicles on the plasma membrane of plant cells. Eur. J. Cell Biol. 41:57-64.
Fowke LC, Gamborg OL (1980) Applications of protoplasts to the study of plant cells. Inter. Rev. Cytol. 68:9-51.
Fowke LC, Rennie PJ, Constabel, (1983) Organelles associated with the plasma membrane of tobacco leaf protoplasts. Plant Cell Rep 2:292-295.
Fowke LC, Tanchak MA, Rennie PJ (1989) Serial section analysis of coated pits and coated vesicles in soybean protoplasts. Cell Biol. Int. Rep. 13: 419-425.
Fowke LC, Tanchak MA, Galway ME (1991) Ultrastructural cytology of the endocytotic pathway in plants. *In:* CR Hawes, JOD Coleman, DE Evans, (eds.). Endocytosis, Exocytosis and Vesicle Traffic in Plants. Cambridge Univ. Press, New York, NY, pp. 15-40.
Fowke LC, Griffing LR, Mersey BG, Tanchak MA (1985) Protoplasts for studies of cell organelles. *In:* LC Fowke, F. Constabel (eds.). Plant Protoplasts. CRC Press, Boca Raton, FL, pp. 39-52.
Galway ME, Rennie PJ, Fowke LC (1993) Ultrastructure of the endocytotic pathway in glutaraldehyde-fixed and high pressure frozen/freeze substituted protoplasts of white spruce (*Picea glauca*). J. Cell Sci. 106:847-858.
Goldstein JL, Anderson RGW, Brown MS, (1979) Coated pits, coated vesicles, and receptor-mediated endocytosis. Nature 279:679-685.
Griffing LR, Fowke LC (1985). Cytochemical localization of peroxidase in soybean suspension culture cells and protoplasts: intracellular vacuole differentiation and presence of peroxidase in coated vesicles and multivesicular bodies. Protoplasma 128:2230.
Griffing LR, Mersey BG, Fowke LC (1986) Cell fractionation analysis of glucan synthase I and II distribution and polysaccharide secretion in soybean protoplasts. Evidence for the involvement of coated vesicles in wall biogenesis. Planta 167:175-182.
Griffing, LR, (1991) Comparisons of Golgi structure and dynamics in plant and animal cells. J. Electron Microscopy Tech. 17:179-199.
Harley SM, Beevers L, (1989) Isolation and partial characterization of clathrin coated vesicles from pea (*Pisum sativum*) cotyledons. Protoplasma 150:103-109.
Holstein SEH, Drucker M, Robinson DG (1994) Identification of a type of adaptin in plant clathrin coated vesicles. J. Cell Sci. 107:945-953.
Horn MA, Heinstein PF, Low PS (1990) Biotin-mediated delivery of exogenous macromolecules into soybean cells. Plant Physiol. 93:1492-1496.

Horn MA, Heinstein PF, Low PS (1992) Characterization of parameters influencing receptor mediated endocytosis in cultured soybean cells. Plant Physiol. 98: 673-679.

Joachim S, Robinson DG (1984). Endocytosis of cationic ferritin by bean leaf protoplasts. Eur. J. Cell Biol. 34:212-216.

Lin H-B, Harley SM, Butler JM, Beevers J (1992). Multiplicity of clathrin light-chain-like polypeptides from developing pea (*Pisum sativum*) cotyledons. J. Cell Sci. 103:1127-1137.

Low PS, Chandra S (1994). Endocytosis in plants. Ann. Rev. Plant Physiol. Plant Molec. Biol. 45:609-631.

Marchant HJ, Hines ER (1979). The role of microtubules and cell-wall deposition in elongation of regenerating protoplasts of *Mougeotia*. Planta 146:41-48.

Mersey BG, Fowke LC, Constabel F, Newcomb EH (1982). Preparation of a coated vesicle-enriched fraction from plant cells. Exp. Cell Res. 141:459-463.

Mersey BG, Griffing LR, Rennie PJ, Fowke LC (1985). The isolation of coated vesicles from protoplasts of soybean. Planta 163:317-327.

Newcomb EH (1980). Coated vesicles: their occurrence in different plant cell types. *In:* CJ Ockleford, A Whyte (eds.) Coated Vesicles. Cambridge Univ. Press, Cambridge, UK, pp. 55-68.

Pesacreta TC, Lucas WJ (1985). Presence of a partially coated reticulum in angiosperms. Protoplasma 125:173-184.

Robinson DG, Balusek K, Depta H, Hoh B, Holstein SEH (1991). Isolation and characterization of plant coated vesicles. Semin. Ser. - Soc. Exp. Biol. 45:65-79.

Ryser U (1979). Cotton fibre differentiation: occurrence and distribution of coated and smooth vesicles during primary and secondary wall formation. Protoplasma 98:223-239.

Schmidt WE, Ebel J (1987). Specific binding of a fungal glucan phytoalexin elicitor to membrane fractions from soybean *Glycine max*. Proc. Nat. Acad. Sci., USA, 84:4117-4121.

Steer MW (1985). Vesicle dynamics. *In:* AW Robards, (ed.). Biological Microscopy. Oxford Univ. Press, Oxford, UK, pp. 129-155.

Tanchak MA, Fowke LC (1987). The morphology of multivesicular bodies in soybean protoplasts and their role in endocytosis. Protoplasma 138:173-182.

Tanchak MA, Rennie PJ, Fowke LC (1988). Ultrastructure of the partially coated reticulum and dictyosomes during endocytosis by soybean protoplasts. Planta 175:433-441.

Tanchak MA, Griffing LR, Mersey BG, Fowke LC (1983). Functions of coated vesicles in plant protoplasts: endocytosis of cationized ferritin and transport of peroxidase. J. Cell Biol. 97:177a.

Tanchak MA, Griffing LR, Mersey BG, Fowke LC (1984). Endocytosis of cationized ferritin by coated vesicles of soybean protoplasts. Planta 162:481-486.

Ueda T, Yamaguchi M, Uchimiya H, Nakano A (2001). Ara6, a plant-unique novel type Rab GTPase, functions in the endocytic pathway of *Arabidopsis thaliana*. EMBO J. 20:4730-4741.

Van der Valk P, Rennie PJ, Connolly JA, Fowke LC (1980). Distribution of cortical microtubules in tobacco protoplasts. An immunofluorescence microscopic and ultrastructural study. Protoplasma 105:27-43.

Van der Valk, Fowke LC (1981). Ultrastructural aspects of coated vesicles in tobacco protoplasts. Can. J. Bot. 59: 1307-1313.
Wiedenhoeft RE, Schmidt GW, Palevitz BA (1988). Dissociation and reassembly of soybean clathrin. Plant Physiol. 86:412-416.
Xu H, Mendgen K, 1994. Endocytosis of 1,3—glucan by broad bean cells at the penetration site of the cowpea rust fungus (haploid stage). Planta 195:282-290.

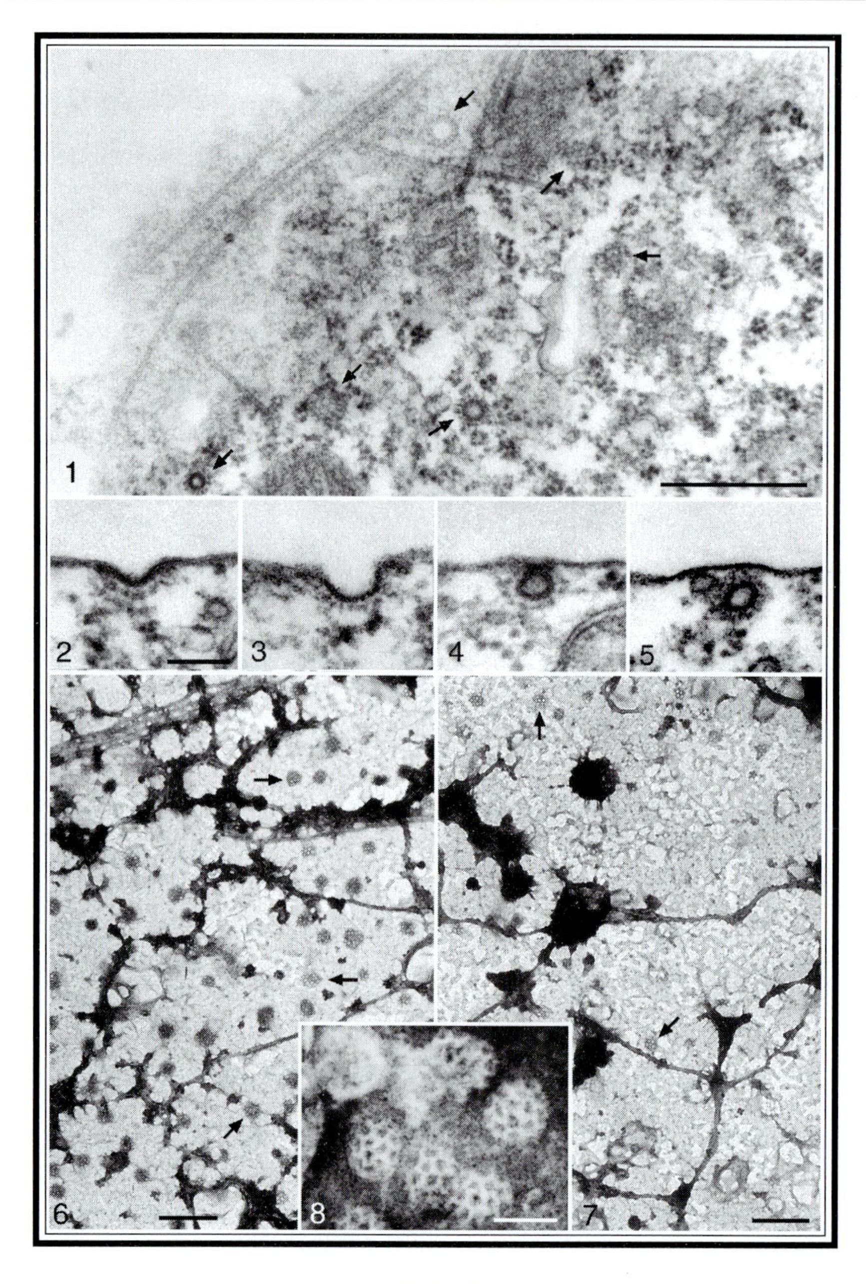

Plate VI

Plate VI

1-Grazing section of tobacco protoplast showing numerous clathrin-coated vesicles (arrows), some sectioned obliquely to show the patterned clathrin coat. (from Van der Valk and Fowke 1981). Bar = 500 nm. 2-5-Cross sections of coated pits and a coated vesicle (5) illustrating the sequential steps of endocytosis at the plasma membrane of a tobacco protoplast. (from Van der Valk and Fowke 1981). Bar = 100 nm. 6-Large fragment of tobacco protoplast plasma membrane showing frequency and distribution of coated pits/coated vesicles (arrows) on the inner surface of the plasma membrane. This protoplast was prepared from a suspension cultured tobacco cell in the process of rapid cell wall regeneration. (from Fowke et al. 1983). Bar = 300 nm. 7-Large fragment of tobacco protoplast plasma membrane showing frequency and distribution of coated pits/coated vesicles (arrows) on the inner surface of the plasma membrane. This protoplast was prepared from a leaf cell which was slow to regenerate new cell wall. (From Fowke et al. 1983). Bar = 300 nm. 8-Negatively stained coated pits on the inner surface of a tobacco plasma membrane fragment, showing the patterned clathrin coats. Bar = 100 nm.

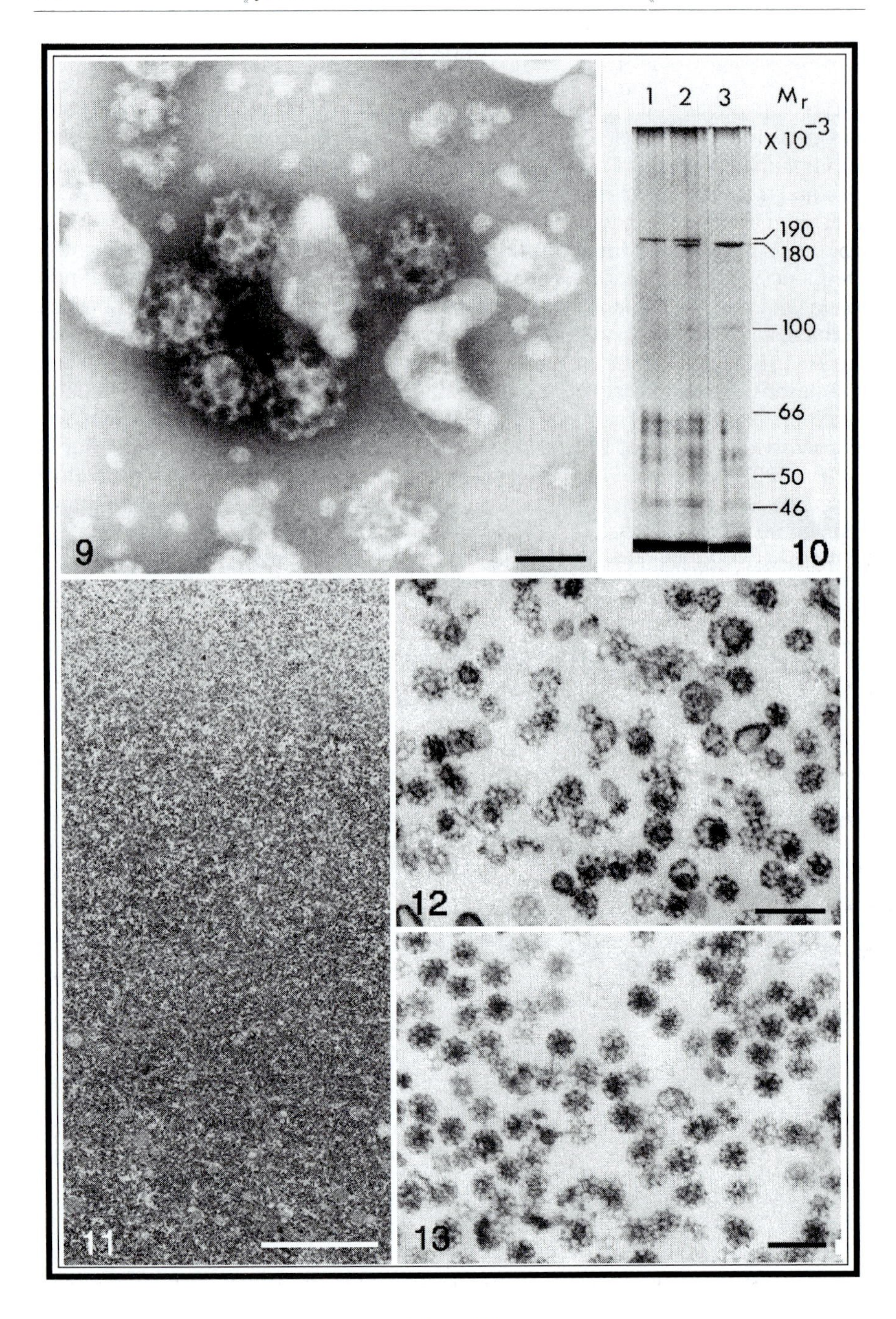

Plate VII

Plate VII

9-Negatively stained coated vesicles isolated from tobacco suspension cultured cells. Bar = 100 nm. 10-SDS polyacrylamide gel electrophoresis of coated vesicles from soybean protoplasts and bovine brain prepared by isopycnic centrifugation followed by rate zonal centrifugation through linear sucrose gradients. Lanes: 1 - soybean coated-vesicles, 2 - soybean plus bovine brain-coated vesicles, 3 - bovine-brain-coated vesicles. Note the difference in molecular mass of the most prominent polypeptide, clathrin, from plant and animal coated vesicles. (from Mersey et al. 1985). 11-Low magnification micrograph of a vertical section through the center of a soybean coated vesicle pellet. Bar = 3 μm 12-Enlargement from this pellet to show coated vesicles. (from Mersey et al. 1985). 13-Bouine brain coated vesicle fraction from rate zonal gradient Bar = 200 nm.

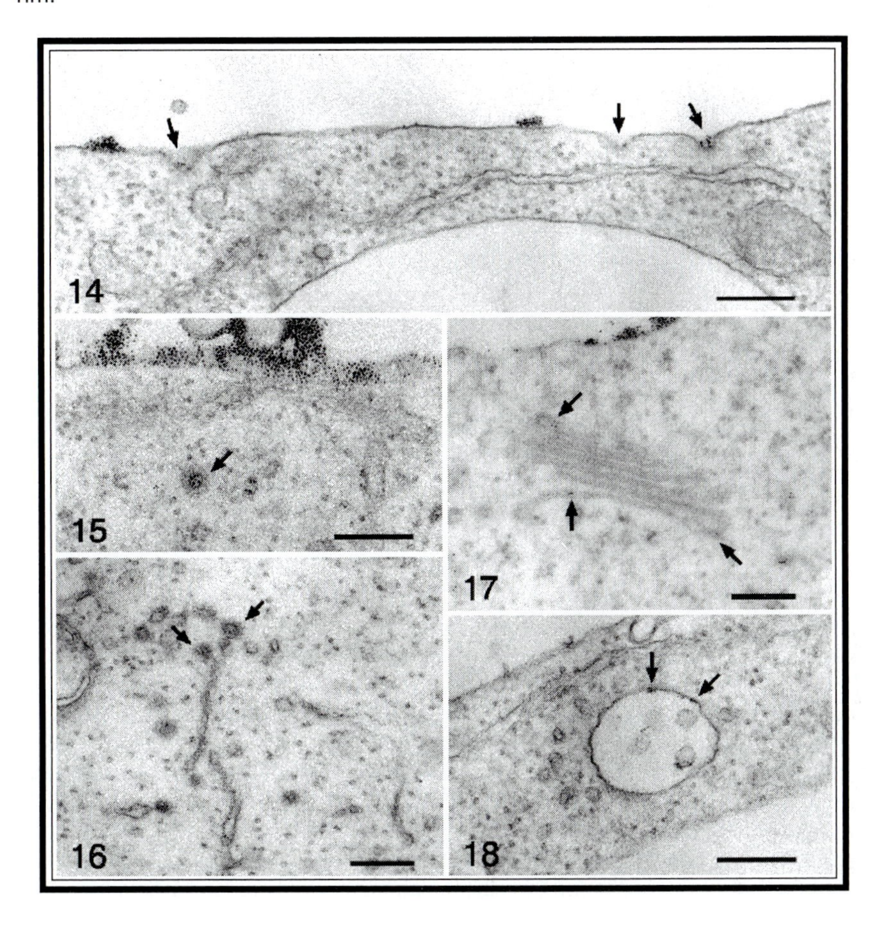

Plate VIII

Plate VIII

14 to 18-Representative micrographs of soybean protoplasts showing endocytosis and subsequent fate of cationized ferritin. All bars = 200 nm. 14-Cationized ferritin in coated pits (arrows) (from Fowke et al. 1985). 15-Cationized ferritin in a coated vesicle (arrow) within the cytoplasm (from Fowke et al. 1985). 16-Cationized ferritin (arrows) in the partially coated reticulum (from Tanchak et al. 1988). 17-Cationized ferritin (arrows) in cisternae of the Golgi (from Tanchak et al. 1988). 18-Cationized ferritin (arrows) in a multivesicular body (from Fowke et al. 1985).

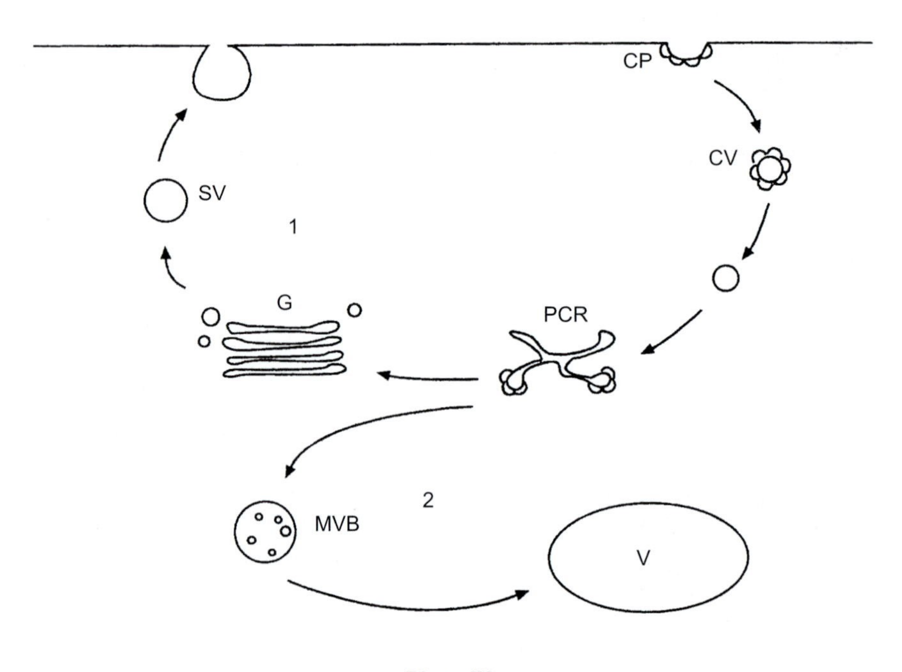

Plate IX

19-Diagram illustrating two possible pathways for molecules internalized by coated membranes, based on time course experiments using cationized ferritin. Both pathways involve uptake by coated pits (CP) to coated vesicles (CV), uncoating of coated vesicles, and delivery to the partially coated reticulum (PCR). Pathway 1 results in subsequent movement to the Golgi (G) and likely reflects plasma membrane recycling to compensate for delivery of matrix polysaccharides from the Golgi to the cell surface via secretory vesicles (SV). Pathway 2 involves transfer of molecules to multivesicular bodies (MVB) and the plant vacuole (V), both thought to comprise the lysosomal compartment of plant cells. This pathway could be used to degrade foreign molecules such as toxins and elicitors.

6

Discovery of Anther Culture Technique of Producing Pollen Haploids

S.C. Maheshwari
International Centre for Genetic Engineering and Biotechnology
Aruna Asaf Ali Marg,
New Delhi 110067, India.
e-mail:maheshwarisc@hotmail.com.

INTRODUCTION

The discovery of the technique of production of haploids by anther and pollen culture has a long history. Interest in haploid plants began right after the rediscovery of Mendel's laws and after Johansen (1903, 1909) enunciated the concept of pure lines. As plant breeders began to apply the knowledge of the new developing field of genetics in their work, it became evident that haploids would be extremely important material for producing pure lines in various crops. The conventional program of producing "pure lines" requires at least 6-7 years of selfing to achieve a satisfactory level of homozygosity. In contrast, one could foresee that with haploids, pure lines could be had in a single generation. The discovery of colchicine-induced polyploidy in plants by Blakeslee and Avery (1937) and Eigsti (1938) also opened a means of restoring diploidy and normal fertility, retaining the advantage of homozygosity.

DISCOVERY OF HAPLOIDS

The first natural haploid plant was found in *Datura stramonium* by Blakeslee (Plate X, 1) and co-workers in 1922 at the Cold Spring Harbor Laboratory in Long Island, New York. This discovery was followed by intensive search for haploids in many other taxa throughout the world. Haploids were thus reported in tobacco (*Nicotiana tabacum*) by Clausen and Mann (1924), wheat (*Triticum compactum*) by Gaines and Aase (1926), *N. glutinosa* by Goodspeed and Avery (1929) and also by Kostoff (1929) in *N. langsdorfii*. In the 1930s more instances of haploids were discovered such as in *Oenothera* by Gates and Goodwin (1930), maize by Randolph (1932), another wheat, *Triticum monococcum*, by Kihara and Katayama (1932), Katayama (1934), rice *Oryza sativa* by Ramaiah et al. (1933), cotton by Harland (1936), rye, *Secale cereale*, by Müntzing (1937) and Nordenskiold (1939), and potato by Lamm (1938). In the 1940s, further reports of haploids followed, for example, in *Lilium* by Cooper (1943) and *Capsicum annuum* by Christensen and Bamford (1943); Stadler (1942) made new observations on haploids in maize. In India, Swaminathan and coworkers made several observations on haploids during the 1950s (see article by Lacadena in monograph by Kasha 1974).

From the foregoing, it is evident that many noteworthy geneticists searched for haploids. And over the last few decades haploids have been found in many flowering plants, including shrubs and trees. An extensive documentation was made by Kimber and Riley (1963) of various kinds of haploids known at that time. The review lists both haploids that had been then obtained spontaneously as well as those in which some kind of experimental intervention was done. In some of the earlier reports, haploids arose following wide hybridization or such manipulations as delayed pollination, or exposure to high and low temperatures so that growth factors could be released (for example by the pollen or pollen-tube) leading to parthenogenesis, yet normal fertilization prevented. But after Muller (1927) discovered the mutagenic effect of X-rays, such treatment became a popular means of inducing haploidy. Thus, in some cases cited above, haploids arose as a result of inactivation of one of the gametes and generally pollen was irradiated, resulting in female parthenogenesis.

ANDROGENETIC HAPLOIDS

In principle, haploids can arise from both an unfertilized egg and from the male gamete. A large number of haploids (the so-called "gynogenetic" haploids) arise no doubt as a result of the stimulation of the unfertilized egg (or occasionally from one of the synergids). Nevertheless, a few examples were discovered of "androgenetic" haploids in which the offspring resembled the male parent and it was thus thought to have originated from the pollen. The first two cases were reported in *Nicotiana* by Clausen and Lammerts (1929) in the USA and by Kostoff (1929) in Bulgaria resulting from attempts to obtain interspecific hybrids (Plate XI,

2, 3). A third case of *Crepis tectorum* was reported by Gerassimova in Russia (1936) following attempts to induce haploidy by X-rays; here irradiation treatment was given to the female organs. Rhoades (1948) and Ehrensberger (1948) also made observations on androgenetic haploids in maize and *Antirrhinum*. Yet another example was reported in pepper (*Capsicum frutescens*) by Morgan and coworkers (Campos and Morgan 1958) again following X-ray irradiation.

EXPERIMENTAL INDUCTION OF HAPLOIDY

Valuable as the early work was with X-rays and other methods of inducing haploidy, a causal relationship could not always be established unequivocally between a treatment and haploidy since the frequency achieved was rather low. Thus, new methods of inducing haploidy were eagerly sought. Following the pioneering work of Haberlandt (1902, 1922) and later the isolation of plant hormones, attempts were made to induce parthenogenesis chemically employing growth substances of various kinds. Since embryos normally arise from the egg, the early attempts were designed to stimulate the egg cell in the ovule. Yasuda (1940) published some observations for *Petunia violacea* claiming some success in inducing the egg cell to undergo one or two divisions in response to certain chemical treatments, but apparently growth was arrested thereafter. Almost simultaneously, in Blakeslee's laboratory attempts were made, employing the tissue culture approach, to induce parthenogenesis of the egg in *Datura*, albeit likewise without success (van Overbeek et al. 1941). During the 1950s and early 1960s, my father, late Professor P. Maheshwari (Plate X, 4), led an active school of research at Delhi University where new efforts were made to induce parthenogenesis in ovaries and ovules cultured *in vitro*. The most extensive work was done in poppy (*Papaver somniferum*) given its advantage of a very large number of ovules for experimentation (Maheshwari 1958). Apart from auxins, cytokinins and gibberellins had also then become available and were tried in various combinations. But at that time only ovules in which fertilization had just taken place could be cultured – even though in later years employing other plants as experimental material, parthenogenesis of the egg could indeed be induced in culture (*see* Rangan 1994; Keller and Korzun 1996).

ANTHER AND POLLEN CULTURE

As evident, the occurrence of androgenesis had long been deduced in a few cases on the basis of circumstantial evidence, i.e., similarity of the offspring to the male parent. Yet no one seriously attempted pollen culture as a route for production of haploid plants, despite the fact that tissue culture had begun in many laboratories across the world. At least in gymnosperms such as *Ginkgo* and *Taxus*, LaRue and his pupil Tulecke in the USA had achieved some success in growing male gameto-phytic tissue from cultured pollen (Tulecke 1953, 1959; Tulecke and Sehgal 1963). Turning to angiosperms, even anther culture studies had begun in some laboratories (Taylor 1950; Sparrow et al. 1955). At Delhi University, Vasil (1958, 1960), my former classmate, conducted studies on anther and pollen culture for his PhD thesis under the supervision of Professor Panchanan Maheshwari. But the underlying aim in all these investigations was more to understand the physiology of anther development.

Although anther or pollen culture studies with the specific aim of obtaining haploids had not begun, the possibility that pollen grains could perhaps be made to develop into plantlets did occur to my father. His ideas were reported in a brief history of the discovery of anther culture technique of raising haploid plants that had appeared earlier (Maheshwari 1996). At the conclusion of the first Tissue Culture symposium organized in India (and in which the late Street, Nitsch, Steward and also Potrykus participated), he said: "There is another question that arises from experiments on anther culture. It is known that at least on some occasions the egg can develop parthenogenetically with an organized mass of cells – the embryo. Is it possible to obtain something similar from pollen grain?" Many of us attended this symposium and these prophetic remarks were later published in a concluding article with one of his pupils (Maheshwari and Rangaswamy 1963). However, because of preoccupation with other researches, the significance of this far-sighted statement did not register in my mind. A little later, the Japanese geneticists, Katayama and Nei (1964), in a review on plant haploidy made a similar remark: "It is to be noted that pollen cultures, if possible, would be a most efficient method of obtaining haploids on a large scale." But again this observation came to my notice only years later. Clearly, these ideas and reports were ignored by many as "wishful thinking". Konar (1963), one of my father's students, did obtain a haploid callus from the gymnosperm *Ephedra*, but the callus from anthers of *Ranunculus sceleratus* (an angiosperm) was sporogenous in nature, originating from the connective (Konar and Nataraja 1965), probably dampening the enthusiasm of many potential workers.

WORK ON *DATURA*

As mentioned earlier the discovery of the induction of haploidy by anther culture was serendipitous and a consequence of our interest in the switch which leads cells from mitosis to meiosis (Maheshwari 1996). However, even the idea of work on the mitotis-to-meiosis switch was an afterthought. When Sipra Guha (later to become Guha-Mukherjee, Plate X, 5) joined my laboratory in 1963 as a post-doctoral fellow, she was seeking some biochemical experience and our principal efforts were directed to exploring the relationship between plant hormones and nucleic acid synthesis—areas in which I had gained some research experience at Yale and Caltech in the USA in the laboratories of A.W. Galston and J. Bonner.

Our parallel interest in anther culture and the discovery of pollen haploids was a consequence of the fact that biochemical facilities still needed to be establised at that time in the Botany Departments in India. Also, because of the high cost of radioisotopes and their erratic supply (to import such chemicals from abroad required hard currency and special permission from the Govern-ment of India), I thought Sipra should have a second research challenge. With late Professor Johri, one of my own teachers and under whom she had also done her PhD work, she had gained experience on tissue culture for which excellent facilities were available in the Department. For work on cell suspensions, we had also acquired a rotating shaker similar to that used by F.C. Steward at Cornell. And now our idea was to see whether these could be utilized to culture anthers. If via anther culture, a callus or a suspension of sporogenous cells could be obtained, this in turn could be a good material for studying the transition from mitosis to meiosis and manipulating this switch (at that time nothing was known about cyclins or cyclin-dependent kinases and obviously my ideas were much too simplistic). I still recall a visit to the modest-sized Botanical Garden of the University right beside the laboratory building and surveying plants then in flower. Cultures in standard glass tubes had been raised of anthers of several different plants as a source for primary callus, but most of the cultures had browned and died. It was during a cleaning operation one afternoon that Sipra called me to the culture room and pointed to a rack of *Datura* cultures in which some tubes contained plantlets emerging from anthers (Plate X, 6).

I have to admit that the possibility that these plantlets were actually pollen plants seemed rather remote to me. Even after many preparations had been made for microscopic examination showing dividing cells within pollen grains, I long remained reluctant to believe that they were of pollen origin and haploid. Although I had some experience not only of biochemical work, but also microtechnique and histological studies (my PhD thesis under Professor Johri concerned the morphology and embryology of Lemnaceae), no cytological work had been ever undertaken by our group. It was only when another student, R.N. Bhatt, working under the late Professor S.L. Tandon (who specialized in cytogenetics) helped us, could we determine the haploid nature of the plantlets. Looking back, it seems foolish that I was so slow in accepting the idea of the origin of the plantlets from microspores, and mastering cytological techniques — thus leading to a gap of two years between the two publications on haploidy (Guha and Maheshwari 1964, 1966)! However, no one apparently tried to duplicate the work meanwhile. And a reason for some satisfaction was that my father was still alive to see the final results compiled in the spring of 1966. Indeed, it is he who edited both of our manuscripts and forwarded them to the editor of Nature, even though unfortunately he did not live to see the second paper demonstrating the haploid nature of pollen culture actually in print.

POST-1966 DEVELOPMENTS

In science, unusual observations and ideas are difficult to accept. We were happy that prominent plant biologists, e.g. Dr M.S. Swaminathan in India and many others were quick to accept our findings (we were flooded with letters of appreciation from all over the world and were hard put to send timely acknowledgments and replies). Their attention and encouragement prompted later achievements. But there were skeptics also, including several of my own colleagues in the Botany Department. In the Department, then headed by my father, in addition to weekly seminars it was customary to hold a "demonstration" of any new work ready for publication, so that people could see actual experimental material and offer criticism and opinion for finalizing manuscripts. It was usual to have many microscopes lined up on benches in a laboratory with different slides and preparations for other workers to examine. The *Datura* preparations were duly arranged for viewing. The embryos were there for everyone to see, but several colleagues proposed that the embryos arose from connective cells which had rounded up, migrated to the anther-sac, and developed exine-like thickening under the influence of hormones and other adjuvants! It was thus a great joy when Nitsch and his colleagues in France published their observations on pollen embryoids in tobacco (Bourgin and Nitsch 1967; Nitsch and Nitsch 1969). This was shortly after I had come back from a visit to Oxford, England in January 1967 and en route stopped by the Nitsch lab at Gif-sur-Yvette to give a brief seminar about our work on flowering in duckweeds as well as *Datura* haploids. Even though our work was not cited in a couple of key publications by the French group in retrospect, it is a matter of immense satisfaction to me that research was undertaken in the Nitsch laboratory on *Nicotiana* and also by Nakata and Tanaka (1968) and Niizeki and Oono (1968) in Japan (the latter work was done on rice). These studies paved the way for larger scale use of microspore haploids for plant breeding and genetical research in laboratories all over the world.

LATER WORK

No account of history would be fair or complete without reference to certain other investigators. Particular mention needs to be made of G. Melchers who headed the Max Planck Institute of Biology in Tubingen in Germany, and was later responsible for establishing another facility for research work on pollen haploids in Ladenburg. He published extensively on haploidy (Melchers 1960; Melchers and Labib 1970; Melchers 1972). Indeed, his interest in haploids was much older than our discovery of the anther culture technique. With Bergmann, he obtained—through tissue culture methods—a haploid in *Antirrhinum* (Melchers and Bergmann 1958). Later, after anther culture was discovered, and after it had been shown that tobacco protoplasts could be grown into adult plants (Takebe et al. 1971), his laboratory obtained one of the first protoplast fusion hybrids. Protoplasts were isolated from haploid parents of *Nicotiana plumbaginifolia*, each defective in a separate step of chlorophyll biosynthesis the fusion event restored normal chlorophyll synthesis, and fertility (Melchers and Labib 1974). A little earlier, Carlson et al. (1972) working at the Brookhaven National Laboratory also employed the anther culture technique for obtaining the first parasexual hybrids in tobacco.

Keen interest in haploids was further shown by Straub's group in Germany (Binding et al. 1970), who later directed the Max Planck Institute for Plant Breeding Research at Cologne and where anther-derived haploid plants were employed for the first demonstration of successful genetic engineering of plants via the Ti plasmids by Schell, Straub and coworkers (Otten et al. 1981). Kohlenbach in Germany was also one of the early researchers on pollen haploids.

Further development of work on pollen haploids in the late 1960s and early 1970s also owes a great deal to Sunderland and Dunwell and Clapham in the UK, Picard and de Buyser (apart from Mrs Nitsch and coworkers in France), Devreux in Italy, Baenziger, Burk, Collins, Schaeffer, Sharp in the USA, Keller in Canada, and Hu Han and others in China, in addition to the Japanese workers cited earlier for their pioneering work. Madam Nitsch made a significant advance by developing the technique of isolated pollen culture (Nitsch 1974) which has been employed by many later workers. In *Brassica*, protocols have been developed whereby > 1000 embryos can be had per anther (Polsoni et al. 1988). Credit is due also to Kasha for organizing the First International Symposium on Haploids in Guelph in Canada in 1974, which brought together many workers from all over the world in this field (Kasha 1974), who also illustrated several uses of haploids other than raising pure lines. Kasha has been a pioneer himself in producing haploids by the chromo-some elimination method, an alternative mode for obtaining such plants (Kasha and Kao 1970). Similarly, the Chinese are applauded not only for the most extensive work on developing superior lines of wheat which reportedly cover hundreds of thousands of acres, but for organizing the second international symposium (Hu and Yang 1986; Maheshwari 1996 and references therein). Finally, mention may be made of the 5-volume work, *In Vitro Haploid Production in Higher Plants* by Jain, Sopory, and Veilleux in which the extensive research on this aspect has been summarized as of 1996 by world experts.

ACKNOWLEDGMENTS

We are very grateful to the authorities of South College, Northampton, MA and Buchanan Library, University of California, Berkeley for photographs reproduced here in Plate X, 2, 3 and Prof. A. Atanassov of Bulgaria for photograph in Plate X, 3. Sincere thanks are also due to Dr. (Mrs.) Nirmala Maheshwari for reviewing the manuscript and Mrs. Shashi Mehta and Mr Satyendra Patwal of Department of Plant Molecular Biology, University of Delhi, South Campus for assistance in typing and final processing of the manuscript.

REFERENCES

Binding H, Binding K, Straub J (1970). Selektion in Gewekekulturen mit haploiden Zellen. Naturwiss. 57:138-139.

Blakeslee AF, Avery AG (1937). Methods of inducing doubling of chromosomes in plants by treatment with colchicine. J. Hered. 28:393-411.

Blakeslee AF, Belling J, Farnham MW, Bergner AD (1922). A haploid mutant in the Jimson weed, *Datura stramonium*. Science 55:1433.

Bourgin JP, Nitsch JP (1967). Obtention de *Nicotiana* haploides a partir d'etamines cultivees *in vitro*. Ann. Physiol. Veg. 9:377-382.

Campos FF, Morgan Jr DT (1958). Haploid pepper from a sperm. J. Hered. 49:134-137.

Carlson PS, Smith HH, Dearing RD (1972). Parasexual interspecific plant hybrids. Proc. Natl. Acad. Sci. (USA) 69:2292-2294.

Christensen HM, Bamford R. (1943). Haploids in twin seedlings of peppers, *Capsicum annuum* L. J. Hered. 34:99-104.

Clausen RE, Mann MC (1924). Inheritance in *Nicotiana tabacum*. V. The occurrence of haploid plants in interspecific progenies. Proc. Natl. Acad. Sci. (USA) 10:121-124.

Clausen RE, Lammerts WE (1929). Interspecific hybridization in *Nicotiana*. X. Haploid and diploid merogony. Amer. Nat. 43:279-282.

Cooper DC (1943). Haploid-diploid twin embryos in *Lilium* and *Nicotiana*. Amer. J. Bot. 30:408-413.

Ehrensberger R (1948). Versuche zur Auslösung von Haploidie bei Blütenpflanzen. Biol. Zentralbl. 67:537-546.

Eigsti O (1938) A cytological study of colchicine effects in the induction of polyploidy in plants. Proc. Natl. Acad. Sci. (USA) 24:56-63.
Gaines EF, Aase UC (1926) A haploid wheat plant. Amer. J. Bot. 13:373-385.
Gates RR, Goodwin KM (1930). A new haploid *Oenothera*, with some considerations on haploidy in plants and animals. J. Genet. 23:123-156.
Gerassimova H, (1936) Experimentele erhaltens haploide Pflanze von *Crepis tectorum* L. Planta 25:696-702.
Goodspeed TH, Avery P (1929) The occurrence of a *Nicotiana glutinosa* haplont. Proc. Natl. Acad. Sci. (USA) 15:502-504.
Guha S, Maheshwari SC (1964) *In vitro* production of embryos from anthers of *Datura*. Nature (Lond.) 204:497.
Guha S, Maheshwari SC (1966) Cell division and differentiation of embryos in the pollen grains of *Datura in vitro*. Nature (Lond.) 212:97-98.
Haberlandt G (1902) Kulturversuche mit isolierten Pflanzenzellen. Sitzungsber. Akad. Wiss. Wien, Math. – Naturwiss. Kl. 1, (111):69-92.
Haberlandt G (1922) Über Zellteilungshormone und ihre Beziehungen zur Wundheilung, Befruchtung, Parthenogenese und Adventivembryonie. Biol. Zentralbl. 42:145-172.
Harland SC (1936) Haploids in polyembryonic seeds of Sea Island cotton. J. Hered. 27:229-231.
Hougas RW, Peloquin SJ (1958) The potential of potato haploids in breeding and genetic research. Amer. Potato J. 35:701-707.
Hu H, Yang H (1986) *Haploids of Higher Plants in Vitro*. China Academic Publishers, Beijing, and Springer-Verlag, Berlin.
Jain SM, Sopory SK, and Veilleux RE (eds.) (1996) *In Vitro Haploid Production in Higher Plants*. Kluwer Academic Publishers, Dordrecht, The Netherlands.
Johansen W (1903) *Über Erblichkeit in Populationen und in reinen Linien*. Gustav Fischer, Jena, Germany.
Johansen W (1909) *Elemente der exakten Erblichkeitslehre*. Gustav Fischer, Jena, Germany.
Kasha KJ (1974) *Haploids in Higher Plants: Advances and Potential, Proc. First Inter. Symp.* Univ. of Guelph, Guelph, Ontario, Canada.
Kasha KJ, Kao KN (1970) High frequency haploid production in barley (*H. vulgare* L.). Nature (Lond.) 225:874-875.
Katayama Y (1934) Haploid formation by X-rays in *Triticum monococcum*. Cytologia 5:234-237.
Katayama Y, Nei M, (1964) Studies on the haploidy in higher plants. Rep. Fac. Agric., Univ. Miyazaki, Japan.
Keller ERJ, Korzun L (1996) Ovary and ovule culture for haploid production. *In:* SM Jain et al. (eds.) *"In Vitro Haploid Production in Higher Plants"*, vol. 1. Kluwer Acad. Publi., Dordrecht, The Netherlands.
Kihara H, Katayama Y (1932) Über das Vorkommen von haploider Pflanzen bei *Triticum monoscoccum*. Kwagaku 2:408-410.
Kimber G, Riley R (1963). Haploid angiosperms. Bot. Rev. 29:480-531.
Konar RN, (1963) A haploid tissue from the pollen of *Ephedra foliata* Boiss. Phytomorphology 13:170-174.
Konar RN, Nataraja K (1965) Production of embryoids from anthers of *Ranunculus scleratus* L. Phytomorphology 15:245.

Kostoff D (1929) An androgenic *Nicotiana* haploid. Zeit. Zellforschg 9:640-642.
Lamm R (1938) Notes on haploid potato hybrid. Hereditas 24:391-396.
Maheshwari N. (1958) *In vitro* culture of excised ovules of *Papaver somniferum*. Science 127:342.
Maheshwari P, Rangaswamy, NS (1963) Plant tissue and organ culture from the viewpoint of an embryologist. *In: Plant Tissue and Organ Culture – A Symposium*, pp. 390-420.
Maheshwari SC (1996) The discovery of anther culture technique for the production of haploids—a personal reflection. *In:* SM Jain, et al. (eds.). *In Vitro Haploid Production in Higher Plants*, Kluwer Acad. Publ., Dordrecht, The Netherlands, Vol. 1.
Melchers G (1960) Haploide Blütenpflanzen als Material der Mutationzuchtung. Züchter 30:129-134.
Melchers G (1972) Haploid higher plants for plant breeding. Z. Pflanzenzuchtg 67:19-32.
Melchers G, Bergmann L (1958) Untersuchungen an Kulturen von haploiden Geweben von *Antirrhinum majus*. Ber. dtsch. bot. Ges. 71:459-473.
Melchers G, Labib G (1970) Die Bedeutung haploider höherer Pflanzen fur Pflanzenphysiologie und Pflanzenzüchtung. Ber. dtsch. bot. Ges. 83:129-150.
Melchers G, Labib G (1974) Somatic hybridization of plants by fusion of protoplasts. I. Selection of light resistant hybrid of haploid light-sensitive varieties of tobacco. Molec. Gen. Genet. 135:277-294.
Muller HJ (1927) Artificial transmutation of the genes. Science 66:84-87.
Müntzing A (1937) Note on a haploid rye plant. Hereditas 23:401.
Nakata K, Tanaka M (1968) Differentiation of embryoids from developing germ cells in anther culture of tobacco. Jpn. J. Genet. 43:65-71.
Nei M (1963) The efficiency of haploid method of plant breeding. Heredity 18:95-100.
Niizeki H, Oono K (1968) Induction of haploid rice plant from anther culture. Proc. Jpn. Acad. 44:554-557.
Nitsch C (1974) La culture de pollen isolé sur milieu synthétique. C.R. Acad. Sci. Paris 278:1031-1034.
Nitsch JP, Nitsch C (1969) Haploid plants from pollen grains. Science 163:85-87.
Nokdenskiold H (1939) Studies of a haploid rye plant. Hereditas 25:204-210.
Otten L, De Greve H, Hernalsteens, JP, van Montagu, M, Schieder O, Straub J, Schell, J. (1981). Mendelian transmission of genes introduced into plants by T1 plasmid of *Agrobacterium tumefaciens*. Molec. Gen. Genet. 183:209-213.
Polsoni L, Kott LS, Beversdorf WD, (1988) Large-scale microspore culture technique for mutation selection studies in *Brassica napus*. Can. J. Bot. 66:1681-1685.
Ramaiah H, Parthasarathy H, and Ramanujam S (1933) Haploid plant in rice, *Oryza sativa*. Curr. Sci. 1:277-278.
Randolph LF (1932) Some effects of high temperature on polyploidy and other variations in maize. Proc. Natl. Acad. Sci. (USA) 18:222-229.
Randolph LF (1938) Note on haploid frequencies. Maize Genet. Coop. Newsl. 12:12.
Rangan TS (1984) Culture of ovules. *In:* IK Vasil (ed.). Cell Culture and Somatic Cell Genetics of Plants. Acad. Press, New York, NY, Vol. 1.

Rhoades MM (1948) Androgenesis. Maize Genet. Coop. Newsl. 22:10.
Sparrow AH, Pond V, Kojan, S (1955) Microsporogenesis in excised anthers of *Trillium erectum* grown in sterile culture. Amer. J. Bot. 42:384-394.
Stadler LJ (1931) The experimental modification of heredity in crop plants. L. Induced chromosomal irregularities. Sci. Agric. 11:557-572.
Stadler LJ (1940) Note on haploids. Maize Genetics Coop Newsletter 14:27.
Stadler LJ (1942) Frequency of haploids. Ann. Rep. Coop. Corn. Invest. Missouri Ag. Expt. Sta. Columbia, MO.
Takebe I, Labib G, Melchers G (1971) Regeneration of whole plants from isolated mesophyll protoplasts of tobacco. Naturwiss. 58:318-320.
Tanaka M, and Nakata K (1969) Tobacco plants obtained by anther culture and experiments to get diploid seeds from haploids. Jpn. J. Genet. 44:47-54.
Taylor JH (1950) The duration of differentiation in excised anthers . Amer. J. Bot. 37:137-143.
Tulecke WR (1953) A tissue derived from the pollen of *Ginkgo biloba*. Science 117:599-600.
Tulecke WR (1959) The pollen cultures of C.D. LaRue: a tissue from the pollen of *Taxus*. Bull. Torry Bot. Club. 86:283-289.
Tulecke WR, Sehgal N. (1963) Cell proliferation from the pollen of *Torreya nucifera*. Contrib. Boyce Thompson Inst. 22:153-163.
van Overbeek J, Conklin ME, and Blakeslee AF (1941) Chemical stimulation of ovule development and its possible relation to parthenogenesis. Amer. J. Bot. 28:647-656.
Vasil IK (1958) PhD Thesis. University of Delhi, Delhi.
Vasil IK (1960) Physiology of anthers. *In:* P Maheshwari, et al. (eds.). Proc. Summer School of Botany, Darjeeling. Ministry of Scientific Research and Cultural Affairs, Govt. of India, New Delhi.
Yasuda S (1940) A preliminary note on the artificial parthenogenesis induced by application of growth promoting substances. Bot. Mag. (Tokyo) 54:506-510.

1

2

3

4

5

6

Plate X

1-Albert F. Blakeslee; (1874-1954)
2-Roy E. Clausen; (1891-1956)
3-Detlev Kostov; (1894-1949)
4-Panchanan Maheshwari; (1904-1966)
5-Sipra Guha-Mukherjee (1938-).
6-An anther of *Datura innoxia* showing androgenic embryos.

7

Plants from Haploid Cells

K.J. Kasha
Department of Plant Agriculture, University of Guelph,
Guelph, ON Canada, N1G 2W1

INTRODUCTION

Haploid plant production in the cereals has been one of the major thrusts of my research for the past 35 years and has kept me so intrigued that it has continued into my retirement. In this review the historical development of this research into two different systems of haploid production is presented and the potential and application of the techniques involved covered briefly.

The story of developing the wide hybridization/chromosome elimination system (sometimes called the "Bulbosum Method") of haploid production in barley (*Hordeum vulgare* L.) is presented first. This was the first system in a crop to result in sufficient numbers of haploids from most genotypes that it was potentially useful as a plant breeding method. It allowed comparison of how doubled haploids would perform in a breeding program relative to other inbreeding systems. Moreover, the genetics of the system are fascinating.

Work on isolated microspore culture in barley and wheat (*Triticum aestivum* L.) is then described, wherein producing a system that could use a completely defined culture medium in order to identify and optimize crtical components was attempted. This was achieved in barley. It has the advantages of producing large numbers of haploid plants from most genotypes and a high frequency of induced doubling of the chromosomes at the embryo induction stage. Doubling chromosomes at the time of induction results in plants that are completely fertile and produce a good quantity of seed per plant. The high response and defined medium have reduced the wide response differences among genotypes so that the system is effective in barley breeding programs. However, in wheat the need to use ovary co-culture with the isolated microspores slowed efforts to develop a defined medium. Recourse was taken in isolating and characterizing factors secreted from ovaries that are beneficial when present in the media.

As with any field of scientific enquiry, numerous graduate students, postdoctoral fellows, visiting scientists and excellent technical assistants were involved over the years. Moreover, all of us gleaned ideas from discoveries and publications made in many other labs that greatly added to our progress, much as our studies have assisted others.

WIDE HYBRIDIZATION FOR HAPLOID PRODUCTION

Interest in wide hybridization between barley (*Hordeum vulgare* L.) and *Hordeum bulbosum* L. originated from a request for collaboration from Dr. Tibor Rajhathy from the Ottawa Research Station of Agriculture Canada. His team had produced hybrids between these two species (Rajhathy 1967) and he wanted to transfer genes from the hybrids into barley by using irradiation to induce chromosome breakage and rearrangement. I had agreed to help in the evaluation of progeny from such X-irradiated plants. However, the hybrids were surprisingly very sensitive to irradiation and subsequently died before they could be crossed or set seed. I later thought this sensitivity might be attributable to the delicate balance of the chromosome elimination phenomenon that was uncovered.

The next important event was a call from Dr. Ray Miller in Olaf Gamborg's lab at NRC in Saskatoon. He had a recent PhD graduate whom they wanted to hire but didn't have the funds for the current year and wondered whether I could support him for a year as a PDF. This was Dr. K.N. Kao who was well known to us as he had completed an MSc. at Guelph with my colleague Dr. Reinbergs. I agreed and decided to have him work on producing hybrids between barley and *Hordeum bulbosum* in an effort to transfer disease resistance genes into barley. Dr. Kao was very skilled at plant hybridization and an experienced 'green thumb' in tissue culture, skills that would be needed for the rescue of the embryos from wide crosses. We had autotetraploid (4*x*) barley lines from the research of Dr. Reinbergs, the barley breeder at Guelph, and a collection of accessions left by Dr. Bryan Harvey of the cross-pollinated species *Hordeum bulbosum*, both diploid (2*x*) and autotetraploid (4*x*).

Crosses and Theory Development

We planned to cross the autotetraploids of both species, as such hybrids should be more fertile and increase the potential of obtaining backcrosses to barley. To our surprise, we obtained a large number of fertile diploid barley plants (VV genomes) from the embryo rescue of these crosses. There were also three rare hybrids that had 27 rather than the expected 28 chromosomes (VVBB). When we crossed diploid barley with tetraploid *H. bulbosum,* we obtained the expected 21 chromosome hybrids (VBB). These results triggered Kasha to speculate that these species crosses tended to produce hybrids when there was a specific balance between the number of genomes of the two species, in this case in a ratio of 2B:1V genomes (Kao and Kasha 1970). Furthermore, based on the genome balance theory, Kasha surmised that if crosses were made between the diploid genotypes of both species, we might expect to obtain haploid barley plants. This idea was successful and we obtained large numbers of haploid plants in barley (Kasha and Kao 1970) (Table 7.1). Furthermore, the frequencies of haploids were too high for parthenogenesis and Kasha proposed that fertilization was followed by **preferential chromosome elimination** that must occur in the early embryo stage. Important to our success and the development of these theories were the skills of Kao in rescuing the embryos and the use of Gamborg's B5 medium (Gamborg et al. 1968) for culturing. This medium provided a more balanced source of nutrients and improved the survival and growth of the young haploid embryos. To improve our production of embryos we later adopted the idea of Larter and Enns (1960) of using 75 ppm GA3, added as drops to the florets or sprayed on the spikes after fertilization, to stimulate embryo development. A key component of this study was the reciprocal crosses between all of these ploidy levels that gave us the same results (Table 7.1). Not only did it help to confirm the genome balance concept, but also that this was chromosome elimination occurring after fertilization. We reasoned that it was not parthenogenesis (stimulation of the egg cell to form an embryo) because of the high frequencies and that haploid plants also developed from the male gamete of barley when used as the pollen source. Previous reports on crosses between these two species had proposed parthenogenesis to explain the rare haploids obtained (Davies 1958).

Table 7.1 Types and frequencies of progenies obtained from interspecific crosses between *Hordeum vulgare* (V) and *H. bulbosum* (B), from Kasha (1974).

Cross	*combination*		*# plants*	*Genotype and chromosome number*					
♀		♂	*obtained*	*V(7)*	*VV(14)*	*VB(14)*	*VBB(21)*	*VVBB(27)*	*VVBB(28)*
VV	x	BB	1544	1517		26	1		
BB	x	VV	35	35					
VV	x	BBBB	87				87		
BBBB	x	VV	6				6		
VVVV	x	BB	4		4				
VVVV	x	BBBB	79		76			3	
BBBB	x	VVVV	34	34					

Confirmation and Elaboration of Theories

Development of the embryos after fertilization was studied in my lab by N.C. Subrahmanyam who demonstrated the gradual elimination of chromosomes in the early development of the embryos (Subrahmanyam and Kasha 1973). Because of the rare 27 chromosome hybrids obtained, we also suspected that a specific chromosome might carry the genes responsible for the balance phenomenon that was being expressed. A graduate student with me, Keh Ming Ho, set up experiments to cross a set of the seven different trisomic ($2x + 1$) chromosome lines of barley with a tetraploid accession of *H. bulbosum* (Table 7.2). This study demonstrated that chromosomes 2 and 3 of barley were involved in the genome balance phenomenon (Ho and Kasha 1975). The n+1 gamete from trisomics for either of these chromosomes caused elimination of *H. bulbosum* chromosomes as they were in a 1:1 ratio in such embryos. The frequency of 22 chromosome hybrids (Table 7.2) was expected to be the same as the usual transmission frequency of the trisomic chromosome unless chromosome elimination had occurred. There was a significant reduction of such hybrids in the progeny from crosses with trisomics for chromosomes 2 and 3, while 8 chromosome haploids were not expected to survive. Also, after crossing monotelotrisomics (half of the chromosome being extra) lines of barley with tetraploid *bulbosum*, we concluded that

Table 7.2 Chromosome numbers observed and expected in progeny from crosses between primary trisomics of diploid *H. vulgare* and tetraploid *H. bulbosum.* (From Ho and Kasha, 1975)

Trisomic type	*Total no. of progeny obtained*	*Frequency and (%) of progeny with:* 21 chromosomes	22 chromosomes	Others	*Expected trisomic trans-mission†*	P^3 value
1	81	69(85.2)	11(13.6)	1(1.2)	15.0	0.06
2	57	55(96.5)	1(1.8)	1(1.8)	22.5	10.73**
3	75	72(96.0)	2(2.7)	1(1.3)	20.3	11.44**
4	39	28(71.8)	10(25.6)	1(2.6)	29.6	0.19
5	60	52(86.7)	8(13.3)	0(0.0)	20.0	1.31
6	99	80(80.0)	18(18.2)	1(1.0)	16.1	0.17
7	95	79(83.2)	14(14.7)	2(2.1)	20.0	0.60
Total	506	435(86.0)	64(12.6)	7(1.4)		

† Trisomic transmission frequencies reported by Yu (1968).
** Indicating significance deviation from expected # of 22 chromosome plants at the 1% level.

both arms of chromosome 2 and the short arm of chromosome 3 contained genetic factors that interacted with factors on homeologous *bulbosum* chromosomes to influence chromosome elimination. This indicated that at least three different genes were involved in the balance phenomenon between barley and *bulbosum* that led to chromosome elimination.

Related Studies

In 1969 at a barley conference in Pullman Washington, we announced our findings regarding genome balance and chromosome elimination theory (Kao and Kasha 1970) that led to the Bulbosum Method of haploid production. It was then that we learned that others were also obtaining haploids from the same crosses but they had not determined the mechanism behind it or its significance to breeding methodology. Symko (1969) at Ottawa had also produced 59 haploid plants from crossing diploid *H. bulbosum* as the female parent and using barley pollen and rescuing the embryos. These haploid plants had the cytoplasm of *bulbosum* and 54 of them exhibited some characteristics, such as later heading, more prostrate growth and hairy leaves of *H. bulbosum*. The other five were more like barley with glabrous leaves, while all haploids had spikes that resembled barley. He speculated that the plants were possibly of hybrid origin followed by somatic reduction or by male parthenogenesis. W. Lange also produced barley haploids from the same crosses in his PhD thesis studies in the Netherlands and later reported on haploids from the elimination of chromosomes (Lange 1971). Again, the original focus for such crosses in these labs had been to obtain hybrids in order to transfer genes into barley and that research continued (Lange and Jochemsen 1976). The major difference from our results and those of Lange, as well as other studies of Symko, were that they produced many more hybrids and they also obtained hybrids with different chromosome numbers that ranged from 14 to 28 when crossing the tetraploid forms. A number of factors could have led to differences in results in different labs. One was the different temperatures at which the plants used for crossing were grown. We later demonstrated that when plants are grown at higher temperatures, fewer hybrids are obtained, presumably due to more extensive elimination of chromosomes at the higher temperature. A second factor was the different accessions of *H. bulbosum* used in crossing with barley. Since *H. bulbosum* is a cross-pollinated species and highly heterozygous and heterogeneous, much more variability in the genes involved in the balance may be expected compared to self-pollinated barley. Thus, selection of *H. bulbosum* lines to be used for haploid production was necessary. Fukuyama and Takahashi (1976) reported crossing various accessions of tetraploid *H. bulbosum* (as the female parent) with tetraploid barley and observed a wide range in the proportions of hybrids and diploid barleys depending on the accession used. They also obtained seeds from the crosses, some of which could be germinated. Again, this could be expected as the endosperm contains the chromosome ratio of 2B:1V and is able to survive better and nourish the developing embryo. They also demonstrated a wide variation in chromosome numbers of resultant plants and suggested that barley chromosomes were also eliminated in some instances. One line of evidence was that they obtained a couple of diploid *H. bulbosum* plants from these crosses. This may be feasible by chromosome elimination but such diploids do not rule out parthenogenesis in these rare events. They also demonstrated a very strong effect of the *bulbosum* cytoplasm on the plant morphology of barley plants obtained as had been observed by a number of other researchers. They clearly demonstrated that the *H. bulbosum* accession used in crosses made a difference in the extent of elimination that occurs and may lead to some elimination of barley chromosomes. R.A. Pickering (pers. comm.), using GISH clearly showed that some haploid plants are a mixture of chromosomes of the two species. (Kasha et al. 1996) we had earlier demonstrated with GISH that it was possible to distinguish chromosomes of *H. bulbosum* from those of barley.

The difficulty in obtaining gene transfers from *H. bulbosum* to barley was surprising in that there was quite good homology between their genomes. Crosses between the interspecific hybrids and barley consistently gave barley haploids (Fukuyama and Takahashi 1976 and many other reports). Kasha and Sadasivaiah (1971) studied the meiotic chromosome pairing in diploid, triploid, and tetraploid hybrids between barley and *H. bulbosum*. In diploid hybrids, only 40% of the cells at metaphase had 14 chromosomes but those cells with 14 chromosomes had a mean of 5 bivalents with some cells having maximum of 7 bivalents. This indicates quite good homology between the species chromosomes because the barley haploids rarely showed bivalents (Sadasivaiah and Kasha 1971). Noda and Kasha (1981) examined chromosome stability in various tissues such as spike primordia and root and leaf meristems in three different diploid hybrids. They observed that root and leaf primordia were quite stable with only 2-3% cells showing variation in chromosome numbers, while the spike primordia had 40% cells with fewer than 14 chromosomes. The latter is consistent with the studies mentioned above on the instability at meiosis in these hybrids and so preferential elimination of chromosomes could be an explanation for the difficulty in obtaining gene transfers.

Some years later, a graduate student with me at Guelph, Xu Jie, proposed an idea for transferring the genes from hybrids into barley and succeeded in transferring powdery mildew resistance (Xu and Kasha 1992). He used triploid hybrids that produced some pollen and had a high frequency of pairing between the barley and *bulbosum* chromosomes. When he crossed them onto diploid barley he obtained a few seeds. Three resultant plants had bulbous DNA and were screened for disease resistance. It was later discovered that one of the selected barley plants also had yellow mosaic virus resistance transferred from *H. bulbosum*. Lines containing the yellow mosaic virus resistance are being used in cultivars in China. Pickering (2000) has also transferred genes from *bulbosum* to barley through finding pollen fertility in triploid (VBB) interspecific hybrids and using it to pollinate barley. Thus, the key to obtaining gene transfers from the interspecific hybrids would appear to be finding hybrids with good chromosome pairing and pollen fertility so that the chromosome elimination phenomenon and certation is reduced.

Chromosome Doubling and Other Improvements

Because the resultant haploid barley plants from wide crosses are sterile with very little spontaneous doubling (1 to 3% of plants), an improved simple and rapid method of chromosome doubling was needed to make the system efficient for barley breeding programs. Previously published procedures wherein colchicine was applied to plants in a paste for two or three days or was taken up by a wick placed through the crown of the seedling were too tedious and time consuming, and not suitable for our needs. I had previously worked on chromosome doubling in alfalfa and used a short 5-h colchicine treatment with success. We examined both nitrous oxide and different ways of applying colchicine to the haploid barley seedlings (Subrahmanyam and Kasha 1975; Thiebaut and Kasha 1978; Thiebaut *et al.* 1979). C.J. Jensen in Denmark was a close collaborator in developing the Bulbosum Method and contributed to chromosome doubling (Jensen 1974, 1976). The system that evolved was a short 5-h colchicine treatment at room temperature under bright lights, which caused the plants to respire and take up the colchicine. It involved cutting back the roots, using wetting agents and a carrier (DMSO) for the colchicine, and placing the shortened roots and base of the seedling in the solution. After 5-h we rinsed the seedlings in water and stimulated rapid plant recovery by treatment with hormones such as GA_3. This system of chromosome doubling gave less plants loss and was effective in doubling sectors on 70-80% of the haploid plants. This method of treatment is widely used today for chromosome doubling of haploids in many species.

Many other researchers have also worked on these species hybrids to improve the Bulbosum Method of haploid barley production initiated in our lab. Hayes and Chen (1989) reported the highest frequency of haploids produced as 48/100 florets by modifying the culture procedures. R.A. Pickering has studied many aspects of producing haploids or hybrids for gene transfer in Wales and New Zealand. These include selection of *bulbosum* lines, chromosome pairing, and plant growth conditions; he has also produced barley plants with extra DNA (recombinant lines) and with substitution chromosomes from *bulbosum* (Pickering *et al.* 1994; Pickering 2000: Johnston and Pickering 2002). M. D. Bennett at PBI in Cambridge was also intrigued by how the chromosomes could be eliminated and from cytological studies concluded that elimination related to chromosome arrangement within the nucleus and centromere attachment to spindles during cell division (Bennett *et al.* 1976; Finch and Bennett 1983).

Haploids Produced from Other Wide Hybridizations

The discovery of the principle of preferential chromosome elimination in crosses between barley and *H. bulbosum* stimulated researchers to look at other wide crosses for the production of haploids in other crops. The key to commercialization was to find a good cross-pollinator such as *H. bulbosum* that would produce an abundant amount of pollen for crossing and would stimulate haploid production from most genotypes. Barclay (1975) demonstrated that crosses between the cv. Chinese Spring wheat and diploid or tetraploid *H. bulbosum* also produced haploid wheat plants. However, successful crossing with wheat was restricted to wheat genotypes that carried the recessive genes for cross incompatability (Snape *et al.* 1979; Sitch *et al.* 1985; Falk and Kasha 1983). Some time later, Laurie and Bennett (1986) reported that pollination of wheat with corn (*Zea mays* L.) yielded embryos that could be rescued to produce haploid wheat plants. This system has been improved and is now widely used in wheat breeding and genetic studies and works well across wheat genotypes. After pollination, the wheat spikes are usually sprayed with auxin 2,4-D to stimulate embryo development. Corn pollen can also be used on oats (*Avena sativa* L.) and Rines et al. (1997) have also been able to produce oat plants with extra chromosomes from corn. Other pollinators that work on wheat are sorghum and millets (Ahmad and Comeau 1989; Laurie *et al.* 1990) and Job's tears (Mochida and Tsujimoto 2001). The chromosome elimination phenomenon is quite prevalent among wide crosses between Hordeum species (Rajhathy and Symko 1974; Jørgensen and von Bothmer 1988; Subrahmanyam 1982) and is now recognized as a phenomenon in wide crosses.

Applications of Haploids from Wide Crosses

Since this wide hybridization technique appeared to work well with most genotypes of barley, it was of interest to know whether a random sample of gametes could be obtained from a F_1 barley hybrid. Also, we wished to know how these genetically homozygous lines would perform across different environments as some breeders questioned whether complete homozygosity would be too inflexible for adaptation. With the enthusiastic collaboration and guidance of Dr. E. Reinbergs and working with a number of excellent PDFs and graduate students, populations were developed from the same F_1's by different breeding methods, namely pedigree, single seed descent and bulk, as well as from doubled haploids. It was found that the best performing doubled haploid lines were equal to the best lines from other methods of inbreeding and were just as stable across different environments. The sample of gametes appearing as haploid plants from an F_1 appeared to be random based upon gene segregations in such populations. We published many papers on the applications of doubled haploids in barley breeding and genetic studies, as reviewed by Choo et al. (1985). Using doubled haploids, there were dominant genetic effects because such plants are genetically homozygous and our studies showed that additive genetic effects were the major components of inheritance of agronomic traits in barley. Snape and Simpson (1981) have laid out genetic expectations of doubled haploids derived from different filial generations. Thus, doubled haploids are well suited as a breeding tool in self-pollinated crops and a means of developing inbreds for hybrid production in cross-pollinated crops.

A recent summary by Thomas *et al.* (2002) lists 58 new barley cultivars produced world wide by the Bulbosum Method. The first cultivar produced was named Mingo after Dr. Keh Ming Ho who developed it and had it evaluated. It was registered for release five years after initiating a new commercial breeding program with this haploid system for Stewarts Seeds in Canada (Ho and Jones 1980). Normally, using other barley breeding methods it would take at least 8 to 10 years to reach the registration stage. While this company has changed names, the breeding program is still in operation at the same location and has produced at least 20 registered cultivars using the Bulbosum Method. Because the haploid system produced a random sample of gametes it was used to produce doubled haploid populations for the North American Barley Genome Mapping Project that provided molecular marker maps for barley (Kleinhofs et al. 1993; Kasha et al. 1995). The random sample of doubled haploids allowed researchers at many locations to collaborate, knowing that they were mapping on identical genotypes. These maps have been extensively used to identify and tag genes and QTL for many traits of agronomic value in barley (Tinker *et al.* 1996; Hayes et al. 1993; and many others). Doubled haploid lines have also proven to be valuable in mutation studies (Maluszynski *et al.* 1996). A number of reviews of haploids in barley provide a perspective on methods of production and applications (Kasha *et al.* 1990; Pickering and Devaux 1992) while microspore culture of cereals was reviewed by Jähne and Lörz (1995).

ISOLATED MICROSPORE CULTURE IN CEREALS

Since the wide hybridization of wheat with *H. bulbosum* was limited and not well suited to producing haploids for breeding programs, we decided to improve anther and isolated microspore culture as an alternative system for wheat. It was expected to be a long-term project as many excellent researchers had worked on these culture systems for some time and, although great progress had been made in some species, there were still strong genotype differences in response in anther and microspore culture. Also, the frequencies were too low for large-scale cereal breeding programs. Although at the time we were starting cultures, Devaux (1987) had compared anther culture in barley with the Bulbosum Method and found no significant difference in the frequencies of haploids produced in winter barley, still the large number of microspores per floret made the potential for such a system was much greater than wide hybridization (one egg cell per floret).

In microspore culture there are so many factors to consider, such as growth of donor plants, pretreatment for induction of embryogenesis, isolation of microspores, induction media components, regeneration media and culture conditions (Fig. 7.1). Each species appeared to have its own media and environmental requirements and exhibited strong genotype differences in response. Most often there are coculture requirements or conditioned media involved or unknown nutrients such as from potato extract or coconut milk.

- Optimize donor plant growth conditions and fertilize weekly.
- Collect tillers when oldest florets are at mid to late uninucleate stage.
- Sterilize spikes with 70% ethanol or also with 10% bleach if spikes are exposed.
- Place spikes in pretreatment—preferably cold + mannitol for 4-7 days.
- Isolate microspores by blender, filter and wash in cold mannitol, centrifuge.
- Plate microspores on filter paper on solid media or put in liquid induction media.
- Place cultures in darkness at recommended temperature, 25-30°C, replenish media as needed over 4-5 weeks.
- Transfer 1-2 mm embryos to differentiation media for 1-2 weeks.
- Transfer to regeneration media for 1-2 weeks and then into tubes.
- At three-leaf stage and when well rooted, transfer to mixture of peat moss and soil.

Fig. 7.1 Steps in isolated microspore culture in cereals [See Kasha et al. (2001a) for details].

Objectives

With all these factors in mind, we asked ourselves: "Where does one start?" We decided to try:

(i) to develop a completely defined induction medium

(ii) to work on both barley and wheat isolated microspore culture
(iii) to select a model genotype with which to work

Igri, a two-row winter barley of European origin, was a model genotype in barley androgenesis as it responded better in cultures than other genotypes. Once isolated microspore culture had been improved with the model genotype, we would then determine how the improvements worked on other genotypes. There was no clear superior model genotype for wheat at that time and we would have to search for the best one to use.

BARLEY

Staging of microspores: When we started, we saw the need for some clear and quick way to identify the uninucleate stages. George Wheatley, a graduate student, undertook to describe the uninucleate stages based upon the position of the nucleus relative to the cell wall pore and the size of the vacuole in barley (Wheatley et al. 1986). It was found that in the early uninucleate stage the microspore was smaller, with a small vacuole, and the nucleus located beside the pore in the microspore wall. With time, the nucleus migrated around the wall until it reached a position opposite the pore, which was described as late uninucleate (Fig. 7.2a). The microspore was larger and the vacuole so large that the nucleus was somewhat flattened against the cell wall. When nuclear migration was halfway round the wall, the stage was described as mid-uninucleate. Recent studies in our lab have shown that the nucleus in the early uninucleate stage is in G1 of the cell cycle and that DNA synthesis (S stage) starts from mid to late uninucleate so that the late stage usually is in G2 (Shim and Kasha 2003). In our experience across genotypes, the mid-to late-uninucleate stage is best for induction pretreatments for barley microspores. For wheat, the early uninucleate stage is definitely too young for pretreatments and isolation whereas it can still succeed in barley. We observed that cereal microspores can become too old for good induction. When they are binucleate and start to form starch, response is reduced and culture repeatability is decreased. The stage of the uninucleate microspore in which induction could occur was known to be narrow (Pechan and Keller 1988). Sangwan and Sangwan-Norreel (1996) have recently reviewed this phenomenon and noted that the induction stages feasible varied with species and ranged from the early uninucleate stage in barley to the early binucleate stage in *Brassica* species and tobacco. The optimum stage for other species falls within this range. More recently, Indrianto *et al.* (2001) cytologically tracked, using time-lapse photography, the individually induced wheat microspores and demonstrated the progression of stages from induced microspores in developing structures.

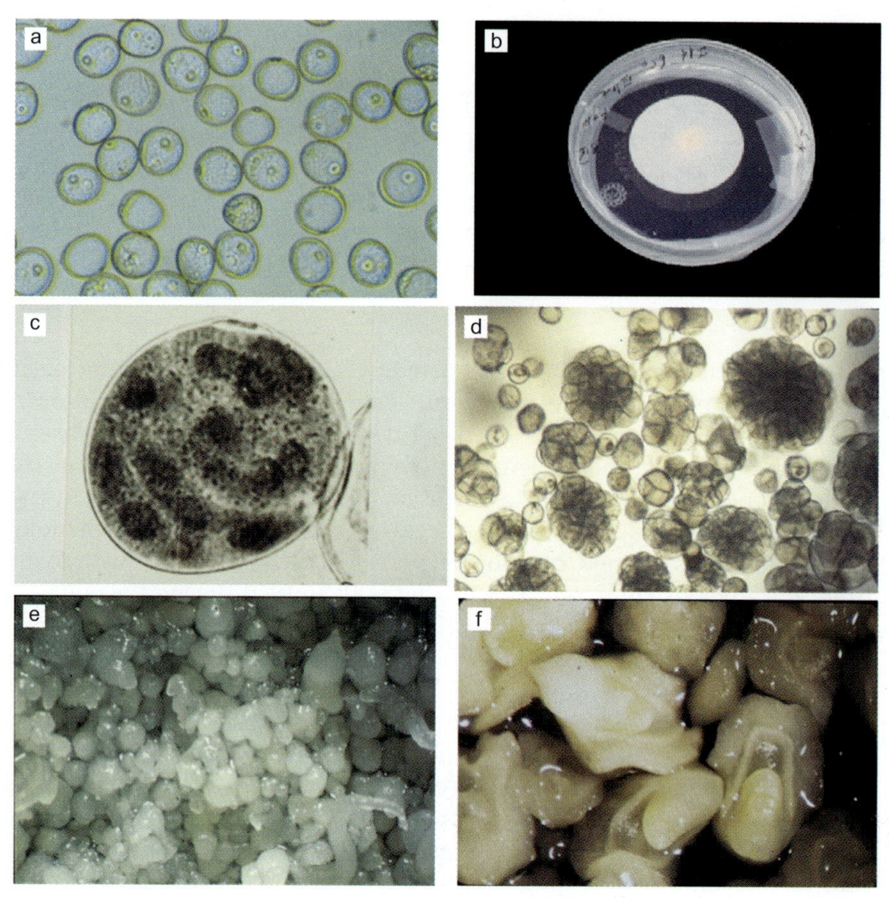

Fig. 7.2 Stages of development in barley isolated microspore culture. a. Freshly isolated barley microspores at the mid-to late-uninucleate stage. b. Isolated microspores transferred to filter paper on top of media. c. Multicellular microspore still within the microspore wall. d. Multicellular structures at d 10-12 exhibiting different sizes. e. Microspore culture at 21 d on filter paper. f. Microspore culture at 25-28 d in culture.

Pretreatments for embryo induction: Pretreatment for the induction of microspore embryogenesis is another area that has received a great deal of attention. The traditional pretreatment developed for barley was placing spikes on cold storage at 4°C for 28 d. In 1990, Roberts-Oehlschlager and Dunwell found that a short 4-d mannitol pretreatment also stimulated production of microspore-derived embryos. This was thought to be an osmotic stress that switched the development from pollen grains into cell division and structure development in culture. Touraev et al. (1996, 1997) proposed that stresses of various types were the factors causing the induction of embryogenesis and hence developed a heat pretreatment that worked well for tobacco and wheat microspores. The lab of Konzak also used heat pretreatment with the addition of various chemicals (2-HNA for one) to provide good induction of wheat microspores (Liu et al. 2002). We found that a combination of cold plus mannitol for a short period of time (4 d for barley or 5-7 d for wheat) was very effective for induction (Kasha et al. 2001a). With this treatment in barley, the microspores collected at the early to mid-uninucleate stages are actually held at those stages during the 4-d pretreatment (Shim and Kasha 2003) and a more synchronized population of microspores may be obtained.

Anthers are known to have an effect on response of microspores in culture and may be a source of genotype response variability. Cho and Kasha (1989) observed wide variation in the release of ethylene from various barley genotypes at the beginning of anther culture and concluded that there appeared to be an optimum early level of ethylene produced in the best responding genotypes. This may be related to the senescence of the anther and contents of the medium used. Thus, it strengthened our resolve to use isolated microspores to develop a repeatable system.

Microspore isolation: The next problem was how to isolate the microspores. The system most often used was a form of mortar and pestle grinding the anthers. Ziauddin et al. (1990) observed that microspores were shed into a medium when placed in a magnetic stirrer after pretreating the anthers for 4 days in mannitol at room temperature and that viability and response were improved by this isolation. However, our system of choice now is to use a blender on cut up spikes as adapted to barley by Olsen (1991). The blender is much faster and can be used with any pretreatment at any stage of the system, such as before or after the pretreatment to induce embryogenesis. However, experience has taught us that the speed and time in the blender are critical. It may need to be varied depending upon the toughness of the spike and where the plants are grown. High speed for too long will result in death of most of the microspores although donor plants grown at higher temperatures require longer blending. It is important to blend at cold temperatures, and do the isolation and washing steps of the microspores as quickly as possible under cold conditions to prevent damage (Kasha et al. 2001a).

Culture media: Finding the appropriate culture media, despite the fact that there have been numerous advances made in this area, was a major challenge. Each lab has tended to use its own modifications. The work of Köhler and Wenzel (1985) with barley anther culture in which they reported finding a factor in anthers that stimulated response in culture was important to us. While unidentified, they determined that it had a phenyl ring. While searching the literature, my graduate student, Albert Marsolais, found a little known auxin in plants that might be the stimulating chemical. It was phenyl acetic acid (PAA) and was reported to be quite abundant and stable in this or related forms in plants (Leuba et al. 1989). When we tested it, the response in number of structures with PAA in barley microspore cultures did not differ at that time from 2,4-D but the PAA appeared to produce better embryo structures and less calli so that regeneration was improved (Ziauddin *et al.* 1992). On the basis of these findings our program to develop a defined induction medium for barley isolated microspore culture was launched. We put together and adapted findings from two other researchers to develop our medium. Hunter (1988) determined that the sugar source was very critical for barley and that maltose or some other sugars were much better than sucrose, the traditional sugar source in tissue culture. Maltose has now become the preferred sugar for anther/microspore culture in most cereals. Mordhorst and Lörz (1993) observed that the type and balance of nitrogen sources was important for barley isolated microspore culture. Putting these findings together with amino acid modifications, we developed a modified version of Hunter's (1988) FHG medium for barley isolated microspore culture that works well across genotypes with high frequencies of embryos (Fig. 7.2) and plants (Kasha et al. 2001a). Visitors such as I. Szarejko from Poland and L. Cistue from Spain made contributions in developing some of the criteria for good response.

Chromosome doubling in microspore culture: Besides the potential number of embryos produced per floret, the frequency of "spontaneous" chromosome doubling with anther and microspore culture is higher than with the Bulbosum Method. This doubling usually occurs at the start of cell division in the microspore and is likely related to pretreatments for induction and the stage of microspore during this pretreatment. Thus "spontaneous" may not be truly spontaneous in this situation as it can be induced. Sunderland and his colleagues proposed that different pathways of development from the uninucleate microspores exist (reviewed in Sunderland 1974) so that different pathways to doubling may be feasible. The pathways depend upon the 1st mitotic division in the microspore. Pathways A and C resulted from an asymmetric 1st division that is typical of normal pollen development while pathway B produced two symmetric nuclei of the larger vegetative nucleus type. In these different pathways for development from the uninucleate microspore, chromosome doubling might occur by endomitosis or nuclear fusion. Sunderland (1974) clearly showed photos of endomitosis while the fusion was hypothesized. A visitor to my lab from Taiwan, Dr. C.C. Chen, undertook EM studies of nuclear division and initial development in barley microspores. He observed what appeared to be coalescence of nuclear membranes from symmetrical first division nuclei lying close together (Chen *et al.* 1984).

The frequency of this nuclear fusion was not determined but was later also observed in rice. More recently in the study of mannitol pretreatments of barley microspores we provided the first clear evidence that binucleate microspores became uninucleate in high frequencies with the accom-panying increase in relative DNA amounts (Kasha et al. 2001b). If there had been simultaneous division on a common spindle or on parallel spindles, two nuclei would have formed. This reversion to a single nucleus provides clear evidence for doubling of chromosomes by nuclear fusion after the first nuclear division in the microspores, leading to completely homozygous fertile plants. The large quantities of seed obtained are often sufficient to avoid the need to grow a seed increase generation before initial field trials. The complete fertility of the plants is an indication that chromosome doubling most likely occurs at the time of pretreatment for the induction of embryogenesis. The frequency of doubling and fertile doubled haploids from barley microspores appears to be higher than frequencies observed in other species. This high frequency in barley may correlate with the early uninucleate stage that responds to pretreatments. When the barley spikes and florets are selected so that the uninucleate stages cultured are predominantly synchronized for transformation, this doubling can be as high as 90% (Shim, unpublished). Other species may require application of colchicine or other doubling agents at the time of culture (Zhao *et al.* 1996) or after seedlings are formed.

Wheat

The system developed for barley isolated microspore culture was not transferable to wheat and required modification of several factors.

Donor plant growth and staging: We observed that some spring wheat genotypes (Chris and Pavon) responded quite well in microspore culture. However, wheat required the use of ovary coculture and hence a defined medium was not achieved. The range of microspore stages in a wheat spike is much wider than barley as there are multiple florets per spikelet in wheat versus a single floret in barley spikelets. In staging, the microspores should be at mid- to late-uninucleate stages for best pretreatment response. The vigorous growth of the donor plants was better at a higher temperature than barley, such as 15-16°C in darkness and 18-20°C with the lights on for at least 16 h d^{-1}. However, the Konzak group successfully grew plants in the greenhouse at temperatures as high as 27°C (Liu *et al.* 2002). Under our conditions, the wheat spikes had partially emerged from the leaf sheath when the right microspore stage was reached. Therefore, sterilization of the spikes with bleach in addition to alcohol was necessary. Also, in donor plants grown at a higher temperature, the spikes were harder and required a longer blending time, possibly leading to more damage to the microspores.

Media development for wheat: We developed for wheat a modified MS medium (MMS4) with lower NH4 and higher organic N, such as 900 mg L^{-1} glutamine. The auxin PAA also worked well with wheat microspore culture. The main problem was the need for ovary coculture to obtain a good response. Also, the condition of the ovaries collected can vary and the genotype of the ovary may be a factor. Some ovaries turned brown during coculture, exerting a negative impact on microspore response. Thus, we needed to find some alternative to ovary coculture in order to obtain a defined medium.

While attending a plant biology conference, I was interested in a paper on arabinogalactan protein (AGP) that was helpful in preventing apoptosis or programmed cell death. It brought two things to mind. First, in microspore cultures some culture plates showed no response while others from the same batch of microspores gave a good response. Could this be attributed to programmed cell death? Second, a graduate student with me, P.L. Vrinten, had isolated and characterized some cDNAs of genes induced to be expressed in the first three days of embryogenesis in barley isolated microspore culture. One of these was an unknown that appeared to be related to AGP (Vrinten *et al.* 1999). Thus, we decided to try adding AGPs to the induction culture media for wheat microspores. In searching for a source of AGP, we found a commercial one designated as AGP by Sigma and called Larcol. By adding it to wheat microspore culture media (10-25 mg L^{-1}) we could greatly increase the survival of the microspores and thus get much better embryo formation (Kasha and Simion 2001). With higher levels of AGP we obtained good embryo formation without the use of ovary cocultures. However, our best response was with the combination of ovaries and AGP. This is an area that we are still researching and have shown similar responses with Gum Arabic and AGP isolated from wheat seeds and ovaries (Unpub). While our induction of embryos in now approaching the high levels we get in barley, regeneration from embryos is still much lower than in barley microspore culture and genotype differences are stronger. We hope that with a defined medium the response differences of genotypes can be reduced, as done in barley.

Breeding applications of haploids from isolated microspore culture: Working across a large and varied number of barley genotypes (spring, winter, 2-row, and 6-row) we produce on average about 10,000 structures per culture plate. These are mostly embryo-like structures (els) and show a wide range of size variation. On examining the regeneration of the first 500 embryos from a number of genotypes, regeneration ranged from 36 to 97% (Kasha *et al.* 2001a). One culture plate would initially contain about 400,000 microspores. These numbers could be extracted from ten 2-row spikes or four 6-row spikes, indicating that about 2.5% of the microspores developed into els. Divisions were initially observed in over 50% of the microspores in a plate during the first few days, indicating high levels of viability and induction. Other labs have developed similar media and protocols but find this system to be very effective for barley breeding programs and new cultivars have been released (A. Jensen, Denmark, pers. comm.). Thomas et al. (2003) attempted to find and summarize the numbers of cultivars produced in various crops using haploid methods and found over 200, of which 96 are in barley. The barley lines were produced by different methods, 58 by the Bulbosum Method, 35 by anther culture, and 3 from isolated microspore culture. The cultivars were produced in 9 different countries by the Bulbosum Method, 8 different countries by anther culture, and 2 different countries by microspore culture. There were also 20 cultivars from haploid methods in wheat, 14 by anther culture, and 6 from wide hybridization with corn pollen. In future, we expect to see more cereal cultivars being produced from isolated microspore cultures as the procedures for producing large numbers of haploids become more routine and genotype independent.

Like the Bulbosum Method, doubled haploids from anther or microspore culture have been used for molecular marker mapping although the randomness of the sample of gametes obtained may not be quite as good. Many earlier studies had shown that certain genes influenced success in anther culture and that these could cause segregation distortion in the population from a cross (Thompson et al. 1991; Foisset et al. 1997). The concept is that gametes carrying the alleles for improved response would preferentially survive into haploid plants, giving rise to preferential transmission of alleles of those genes and to those genes linked to them. This did not appear to be a problem with the Bulbosum Method and now needs to be tested with improved isolated microspore culture procedures. In wheat, we have achieved upto 300 green plants per spike with the higher responding genotypes Chris and Bob White. This frequency would be suitable for breeding but the numbers are much lower across other genotypes. In the other genotypes, the initial induction of embryos is almost as high as barley but the regeneration rate is very low. At present, the frequencies with genotypes in a wheat breeding program are only slightly better than those using the wide hybridization method with corn pollen. With better regeneration of embryos, isolated microspore should become more efficient in producing doubled haploids for wheat breeding.

Other applications and potential uses of haploids: The major applications of haploids to date have taken advantage of the reduction in time to obtain homozygous lines compared to traditional methods of inbreeding, and facilitation of selection of pure lines versus individual plants. However, many other applications are practiced or have great potential. They are quite extensively used in genetic studies (Pauls 1996). Evaluation of traits and subsequent selection would be advantageous in many types of crops such as vegetables, pharmaceutical or herbal producing crops, ornamentals, long-lived perennials such as forest or fruit trees, and so forth. Use of doubled haploids to sample and select from the wide variability in some species would be useful for pharmaceutical production or new strains of ornamentals. The genes for the trait could be fixed in the homozygous state through doubled haploids so that plants could be readily selected and evaluated for the optimum levels of traits. This may not require high frequencies of haploids (Saxena, Pers. comm.). Haploidy has also been developed in asparagus where male and female plants occur. The production of a doubled haploid from the male plant through anther or microspore culture produces a super male that, crossed with a female, will produce all male progeny which are higher yielding than the female (Tsay 1997). The production of haploids from ovary or ovule culture has been used in crops such as sugar-beets (*Beta vulgaris*) and onion (*Allium cepa*) with good success where androgenesis has not succeeded (Keller and Korzun 1996). Use of haploids in mutation and selection has been reviewed by Maluszynski *et al.* (1996). The target could be the male gametophytes, haploid protoplasts, haploid cell cultures or haploid plants. As for breeding, the main advantage would be directly observing and selecting the mutations, either dominant or recessive, on the mutagenized haploid cells in culture or on the haploid or doubled haploid plants (Castillo *et al.* 2001). Production of doubled haploids from the gametes of M_1 plants could also aid in detection and fixation of the mutants as pure lines. Development of doubled haploid lines, whether for selecting new mutations or variability in nonmutagenized populations, would allow for easier selection of simple or quantitative traits.

In ways similar to mutation, haploids would be advantageous for transformation in that recessive traits or markers could be identified in the haploid and become homozygous by subsequent chromosome doubling of the haploid plants. Isolated microspores provide thousands of target haploid cells and the potential to directly obtain plants homozygous for the transgenes. Jähne *et al.* (1994), Yao *et al.* (1997), and Carlson *et al.* (2001) have produced transgenic barley plants from biolistic bombardment of isolated microspores induced to form embryos. Only Jähne *et al.* (1994) obtained homozygous transgenics while the others obtained heterozygous transgenic plants. Our current research is to determine when DNA synthesis occurs in the uninucleate microspore in order to plan the pretreatments and time of bombardment to achieve homozygous transgenic plants (Shim and Kasha 2003; Shim unpubl.). Others have tried to bombard the immature pollen to incorporate the genes and then develop mature pollen to be used in pollinations (Støger *et al.* 1992). Transformation of isolated microspores by *Agrobacterium* has been attempted but is more difficult than biolistic bombardment due to the problem of subsequently removing the *Agrobacterium.* However, this has been achieved in *Brassica napus* (Dormann et al. 2001) but not as yet in cereals. Transformation of immature haploid embryos produced by any haploidy method would produce plants chimeric for the transgenes or markers so that selection could be achieved with the resultant plants (Folling and Olesen 2002). The closer to the haploid single cell stage one can transform the target, the greater the potential for doubled sectors and thereby the amount of seed they produced.

The applications and potential uses of haploids and doubled haploids are only briefly covered here. Numerous reviews have been published. just two methods of haploid production have been described. The advantages and potential of other procedures have not been touched. I would conclude that the potential applications of haploids are limited solely by the imagination of the researchers.

REFERENCES

Ahmad F, Comeau A (1989) Wheat × pearl millet hybridization: consequences and potential. Euphytica 50:181-190.

Barclay IR (1975) High frequencies of haploid production in wheat (*Triticum aestivum* L.) by chromosome elimination. Nature (London) 256:410-411.

Bennett MD, Finch RA, Barclay IR (1976) The time, rate and mechanism of chromosome elimination in *Hordeum* hybrids. Chromosoma 54:175-200.

Carlson AR, Letarte J, Chen J, Kasha KJ (2001) Visual screening of microspore derived transgenic barley (*Hordeum vulgare* L.) with green-fluorescent protein. Plant Cell Rep. 20:331-337.

Castillo AM, Cistue L, Valles MP, Sanz JM, Romagosa I, Molina Cano JL (2001) Efficient production of androgenic doubled-haploid mutants in barley by the application of sodium azide to anther and microspore cultures. Plant Cell Rep. 20:105-111.

Chen CC, Howarth MR, Peterson RL, Kasha KJ (1984) Ultrastructure of androgenic microspores of barley during the early stages of anther cultures. Can. J. Genet Cytol. 26:484-491.

Cho U-H, Kasha KJ (1989) Ethylene production and embryogenesis from anther cultures of barley (*Hordeum vulgare* L.). Plant Cell Rep. 8:415-417.

Choo TM, Reinbergs E, Kasha KJ (1985) Use of haploids in barley breeding. Plant Breed. Rev. 3:219-252.

Davies DR (1958) Male parthenogenesis in barley. Heredity 12:493-498.

Devaux P (1987) Comparison of anther culture and *Hordeum bulbosum* method for the production of doubled haploid in winter barley. I. Production of green plants. Plant Breed. 98:215-219.

Dormann M, Wang H-M, Oelck M (2001) Transformed embryogenic microspores for the generation of fertile homozygous plants. US Patent # 6316694, Gazette of US Patent and Trademarks 1252(2), issued Nov. 13, 2001.

Falk DE, Kasha KJ (1983) Genetic studies of the crossability of hexaploid wheat with rye and *Hordeum bulbosum.* Theor. Appl. Genet. 64:303-307.

Finch RA, Bennett MD (1983) The mechanism of somatic chromosome elimination in *Hordeum. In:* PE Brandham, MD Bennett (eds.). Kew Chromosome Conference II, Allen and Unwin, London, pp. 147-154.

Foisset N, Delourme R, Lucas M-O, Renard M (1997) *In vitro* androgenesis and segregation distortion in *Brassica napus* L.: spontaneous versus colchicine-doubled lines. Plant Cell Rep. 16:464-468.

Folling L, Olesen A (2002) Transformation of wheat (*Triticum aestivum* L.) microspore-derived callus and microspores by particle bombardment. Plant Cell Rep. 20:1098-1105.

Fukuyama T, Takahashi R (1976) A study of the interspecific hybrid, *Hordeum bulbosum* (4*x*) × *Hordeum vulgare* (4*x*), with special reference to dihaploidy frequency. *In:* Barley Genetics III, Karl Thiemig, Mnchen, pp. 351-360.

Gamborg OL, Miller RA, Ojima K(1968) Nutrient requirements of suspension cultures of soybean root cells. Exp. Cell. Res. 50:151-158.

Hayes PM, Chen FQ (1989) Genotypic variation for *Hordeum bulbosum* L.-mediated haploid production in winter and facultative barley. Crop. Sci. 29:1184-1188.

Hayes PM, Liu BH, Knapp SJ, Chen F, et al. (1993) Quantitative trait locus effects and environmental interaction in a sample of North American barley germplasm. Theor. Appl. Genet. 87:392-401.

Ho KM, Jones GE (1980) Mingo barley. Can. J. Plant Sci. 60:279-280.

Ho KM, Kasha KJ (1975) Genetic control of chromosome elimination during haploid formation in barley. Genetics 81:263-275.

Hunter CP (1988) Plant regeneration from microspores of barley, *Hordeum vulgare* L. PhD thesis. Wye College, Univ. of London, London, UK.

Indrianto A, Barinova I, Toutaev A, Heberle-Bors E (2001) Tracking individual wheat microspores in vitro: identification of embryogenic microspores and body axis formation in the embryo. Planta 212:163-174.

Jähne A, Lörz H (1995) Cereal microspore culture. Plant Sci 109:1-12.

Jähne A, Ecker D, Brettschneider R, Lörz H (1994) Regeneration of transgenic microspore-derived fertile barley. Theor. Appl. Genet. 89:525-533.

Jensen CJ (1974) Chromosome doubling techniques in haploids. *In:* KJ Kasha (ed.). Haploids in Higher Plants: Advances and Potential. Univ. of Guelph, Guelph, Ontario, canada, pp. 153-190.

Jensen CJ (1976) Barley monoploids and doubled monoploids: Techniques and experience. *In:* Barley Genetics III. Karl Thiemig, Mnchen. pp. 316-345.

Johnston PA, Pickering RA (2002) PCR detection of *Hordeum bulbosum* introgressions in an *H. vulgare* background using retrotransposon-like sequence. Theor. Appl. Genet. 104:720-726.

Jørgensen RB, Bothmer R von (1988) Haploids of *Hordeum vulgare* and *H. marinum* from crosses between the two species. Hereditas 108:207-212.

Kao KN, Kasha KJ (1970) Haploidy from interspecific crosses with tetraploid barley. *In:* RA Nilan (ed.), Barley Genetics II. Wash. State University, Pullman, WA, pp. 82-88.

Kasha KJ (1974) Haploids from somatic cells. *In:* KJ Kasha (ed.). Haploids In Higher Plants, Advances and Potential. University of Guelph, Guelph, Ontario, Canada, pp. 67-87.
Kasha KJ, Kao KN (1970) High frequency haploid production in barley (*Hordeum vulgare* L.) Nature 225:874-876.
Kasha KJ, Sadasivaiah RS (1971) Genome relationships between *Hordeum vulgare* L. and *H. bulbosum* L. Chromosoma 35:264-287.
Kasha KJ, Simion E (2001) Embryogenesis and plant regeneration from microspores. Int. Patent Pub. No. W0 01/41557. Electronic June 14, 2001.
Kasha KJ, Ziauddin A, Cho U-H (1990) Haploids in cereal improvement: Anther and microspore culture. *In:* JP Gustafson (ed.), Gene Manipulation in Plant Improvement II. Plenum Press, New York, NY, pp. 213-235.
Kasha KJ, Simion E, Oro R, Yao QA, Hu TC, Carlson AR (2001a) An improved *in vitro* technique for isolated microspore culture of barley. Euphytica 120:379-385.
Kasha KJ, Hu TC, Oro R, Simion E, Shim YS (2001b) Nuclear fusion leads to chromosome doubling during mannitol pretreatment of barley (*Hordeum vulgare* L.) microspores. J. Exp. Bot. 52:1227-1238.
Kasha KJ, Kleinhofs A, Kilian A, Saghai Marrof M, et al. (1995) The North American barley genome map on the cross HT and its comparison to the map on cross SM. *In:* K Tsunewaki (ed.), Kodansha Science Inc., Tokyo, pp. 73-88.
Kasha KJ, Pickering RA, William HM, Hill A, Oro R, Reeder S, Snape JW (1996) GISH and RFLP facilitated identification of a barley chromosome carrying powdery mildew resistance from Hordeum bulbosum. In Proc V. 10 C + VII IBGS Shinkard A, Scolesg, Rossnage B (eds) Univ. of Saskatchewan pp. 238-240.
Keller ERJ, Korzun L (1996) Ovary and ovule culture for haploid production. *In:* SM Jain, SK Sopory, RE Veilleux (eds.), *In Vitro* Haploid Production in Higher Plants. Kluwer Acad. Publ., Dordrecht, The Netherlands, Vol. 1. pp. 217-235.
Kleinhofs A, Kilian A, Saghi Maroof MA, Biyashev RM, *et al.* (1993) A molecular, isozyme and morphological map of the barley (*Hordeum vulgare*) genome. Theor. Appl. Genet. 86:705-712.
Köhler F, Wenzel G (1985) Regeneration of isolated barley microspores in conditioned media and trials to characterize the responsible factor. J. Plant Physiol. 121:181-191.
Lange W (1971) Crosses between *Hordeum vulgare* L. and *H. bulsosum* L. I. Production, morphology and meiosis of hybrids, haploids and dihaploids. Euphytica 20:14-29.
Lange W, Jochemsen G (1976) The offspring of diploid, triploid and tetraploid hybrids between *Hordeum vulgare* and *H. bulbosum*. *In:* Barley Genetics III, Karl Thiemig, Mûnchen. pp. 252-259.
Larter EN, Enns H (1960) The influence of gibberellic acid on the development of hybrid barley ovules *in vitro*. Can. J. Genet. Cytol. 2:435-441.
Laurie DA, Bennett MD (1986) Wheat by maize hybridization. Can. J. Genet. Cytol. 28:313-316.
Laurie DA, O'Donoughue LS, Bennett MD (1990) Wheat × maize and other sexual hybrids: their potential for genetic manipulation and crop improvement. *In:* JP

Gustafson (ed.). Gene Manipulation in Plant Improvement II. Plenum Press, New York, NY, pp. 95-126.

Leuba V, Le Tourneau D, Oliver D (1989) Stability of Phenylacetic acid in liquid media. J. Plant Growth Reg. 8:163-165.

Liu W, Zheng MY, Polle A, Konzak CF (2002) Highly efficient doubled-haploid production in wheat (*Triticum aestivum* L,) via induced microspore embryogenesis. Crop. Sci. 42:686-692.

Maluszynski M, Szarejko I, Sigurbjörnsson B (1996) Haploidy and mutation techniques. *In:* SM Jain, SK Sopory, RE Villeux (eds.). *In Vitro* Haploid Production in Higher Plants. Kluwer Acad. Publi., Dordrecht, Netherlands, Vol. 1, pp. 67-93.

Mochida K, Tsujimoto H (2001) Production of wheat doubled haploids by pollination with Job's Tears (*Coix lachryma-jobi* L.). J. Hered. 92:81-83.

Mordhorst AP, Lörz H (1993) Embryogenesis and development of isolated barley (*Hordeum vulgare* L.) microspores are influenced by amount and composition of nitrogen sources in culture media. J. Plant Physiol. 142:485-492.

Noda K, Kasha KJ (1981) Chromosome elimination in different meristematic regions of hybrids between *Hordeum vulgare* L. and *H. bulbosum* L. Jpn. J. Genet. 56:193-204.

Olsen FL (1991) Isolation and cultivation of embryogenic microspores from barley (*Hordeum vulgare* L.). Hereditas 115:2255-266.

Pauls KP (1996) The utility of doubled haploid populations for studying the genetic control of traits determined by recessive alleles. *In:* SM Jain, SK Sopory, RE Veilleux (eds.). *In Vitro* Haploid Production in Higher Plants. Kluwer Acad. Publ., Dordrecht, Netherlands, Vol. 1, pp. 125-144.

Pechan PM, Keller WA (1988) Identification of potentially embryogenic microspores in *Brassica napus*. Physiol. Plant 74:377-384.

Pickering R (2000) Do the wild relatives of cultivated barley have a place in barley improvement? *In:* S Logue (ed.). Barley Genetics VIII. Adelaide Univ., Glen. Osmond, Australia, Vol. 1, pp. 223-230.

Pickering RA, Devaux P (1992) Haploid production: Approaches and use in plant breeding. *In:* PR Shewry (ed.). Barley: Genetics, Biochemistry, Molecular Biology and Biotechnology. CAB Int. Wallingford, UK, pp. 519-547.

Pickering RA, Timmerman GM, Cromey MG, Melz G (1994) Characterization of progeny from backcrosses of triploid hybrids between *Hordeum vulgare* L. (2*x*) and *H. bulbosum* L. (4*x*) to *H. vulgare*. Theor. Appl. Genet. 88:460-464.

Rajhathy T (1967) Notes on some interspecific *Hordeum* hybrids. Barley Newsletter 10:69-70.

Rajhathy T, Symko S (1974) High frequency of haploids from crosses of *Hordeum lechleri* (6*x*) × *H. vulgare* (2*x*) and *H jubatum* (4*x*) × *H. bulbosum* (2*x*). Can. J. Genet. Cytol. 16:468-472.

Rines HW, Riera-Lizarza O, Nunez VM, Davis DW, Phillips RL (1997) Oat haploids from anther culture and wide hybridization. *In:* SM Jain, SK Sopory, RE Veilleux (eds.). Kluwer Acad. Publi., Dordrecht, The Netherlands, Vol. 4, pp. 205-221.

Roberts-Oehlschlager SL, Dunwell JM (1990) Barley anther culture pretreatment on mannitol stimulates production of microspore-derived embryos. Plant Cell Tiss. Org. Cult. 20:235-240.

Sadasivaiah RS, Kasha KJ (1971) Meiosis in haploid barley—an interpretation of non-homologous chromosome associations. Chromosoma 35:247-263.

Sangwan RS, Sangwan-Norreel BS (1996) Cytological and biochemical aspects of *in vitro* androgenesis in higher plants. *In:* SM Jain, SK Sopory, RE Veilleux (eds.). Kluwer Acad. Publi., Dordrecht, The Netherlands, Vol. 1. pp. 95-109.

Shim YS, Kasha KJ (2003) The influence of pretreatment time on DNA synthesis in barley (*Hordeum vulgare* L.) uninucleate microspores. Plant Cell Rep. 21: 1065-1071.

Sitch LA, Snape JW, Firman SJ (1985) Interchromosomal mapping of crossability genes in wheat (*Triticum aestivum* L.). Theor. Appl. Genet. 70:309-314.

Snape JW, Simpson E (1981) The genetical expectations of doubled haploid lines derived from different filial generations. Heredity 42:291-298.

Snape JW, Chapman V, Moss J, Blanchard CE, Miller TE (1979) The crossability of wheat varieties with *Hordeum bulbosum.* Z. Pflanzenzüchtg. 85:200-204.

Støger E, Moreno RMB,Ylstra B, Vincente O, Heberle-Bors E (1992) Comparison of different techniques for gene transfer into mature and immature tobacco pollen. Transgenic Res. 1:71-78.

Subrahmanyam NC (1982) Species dominance in chromosome elimination in barley hybrids. Current Sci. 51:28-31.

Subrahmanyam NC, Kasha KJ (1973) Selective chromosomal elimination during haploid formation in barley following interspecific hybridization. Chromosoma (Berl) 42:111-125.

Subrahmanyam NC, Kasha KJ (1975) Chromosome doubling of barley haploids by nitrous oxide and colchicine treatments. Can. J. Genet. Cytol. 17:573-583.

Sunderland N (1974) Anther culture as a means of haploid induction. *In:* KJ Kasha (ed.). Haploids in Higher Plants: Advances and Potential. University of Guelph, Guelph, Ontario, Canada, pp. 91-122.

Symko S (1969) Haploid barley from crosses of *Hordeum bulbosum* (2*x*) × *H. vulgare* (2*x*). Can. J. Genet. Cytol. 11:602-608.

Thiebaut J, Kasha KJ (1978) Modification of the colchicine technique for chromosome doubling of barley haploids. Can. J. Genet. Cytol. 20:513-521.

Thiebaut J, Tsai A, Kasha KJ (1979) Influence of plant development stage, temperature, and plant hormones on chromosome doubling of barley haploids using colchicine. Can. J. Bot. 57:480-483.

Thomas WTB, Forster BP, Gertsson B (2003) Doubled haploids in breeding. *In:* M Maluszynski, KJ Kasha, BP Forster, I Szarejko (eds.). Doubled Haploid Production in Crop Plants: A Manual. IAEA, Vienna. pp. 337-349.

Thompson DM, Chalmers K, Waugh R, Forster BP, Thomas WTB, Caligari PDS (1991) The inheritance of genetic markers in microspore derived plants of barley (*Hordeum vulgare* L.). Theor. Appl. Genet. 81:487-492.

Tinker NA, Mather DE, Rossnagel BG, Kasha KJ, et al. (1996) Regions of the Genome that affect agronomic performance in two-row barley. Crop. Sci. 36:1053-1062.

Touraev A, Vincente O, Heberle-Bors E (1997) Initiation of microspore embryogenesis by stress. Trends Plant Sci. 2:297-302.

Touraev A, Pfosser M, Vincente O, Heberle-Bors E (1996) Stress as the major signal controlling the developmental fate of tobacco microspores: towards a unified model of induction of microspore/pollen embryogenesis. Planta 200:144-152.

Tsay H-S (1997) Haploidy in asparagus by anther culture. *In:* SM Jain, SK Sopory, RE Veilleux (eds.). *In Vitro* Haploid Production in Higher Plants. Kluwer Acad. Publ. Dordrecht, Netherlands, Vol. 5, pp. 109-134.

Vrinten PL, Nakamura T, Kasha KJ (1999) Characterization of cDNAs expressed in the early stages of microspore embryogenesis in barley (*Hordeum vulgare* L.). Plant Molec. Biol. 41:455-463.

Wheatley WG, Marsolais AA, Kasha KJ (1986) Microspore growth and staging for barley anther cultures. Plant Cell Rep. 5:47-49.

Xu J, Kasha KJ (1992) Transfer of a dominant gene for powdery mildew and DNA from *Hordeum bulbosum* into cultivated barley (*H. vulgare*). Theor. Appl. Genet. 84:771-777.

Yao QA, Simion E, William M, Krochko J, Kasha KJ (1997) Biolistic transformation of haploid isolated microspores of barley (*Hordeum vulgare* L.). Genome 40:570-581.

Yu RCP (1968) Derivation and study of primary trisomics of common barley, *Hordeum vulgare* L. PhD thesis, University of Manitoba, Winnipeg, Manitoba, Canada.

Zhao J-P, Simmonds DH, Newcombe W (1996) Induction of embryogenesis with colchicine instead of heat in microspores of *Brassica napus* L. cv Topas. Planta 198:433-439.

Ziauddin A, Simion E, Kasha KJ (1990) Improved plant regeneration from shed microspore culture in barley (*Hordeum vulgare* L.). Plant Cell Rep. 9:69-72.

Ziauddin A, Marsolais AA, Kasha KJ (1992) Improved plant regeneration from wheat anther and barley microspore culture using phenlyacetic acid (PAA). Plant Cell Rep. 11:4898-498.

8

Skoog and Miller Legacy: 45 Years of Manipulating Plant Growth

Susan J. Murch[1] and Praveen K. Saxena[2]
[1]Institute for Ethnobotany, National Tropical Botanical Garden, 3530 Papalina Road, Kalaheo, Hawaii, 96741.
e-mail: smurch@ntbg.org
[2]Department of Plant Agriculture, University of Guelph, Guelph, Ontario, Canada, N1G 2W1.
e-mail: psaxena@uoguelph.ca

INTRODUCTION

No volume on the regeneration of plants from a single cell would be complete without a review of the pivotal work of Skoog and Miller (1957) and a commentary on the research conducted in the last four decades. So this chapter gives an overview of the original hypothesized role of different plant growth regulators in plant regeneration and uses exceptions to illustrate some of the research that remains incomplete. Skoog and Miller published their hypothesis as a contribution to a symposium on the Biological Activity of Growth Substances but the impact of this manuscript to the science of plant growth regulation has far exceeded any expectation. The Skoog and Miller paper forms the basis for experiments to redirect plant growth and developmental processes *in vitro*. The impact of this work is seen in the consistently high citation of the hypothesis in plant regeneration papers. To date, more than 1000 authors have cited Skoog and Miller as the inspiration for their work and the rate of citation has remained relatively constant over almost 5 decades, further proof of the fundamental nature and importance of these observations (Fig. 8.1). This chapter examines the ruggedness of the hypothesis as tested over more than 45 years, the impact of Skoog and Miller on younger scientists, and the evolution of their hypothesis with the ever-growing body of literature.

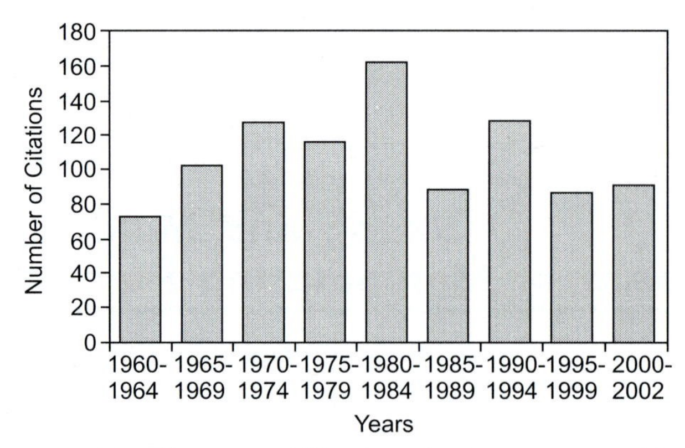

Fig. 8.1 Impact of the Skoog and Miller hypothesis as assessed by the number of citations in the last 45 years.

SKOOG AND MILLER HYPOTHESIS

In its simplest form, Skoog and Miller proposed uniformity in growth factor requirements and regulatory mechanisms for all types of plant growth. Further, they described the interaction between auxin and a cytokinin as the essential requirement with "both types of chemicals required for growth. Low levels of one with high levels of the other and vice versa lead to opposite morphological end-results." An example of the application of this principle is seen in Fig. 8.2. Incorporation of auxin into a culture medium in the absence of exogenous cytokinin results in a proliferation of roots. Alternately, supplementation of the medium with cytokinin promotes shoot organogenesis. The concept that the relative ratio of auxin to cytokinin dictates the route of plant regeneration set into motion a revolution in plant morphogenesis and facilitated modern plant biotechnology.

Historical Context

In 1957, several possibilities still existed to explain the complexities of plant growth. It was possible that plant growth was regulated simply by the availability of metabolic energy and carbon structures produced by photosynthesis and respiration. The theory of limited resources was based in the quantification of the outputs of primary metabolic cycles as a predictor of growth potential but several experimental systems were demonstrating that even moderate declines in respiratory rate could completely eliminate growth and therefore this possibility was reaching the limits of application. Alternatively, plants may have a succession of growth factors that regulate different phases of development. This approach continues to prompt the search for a "florigen", a hormone that will trigger plant flower formation. Investigations in this area have compared cell elongation and division in different tissues and organs for relative changes. Other thoughts include the possibility of specific plant growth regulating chemicals and the potential for interaction with organic and inorganic nutrients.

State of the Art

There are literally tens of thousands of reports of application of the Skoog and Miller hypothesis for the regeneration of tissues and organs and the manipulation of regeneration systems forms the basis of modern agriculture and horticulture. Genetic transformation systems rely on the capacity to regenerate transformed plants, mass propagation systems allow for large scale multiplication of elite germplasm, and even virus-free stocks require the capacity to regenerate shoots from the meristem for proliferation. Each of these innumerable practical applications must include a phase of regulated growth patterns originally described by Skoog and Miller.

Auxins

Exposure of excised plant tissues results in *de novo* root formation both *in vivo* and *in vitro*. This principle has been demonstrated innumerable times and the phenomena form the basis of modern asexual propagation industries wherein cuttings are exposed to commercial auxin preparations and induced to root in artificial media. In addition, experimental systems have demonstrated that auxin, alone or in combination with other plant growth regulators, regulates cellular processes such as cell division, elongation and differentiation. The term auxin is derived from the Greek word "auxine" which literally means "to increase" and auxin activity was first described as promotion of cell elongation (Went and Thimann 1937). In more recent literature auxin has been used in combination with cytokinin for plant regeneration, in conjunction with gibberellic acid for stem elongation (Ross et al. 2000), with ethylene for regulation of root hair differentation (Masucci and Schiefelbein 1994), and the culture of undifferentiated cells in the presence of high auxin followed by subculture onto a medium devoid of auxin has been much investigated as the classic system for induction of somatic embryogenesis (Steward et al. 1958; Reinert 1959). Auxin has a direct impact on the capacity of a cell to divide and was shown to increase the concentration of a cyclin-dependent kinase (CDK) important for transition through the cell cycle but cytokinin was required for activation of the CDK (den Boer and Murray 2000). The most important cell cycle control is the transition points as cells move from G1 to S phase and from G2 to M phase and it is in these phases that auxin has been demonstrated to be essential (den Boer and Murray 2000).

Cytokinins

The possibility of cytokinins, initially defined as a class of compounds that prompted cell division (cytokinesis) was initially proposed by Haberlandt (1921). Later researchers observed that cytokinins are compounds that, *when adequate auxin is present,* promote cell division (Skoog and Miller, 1957) and demonstrated with isolated compounds from complex mixtures, natural products and synthetic derivatives (Jablonski and Skoog 1954). The first identified cytokinin, kinetin, was isolated from a commercial DNA preparation and subsequently synthetic synthesis protocols were developed (Fig. 8.2; Miller et al. 1956). This research provided the means for the detailed investigations of cytokinin-regulated plant responses in tobacco cell cultures (Skoog and Miller 1957). Subsequent researchers identified a range of naturally occurring cytokinins that are primarily N^6-substituted purine derivatives and include isopentenyladenine, zeatin, dihydrozeatin, and their derivatives (Fig. 8.2; Letham 1991). In essence, exposure of plant tissues to cytokinin in the presence of adequate auxin, either endogenous or exogenously provided, results in shoot formation.

Molecular approaches have provided some detailed information describing the biosynthesis and metabolism of cytokinins in plants. Cytokinin primary response genes, genes encoding for cytokinin receptors, and a skeletal pathway of elements that mediate signaling have all been identified (Hutchinson and Kieber 2002). Mutants that overproduce cytokinins demonstrate an increased capacity for *de novo* regeneration (Catterou et al. 2002). Some of the most convincing evidence of the veracity of the Skoog and Miller hypothesis was the recent demonstration that genetic modification to reduce endogenous cytokinin concentration effectively altered plant development (Werner et al. 2001). Results of these studies demonstrated an absolute requirement for cytokinin for leaf formation with a slowed formation of new cells, delayed flowering and retarded shoot development observed in cytokinin-deficient plants (Werner et al. 2001). Cytokinins are required for cell cycle progression into the S phase and therefore cytokinin-deficient tissues have limited cell division and differentiation (Jacqmard et al. 1994).

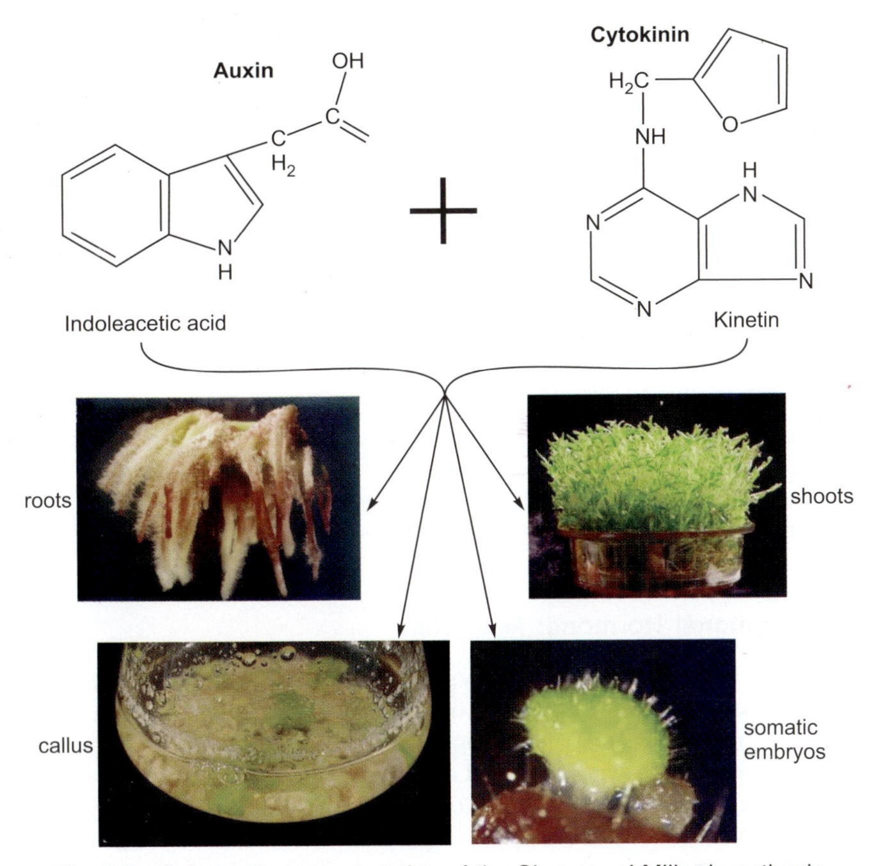

Fig. 8.2 Schematic representation of the Skoog and Miller hypothesis.

EXCEPTIONS TO THE RULE

The vast diversity of plant species and chemical regulators has increased the likelihood of exceptions to any hypothesis that describes these interactions. The Skoog and Miller hypothesis presents us with a more complex pattern of questions since the redirection of plant growth requires several compounds. These phytochemicals may be exogenous or endogenous, applied concurrently or in subsequent phases, identified or completely novel, plant-based or synthetic. Further complicating the interpretation of data is the research approach. The majority of the conclusions concerning the role of specific growth regulators in plant development have been based on exogenous application and observation. In addition, experiments are usually done with undifferentiated or excised tissues that may have altered physiological processes to cope with the stresses of *in vitro* cultures.

As a result, questions remain about the specific mode of action of these compounds. Possible explanations of plant responses include:

(a) The compound may have a specific single activity, including binding to a receptor, activation of a gene or inhibition of an enzyme.
(b) The compound may induce a more generalized physiological response such as a stress response or secondary metabolism cascade.
(c) The compound may act in synergy with other plant metabolites to simultaneously alter the physiological identity of the cells.
(d) The compound may trigger adaptive evolutionary mechanisms for survival in adverse conditions.
(e) The compound may be metabolized to produce the active trigger for morphogenesis.

Exceptions to the Skoog and Miller rule provide tools for addressing some of these questions and more fully investigating plant regeneration responses. Concomitantly, some exceptions to the Skoog and Miller hypothesis raise troubling questions about the assumptions made in regeneration experiments.

An Exceptional Hormone: Melatonin

Melatonin (N-acetyl-5-methoxytryptamine) is synthesized in the pineal gland of mammals and the circulation of microquantities of the hormone regulate the most basic physiological processes (Yu and Reiter, 1993). Recently melatonin was identified and quantified in higher plant species as well (Murch et al. 1997; Murch et al. 2002). Melatonin is derived from tryptophan in a biosynthetic pathway that functions in concert with the biosynthesis of auxin (Fig. 8.3; Murch et al. 2000). Although there is no known role for melatonin in higher plants, it might possibly serve as a chemical signal of darkness and age in entirely different, phylogenetically very distant organisms such as unicellular dinoflagellates, higher plants, and mammals (Murch and Saxena 2002). One of the problems inherent in the Skoog and Miller model describing plant regeneration processes may be the attribution of plant responses to the exogenous application of a plant growth regulator and the failure to incorporate the metabolism of the molecule into the model. In this scenario, it is possible that a portion of the plant responses attributed to auxin may actually be responses of plant tissue to auxin-derived metabolites such as melatonin. To investigate this possibility, we reexamined the process of auxin-induced root formation in St. John's wort (*Hypericum perforatum* L.) and found that alterations in the relative ratio of auxin and melatonin affected root formation (Murch et al. 2001). Elevated concentrations of auxin in the absence of elevated melatonin were not found to result in *de novo* root organogenesis and further, elevated concentrations of the melatonin precursor serotonin

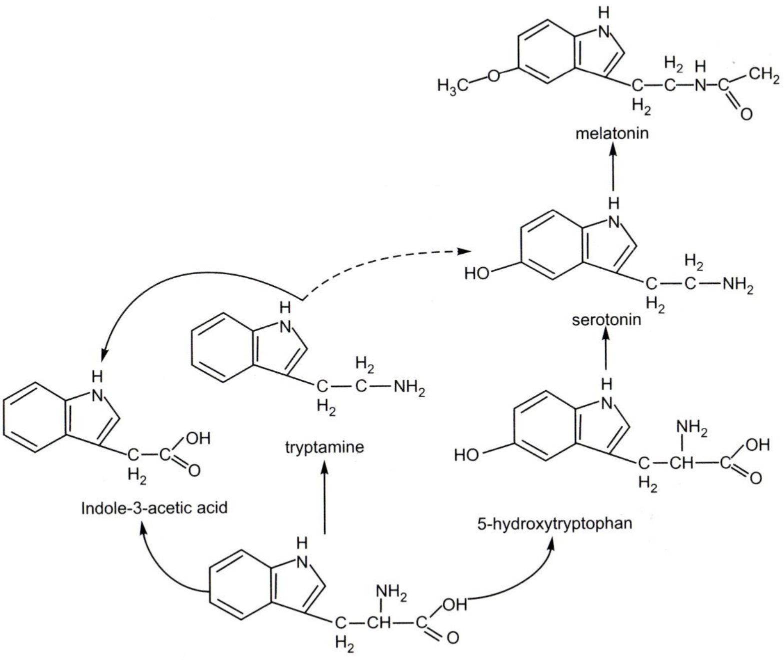

Fig. 8.3 Biosynthetic pathway for the mammalian neurohormones melatonin and serotonin. Characterization of this pathway in *Hypericum perforatum* L. (St. John's wort) provides an early indication of potentially active auxin metabolites in plant tissues.

coupled with decreases in melatonin resulted in *de novo* shoot proliferation on the explants (Murch et al. 2001). Together these data provided early evidence of a role for auxin metabolites in auxin-regulated responses and established a direction for future research. A reexamination of the classic systems of plant physiology may provide new clues to the role of melatonin that may be independent of other growth regulators, may mediate other growth regulators, or may be involved in the expression of responses attributed to other compounds (Murch and Saxena 2002).

An Exceptional Synthetic: Thidiazuron

One of the most interesting synthetic plant growth regulators for evaluation of the Skoog and Miller model is the diphenylurea, N-phenyl-N'-1,2,3-thidiazol-5-ylurea (thidiazuron; TDZ). The physiological responses of plant tissues to this chemical are diverse and have made it difficult to place TDZ in a growth regulator category. Additionally, the chemical structure of TDZ is quite different from either auxin or cytokinin-type plant growth regulators, compounding the difficulty in classification

(Fig. 8.4). Forchlorfenuron (N-(2-chloro-4-pyridyl)-N′-phenylurea), a lesser known phenylurea (Fig. 8.4) that is structurally similar to thidiazuron, also elicits cytokinin-like physiological responses and has been shown to substitute for both auxin and cytokinin complement in *in vitro* regeneration (Murthy et al. 1995). TDZ was initially developed as a cotton defoliant and has been used in commercial cotton production since the 1950s. In cotton, application of TDZ results in the abscission of the leaves without the full senescence process and therefore leaves drop while green and turgid (Arndt et al. 1976; Grossman, 1991). Since the late 1980s the potential of TDZ as an inductive stimulus for plant regeneration has been repeatedly demonstrated. TDZ has been used to effectively induce *de novo* shoot organogenesis on a wide range of species (Huetteman and Preece 1993; Lu 1993). From these studies and earlier bioassay data, it was initially concluded that TDZ was a synthetic cytokinin (Mok et al. 1982; 1987). However, application of TDZ in tissue cultures of various species induced a diverse range of responses including callus formation (Kartomysheva et al. 1983; Jayshankar et al. 1991), androgenesis (Perri et al. 1994), gynogenesis (Bohanee et al., 1995), and bulblet formation (Mohammed-Yasseen et al. 1993). In the early to mid-1990s, work in our lab generated results that contradicted earlier conclusions about TDZ and demonstrated the potential for an improved understanding of plant growth regulation by application of this compound. Culture of geranium explants on a medium supplemented with TDZ resulted in somatic embryogenesis (Visser et al.

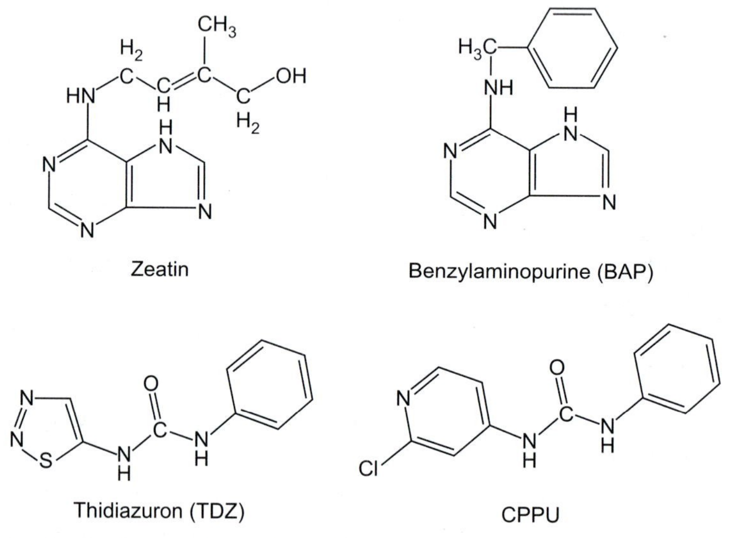

Fig. 8.4 Synthetic growth regulator thidiazuron (TDZ) and CPPU as it compares to other cytokinins.

1992; Gill et al. 1993). Further, studies demonstrated TDZ-induced *in vitro* somatic embryogenesis in a wide range of herbaceous, leguminous, and woody species including: tobacco (Gill and Saxena, 1993), tomato (Gill et al. 1995), peanut (Saxena et al. 1992; Gill and Saxena, 1992; Victor et al. 1999a), chickpea (Murthy et al. 1996), and neem (Murthy and Saxena, 1998). In addition, we made the observation that isolation of an explant was not always required for the TDZ-induced regeneration responses (Malik and Saxena, 1992a, and b; Murthy et al. 1995). Steward (1964) had described the physical isolation of the cells from the maternal tissues as one of the principle requirements for the redirection of plant growth. However, TDZ-induced regeneration proved to be an exception to this rule with profusions of somatic embryos formed at the cotyledonary notch of peanut seedlings germinated *in vitro* (Murthy et al. 1995). These findings raised many tantalizing questions in plant morphogenesis, most obviously, how could one single compound induce such different results in different species and yet induce the same response in such a wide range of species. This seeming contradiction provided the foundation for more than a decade of work to understand plant morphogenesis in TDZ-regulated systems.

Mode of Action of TDZ

The cascade of events in plant tissues following exposure to TDZ involves reprogramming and expression of morphogenetically competent cells to undergo development, thereby leading to morphogenesis. Although the physiological sequences involved in regeneration are not completely understood, studies to date indicate that there are five main possible explanations for the activity of TDZ in induction of plant morphogenesis.

1. TDZ may have a specific, as yet undiscovered, activity unique to the compound and therefore may trigger a specific metabolic event.
2. TDZ may induce morphogenesis as an evolved adaptation to a physiological stress.
3. TDZ may increase the concentration or alter the relative ratio of endogenous plant growth regulators.
4. TDZ may alter the capacity of the tissue to respond to endogenous compounds.
5. TDZ may alter the transport of nutrients or metabolites along or between tissues and therefore cause a gradient of concentrations that ultimately results in morphogenesis.

Investigations began in 1993 to quantify changes in endogenous plant growth regulators in TDZ-exposed tissues. Chemical analyses of peanut seedlings germinated on a medium supplemented with TDZ revealed an overall increase in the activity of endogenous auxins and cytokinins

(Murthy et al. 1995). Further, it was observed that different tissues responded differently to TDZ exposure. The greatest perturbations in endogenous plant growth regulators were observed in the cotyledons of intact peanut seedlings with a >50% increase in concentration of both the cytokinin zeatin and the auxin IAA as well as precursors for each of these (Murthy et al. 1995). Interestingly, TDZ was sequestered in the developing cotyledons and was not detected in the other tissues after 7 days. However, the cotyledons were not the site of regeneration; rather the somatic embryos developed at the junction between the cotyledons and the hypocotyl. Therefore, the TDZ-induced signal for regeneration may involve the translocation of endogenous compounds to specific cells with the capacity to regenerate or may involve formation of a gradient of endogenous plant growth regulators between tissues.

Further studies with inhibitors of auxin and cytokinin transport were initiated to investigate these possibilities. For these studies, isolated tissues were used to reduce the number of complicating factors and a system for induction of somatic embryogenesis on etiolated hypocotyls of the diploid geranium cultivar "Scarlet Orbit Improved" was developed (Hutchinson et al. 1996a). This system has several advantages as a model for investigation of plant growth regulator responses. First, the process of somatic embryogenesis has been separated into distinct phases with a 3-day induction period of exposure to TDZ medium followed by an 18-day expression phase with culture on a medium devoid of exogenous plant growth regulators (Hutchinson et al. 1996a). This separation in time allows investigations into the inductive capacity of TDZ, longevity of the TDZ molecule in plant tissues, and the resultant embryogenesis. TDZ-induced elevation in endogenous auxin concentration observed in legumes (Murthy et al. 1995) was also observed in the induction phase of somatic embryogenesis in geranium (Hutchinson et al. 1996b). As well, inclusion of an inhibitor of auxin 2-(p-chlorophenoxy)-2-methylpropionic acid (PCIB) in the induction medium significantly reduced the regenerative response, indicating that the TDZ-induced accumulation of auxin was essential for somatic embryogenesis (Hutchinson et al.,1996b). However, the absolute concentration was not the only requirement for regeneration as it was also observed that inclusion of 2,3,5-triiodobenzoic acid (TIBA) in the induction medium reduced the regenerative response. TIBA is thought to be an inhibitor of auxin transport and receptors and reduced embryogenesis occurred in the absence of any significant reduction in auxin concentration (Hutchinson et al. 1996b). Together these observations indicated that both the accumulation and transport of endogenous auxin are requirements for somatic embryogenesis in geranium. More recently, the transport of radiolabeled auxin in TDZ-exposed geranium hypocotyls was quantified; these studies demonstrated that TDZ-enhanced auxin transport was the most significant effect in the basipedal transport of auxin through the explants (Murch and Saxena 2001). Further, the TDZ-stimulated auxin transport was impaired by TIBA, thereby demonstrating the need for this transport across the tissue during the induction phase of regeneration (Murch and Saxena 2001).

However, effects on auxin metabolism were not sufficient to explain all the diverse regeneration responses induced by TDZ as the sole growth regulator. Initial references to TDZ described the cytokinin activity of the compound including the cytokinin activity of TDZ in bioassays (Thomas and Katterman 1986; Visser et al. 1995) and in isolated cell preparations. TDZ in combination with various auxins has been used during the initial phases of cell wall formation around protoplasts, protoplast division, and in the later phases of protoplast culture and recovery of regenerated plants (reviewed in Murthy et al. 1998). In the TDZ-induced somatic embryogenesis on intact seedlings of peanut, cotyledons were enlarged and the leaves thickened and much darker green in color. Cytokinin metabolism in the cotyledons was significantly affected by TDZ exposure with significant accumulations of zeatin and dihydrozeatin coupled with a significant decrease in the precursor 2-isopentenyladenine (Murthy et al. 1995). The effects of these cytokinins in the regeneration process were demonstrated with inhibitors of cytokinin activity viz. azaadenine, azaguanine and diaminopurine (Victor et al. 1999b). All three purine analogues inhibited the regeneration process in peanut and also inhibited somatic embryogenesis in geranium explants (Hutchinson and Saxena 1996). Together, these data indicate that modulation of cytokinin metabolism may also be an essential component in the TDZ-induced metabolic processes that culminate in plant regeneration.

TDZ AND STRESS

One of the interesting possibile explanations for the encompassing nature of TDZ-induced physiological responses could be activation of a basic evolutionary mechanism and an associated metabolic cascade. The first serendipitous discovery was the observation that seedlings germinated on a medium supplemented with the synthetic cytokinin TDZ formed *de novo* somatic embryos and shoots (Malik and Saxena 1992a). This process was unique to the synthetic cytokinin and was not observed in seedlings cultured on an auxin medium. These findings contradicted earlier researchers who had suggested that somatic embryogenesis was a process controlled by auxin and that excision of an explant from the maternal tissue was required for dedifferentiation and redifferentiation of the cells (Steward et al. 1964). Removal of one or both cotyledons from the germinating seedlings significantly reduced the number of embryos formed on the hypocotyl regions. Therefore, it seemed that metabolism of the different organs was required for the expression of induced morphogenesis (Murthy et al. 1995). In other legumes such as chickpea and bean as well as in tree species such as neem, similar results were found (Malik and Saxena 1992a; Murthy and Saxena 1998). In these diverse species, the responses of whole seedlings to TDZ in the medium varied in the localization of the organogenesis and embryogenesis (Fig. 8.5). In some species regeneration was limited to a region of the hypocotyl while in other species embryos were observed scattered over the entire seedling surface, indicating that a variety of tissues or cells may become induced, perhaps due to an optimal auxin to cytokinin ratio achieved within the site of regeneration. Therefore, the transport of endogenous growth regulators throughout the intact plants was likely associated with the capacity of individual tissues to respond to the TDZ stimulus (Murthy et al. 1995).

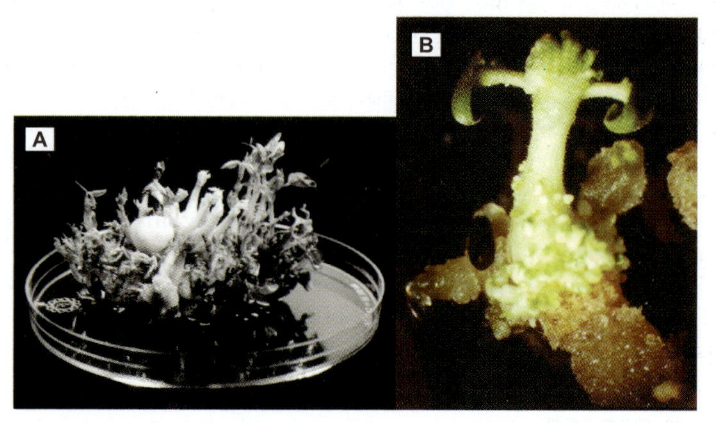

Fig. 8.5 Regeneration on intact seedlings: (A) organogenesis in pea and (B) somatic embryogenesis in a greenhouse.

Our visual observations indicated that the germinating seedlings were in the process of metabolic stress. Seedlings were stunted with thickened leaves and poor root growth (Murthy et al. 1995). Similar studies with peanut revealed the accumulation of proline to 6% of the dry mass of the germinating seedlings cultured on TDZ-supplemented medium, thereby confirming the stress response in the seedlings (Murch et al. 1998). Further, greenhouse grown geraniums watered with a TDZ solution formed regenerative outgrowths on the root tissues. These nodular structures sometimes appeared callused but at other times appeared to have complete shoot apices with green leaves under the soil surface (Murch et al. 1997). Chemical analyses of the tissues indicated an accumulation of stress-related markers, proline, GABA and abscissic acid (Murch et al. 1997). Interestingly both peanut seedlings and geranium plants accumulated significantly more metal ions from the environment following exposure to TDZ. Together these observations led to the hypothesis that TDZ induced a chemical or physical stress on tissues and that to overcome this physiological stress, the plant tissues modified their metabolic processes and reallocated resources via activation of endogenous plant growth regulators, thereby initiating a cascade of metabolic reactions that ultimately resulted in *de novo* regeneration (Murch et al. 1997).

EXCEPTIONS AND THE SKOOG AND MILLER HYPOTHESIS

Studies with melatonin and TDZ tell us a great deal about the Skoog and Miller's postulate. TDZ is unique in that a relatively short exposure to a single compound effectively induces a range of various morphogenetic responses. The response may vary by tissue within the same plant, cultivar within the same species, or between species. Pretreatment with TDZ can predispose a tissue to accept other inductive stimuli. Alternatively, exposure to TDZ can commit a tissue to a regenerative route that is expressed even after the inductive stimulus is removed. Therefore, experiments with TDZ both prove and disprove the hypothesized role of auxin and cytokinin in plant regeneration. If TDZ is in fact a cytokinin, then the activity of the compound in the absence of auxin would be in direct contrast to Skoog and Miller's observations. However, the potential for a TDZ-induced accumulation or increase in activity of endogenous auxin may account for the observations. Alternatively, TDZ may trigger a basic survival mechanism in plant tissues that includes asexual reproduction for species survival. The work with melatonin likewise exposes one of the basic flaws in many plant regeneration studies, correlation of exogenously applied phytohormones with physiological responses without consideration of its possibility of metabolism of the stimulant. The diverse range of responses attributed to auxin increases the possibility that metabolism to other hormones may be essential to the process and, in this scenario, other metabolites may be essential components of the Skoog and Miller process.

Skoog and Miller likely did not realize in 1957 the far-reaching impact their work would have nor that the application of this work would have a major impact on modern agricultural practices. Almost 50 years later we are still struggling to define the roles of auxins and cytokinins in plant regeneration. Further work in this area is essential for a more complete understanding of plant growth regulation and morphogenesis.

REFERENCES

Arndt F, Rusch R, Stilfried H (1976) SN 49537, a new cotton defoliant. Plant Physiol. 57:99.

Bohanee B, Jakse M, Ihan A (1995) Studies of gynogenesis in onion: Induction procedures and genetic analysis of regeneration. Plant Sci. 104:215-224.

Catterou M, Dubois F, Smets R, Vaniet S, Kichey T, Van Onckelen H, Sangwan-Norreel BS, Sangwan RS (2002) hoc: an Arabidopsis mutant overproducing cytokinins and expressing high in vitro organiogenic capacity Plant J. 30 (3):273-287.

den Boer BGW, Murray JAH (2000) Triggering the cell cycle in plants. Trends in Cell Biol. 10(6):245-250.

Den Boer BGW, Murray JAH (2000) Triggering the cell cycle in plants. Trends Cell Biol. 10:245-250.

Gill R, Saxena PK (1993) Somatic embryogenesis in *Nicotiana tabacum* L.: induction by thidiazuron of direct embryo differentiation from cultured leaf discs. Plant Cell Repts. 12:154-159.

Gill R, Gerrath JM, Saxena PK (1993) High-frequency direct somatic embryogenesis in thin layer cultures of hybrid seed geranium (*Pelargonium x hortorum*). Can. J. Bot. 71:408-413.

Gill R, Malik KA, Sanago MHM, Saxena PK (1995) Somatic embryogenesis and plant regeneration from seedling cultures of tomato (*Lycopersicon esculentum* Mill.). J. Plant Physiol. 147:273-276.

Gill R, Saxena PK (1992) Direct somatic embryogenesis and regeneration of plants from seedling explants of peanut (*Arachis hypogaea* L.): promotive role of thidiazuron. Can. J. Bot. 70:1186-1192.

Grossmann K (1991) Induction of leaf abscission in cotton is a common effect of urea and adenine-type cytokinins. Plant Physiol. 95:234-237.

Haberlandt G (1921) Beitr. Allg. Bot. 2:1.

Huetteman CA, Preece JE (1993) TDZ: a potent cytokinin for woody plant tissue culture. Plant Cell Tiss. Org. Cult. 33:105-119.

Hutchinson MJ, Saxena PK (1996) Role of purine metabolism in thidiazuron-induced somatic embryogenesis of geranium (*Pelargonium x hortorum* Bailey) hypocotyl cultures. Physiol. Plant 98:517-522.

Hutchinson CE , Kieber JJ (2002) Cytokinin signaling in arabidopsis. Plant Cell. 14:S47-S50.

Hutchinson MJ, KrishnaRaj S, Saxena PK (1996a) Morphological and physiological changes in thidiazuron-induced somatic embryogenesis in geranium (*Pelargonium x hortorum* Bailey) hypocotyl cultures. Int. J. Plant Sci. 157:440-446.

Hutchinson MJ, Murch SJ, Saxena PK (1996b) Morphoregulatory role of thidiazuron: Evidence of the involvement of endogenous auxin in thidiazuron-induced somatic embryogenesis of geranium (*Pelargonium x hortorum* Bailey). J. Plant Physiol. 149:573-579.

Jablonski JR, Skoog F (1954) Physiol. Plant. 7:16.

Jacqmard A, Houssa C, Bernier G (1994) *In:* DWS Mok and MC Mok (eds.). Cytokinins: Chemistry, Activity and Function. CRC Press, Boca Raton, FL, pp. 197-215.

Jayshankar RW, Dani RG, Aripdjanov SA (1991) Studies on TDZ mediated in vitro callus induction in Asiatic cottons. Adv. Plant Sci. 4:138-142.

Kartomysheva OP, Volkova TV, Nikitenki SS (1983) Dropp, a new promising stimulant of callus formation. Sintez Biologicheskaya Aktivnnost'I Primenenie Pestitsidov, pp. 131-135.

Letham DS, Singh S, Wong OC (1991) Mass-spectrometric analysis of cytokinins in plant tissue. 7. Quantification of cytokinin bases by negative-ion mass spectrometry. J. Plant Growth Reg. 10(2): 107-113.

Lu CY (1993) The use of thidiazuron in tissue culture. In Vitro Cell Dev. Biol. Plant. 29(2):92-96.

Malik KA, Saxena PK (1992a) *In vitro* regeneration of plants: A novel approach. Naturwissenschaften 79: 136-137.

Malik KA, Saxena PK (1992b) Thidiazuron induces high-frequency shoot regeneration in intact seedlings of pea (*Pisum sativum*), chickpea (*Cicer arietinum*) and lentil (*Lens culinaris*). Aust. J. Plant Physiol. 19:731-740.

Masucci JD, Schiefelbein J (1994) The RHD6 mutation of *Arabidopsis thaliana* alters root epidermal cell polarity and is rescued by auxin. Dev. Biol. 163 (2):554-554.

Miller CO, Skoog F, Okumura FS, von Saltza MH, Strong FM (1956) J. Amer. Chem. Soc. 78:1375.

Mok MC, Mok DWS, Armstrong DJ (1982) Cytokinin activity of N-phenyl-N'-1,2,3-thidiazol-5-ylurea (TDZ) Phytochemistry 21:1509-1511.

Mohammed-Yasseen Y, Splittstoesser WE, Litz RE (1993) In vitro bulb formation and plant recovery from onion inflorescences. Hort. Sci. 28:1052-1053.

Mok MC, Mok DWS, Turner JE (1987) Biological and Biochemical effects of cytokinin-active phenylurea derivatives in tissue culture systems. Hort. 22:1194-1197.

Murch SJ, Saxena PK (2001) Molecular fate of thidiazuron and its effects on auxin transport in hypocotyls tissues of *Pelargonium x hortorum* Bailey. Plant Growth Reg. 35 (3):269-275.

Murch SJ, Saxena PK (2002) Melatonin: A potential regulator of plant growth and development? In Vitro Cell. Dev. Biol. 38:531-536.

Murch SJ, KrishnaRaj S, Saxena PK (1997) Thidiazuron-induced morphogenesis of Regal geranium (*Pelargonium domesticum*): A potential stress response. Physiol. Plant. 101:183-191.

Murch SJ, Simmons CB, Saxena PK (1997) Melatonin in feverfew and other medicinal plants. Lancet 350:1598-1599.

Murch SJ, KrishnaRaj S, Saxena PK (2000) Tryptophan is a precursor for melatonin and serotonin biosynthesis in *in vitro* regenerated St. John's wort (*Hypericum perforatum* cv. Anthos) plants. Plant Cell Repts. 19:698-704.

Murch SJ, Campbell SSB, Saxena PK (2001) The role of serotonin and melatonin in plant morphogenesis: Regulation of auxin-induced root organogenesis in *in vitro* cultured explants of St. John's wort (*Hypericum perforatum* L.). In Vitro Cell. Dev. Biol. 37:786-793.

Murch SJ, Victor RMJ, KrishnaRaj S, Saxena PK (1998) The role of proline in thidiazuron-induced somatic embryogenesis of peanut (*Arachis hypogaea* L.). In Vitro Cell Dev. Biol. 35:102-105.

Murthy BNS, Saxena PK (1998) Somatic embryogenesis and plant regeneration of Neem (*Azadirachta indica* A. Juss). Plant Cell Repts. 17:469-475.

Murthy BNS, Murch SJ, Saxena PK (1995) Thidiazuron-induced somatic embryogenesis in intact seedlings of peanut (*Arachis hypogaea*): Endogenous growth regulator levels and significance of cotyledons. Physiol. Plant. 94:268-276.

Murthy BNS, Murch SJ, Saxena PK (1998) Thidiazuron: a potent regulator of *in vitro* plant morphogenesis. *In Vitro* Cell Dev. Biol. 34:267-275.

Murthy BNS, Victor J, Singh RP, Fletcher RA, Saxena PK (1996) *In vitro* regeneration of chickpea (*Cicer arietinum* L): Stimulation of direct organogenesis and somatic embryogenesis by thidiazuron. Plant Growth Reg. 19:233-240.

Perri E, Parlati M, Mule R (1994) Attempts to generate haploid plants from In Vitro cultures of *Olea evropaea* L. anthers. Acta Hort. 356:47-50.

Reinert, J. (1958). Untersuschungen über die Morphogenese an Gewebekulturen. Ber. Dtsch. Bot. Ges. 71:15.

Ross JJ, O'Neill DP, Smith JJ, Kerckhoffs LHJ, Elliott RC (2000) Evidence that auxin promotes gibberellin A(1) biosynthesis in pea. Plant Journal 21(6):547-552.

Saxena PK, Malik KA, Gill R (1992) Induction by thidiazuron of somatic embryogenesis in intact seedlings of peanut. Planta 187:421-424.

Skoog F, Miller CO (1957) Chemical regulation of growth and organ formation in plant tissues cultured *in vitro*. Symp. Soc. Exp. Biol. 11:118-131.

Steward FC, Mapes MO, Kent AE, Holsten RD (1964) Growth and development of cultured plant cells. Science 143:20-27.

Steward, FC, Mapes, MO, and Mears, K (1958). Growth and organized development of cultured cells. II. Organization in cultures grown from freely suspended cells. Am. J. Bot. 45:705-708.

Thomas JC, Katterman FR (1986) Cytokinin activity induced by TDZ. Plant Physiol. 81:681-683.

Victor JMR, Murch SJ, KrishnaRaj S, Saxena PK (1999a) Somatic embryogenesis and organogenesis in peanut: The role of thidiazuron and N^6-benzylaminopurine in the induction of plant morphogenesis. Plant Growth Reg. 28:9-15.

Victor JMR, Murthy BNS, Murch SJ, KrishnaRaj S, Saxena PK (1999b) Role of endogenous purine metabolism in thidiazuron-induced somatic embryogenesis of peanut (*Arachis hypogaea* L.) Plant Growth Reg. 28:41-47.

Visser C, Qureshi JA, Gill R, Saxena PK (1992) Morphoregulatory role of thidiazuron: Substitution of auxin-cytokinin requirement of somatic embryogenesis in hypocotyl cultures of geranium. Plant Physiol. 99:1704-1707.

Visser-Tenyenhuis CF, Fletcher RA, Saxena PK (1995) Thidiazuron stimulates expansion and greening in cucumber cotyledons. Physiol. Mol. Biol. Plants 1:21-26.

Went FW, and KV Thimann (1937) Phytohormones. (New York: Macmillan).

Werner T, Motyaka V, Stmad M, Schmulling T (2001) Regulation of plant growth by cytokinin. Proc. Natl. Acad. Sci. USA 98:10487-10492.

Yu HS, Reiter RJ (1993) Melatonin: Biosynthesis, Physiological Effects and Clinical Applications. CRC Press, Boca Raton, FL (USA).

9

Somatic Embryogenesis

V. Raghavan
Department of Plant Cellular and Molecular Biology,
The Ohio State University,
318 West 12th Avenue, Columbus, OH 43210 USA
e-mail: raghavan.1@osu.edu

INTRODUCTION

Unlike animals and most other phyla of the plant kingdom, angiosperms and gymnosperms do not bequeath to the fertilized egg or zygote the monopoly to differentiate into an embryo. It has long been known that embryo formation and production of viable seeds occur in a few angiosperms from a diploid cell of the ovule or a cell of the unreduced megagametophyte, including the egg, in the absence of fertilization – a phenomenon broadly considered under the rubric apomixis. From the 1960s, tissue culture approaches have shown that single somatic cells and pollen grains can give rise to fertile plants, simulating stages strongly reminiscent of normal embryogenesis, by processes known as somatic embryogenesis and pollen embryogenesis, respectively. This raises the interesting point that whereas the zygote passes through cycles of growth and division to differentiate into an embryo, somatic cells and pollen grains follow dedifferentiative pathways to form embryo-like structures.

A historical account of the discovery of somatic embryogenesis is given here, with emphasis on the structural and developmental aspects of this process. Carrot (*Daucus carota*) is the exemplar since much information on the regulatory mechanism of embryogenic transformation of somatic cells has come from the so-called carrot system. General reviews especially on the genetic and molecular aspects of somatic embryogenesis from different perspectives are provided by Rao (1996), Dodeman, Ducreux, and Kreis (1997), Mordhorst, Toonen, and de Vries (1997), Raghavan (1997, 2000), and Thorpe and Stasolla (2001); these should be consulted for copious references to requirements for the induction of somatic embryogenesis in a wide variety of flowering plants and gymnosperms.

CELL CULTURE AND SOMATIC EMBRYOGENESIS

One of the finest endeavors in the history of plant tissue culture has been the cultivation of free cells and cell groups (cell clusters) derived from higher plants, especially flowering plants, in a chemically defined liquid medium as a suspension (suspension culture) and manipulation of the regenerative potencies of cells by changes in the hormonal or nutritional balance in the medium. One type of regeneration frequently observed in suspension cultures is the development of embryo-like structures that recapitulate with a high degree of precision the typical stages of embryogenesis of the fertilized egg cell, while remaining innocent of sex. Since embryos are formed from the sporophytic or somatic cells of the plant, as opposed to gametophytic or germ cells, the phenomenon is referred to as somatic embryogenesis. To emphasize the divergent pathways through which zygotic embryos and somatic embryos have evolved, the term "embryoid" is often used to refer to somatic embryos and the term "embryo" reserved to designate zygotic embryos. The cryptic potentiality of cells of angiosperms and gymnosperms to grow as callus, the callus to dissociate into free cells, and free cells to regenerate whole plants by recapitulating stages of embryogenesis is very widespread and usually considered a general property of these groups of plants. Lack of success in demonstrating somatic embryogenesis by techniques currently in use, is probably attributable to special inhibiting conditions within the system, rather than to its specialized or primitive state.

Totipotency and Somatic Embryogenesis

The discovery of somatic embryogenesis is often traced to a prophesy by the German botanist Haberlandt in the early 1900s that it would be possible to grow facsimiles of embryos from vegetative cells of plants. Implied in this prophesy is the dictum of totipotency—all plant cells, except perhaps those that have undergone irreversible differentiation, lignified cells, and cells that do not possess a functional nucleus at maturity such as the sieve elements and tracheal cells, are capable of regenerating whole new plants in full multicellularity, sexuality, and structure. Although experiments by Haberlandt failed to achieve this goal, the stage was set to demonstrate totipotency in plant cells beginning in the 1930s with the formulation of methods and media for growing plant organs, tissues, and cells under aseptic conditions. Using cultured secondary phloem of domestic carrot, Steward and coworkers (Steward et al. 1958a; 1958b) initiated the pioneering experiments that led to the demonstration of totipotency. The principal observations made in these investigations may be sum-marized as follows: culture of slabs of the secondary phloem of carrot in a solidified medium containing inorganic salts, vitamins, and organic nutrients, supplemented with the liquid endosperm of coconut (*Cocos nucifera*), known as coconut milk or coconut water, typically gives rise to a proliferating callus constituted of parenchymatous cells; transfer of the callus to a liquid medium of the same composition with gentle agitation results in a suspension of single cells and cell clusters, the latter originating by repeated division of cells dissociated from the callus; subsequent growth of the cell suspension in the liquid medium without subculture results in lignification of inner cells of the clump, formation of cambium-like cells, and eventually in the appearance of lateral root primordia; a final transfer of the rooted aggregate to a solid medium devoid of coconut milk results in the regeneration of a normal carrot plant in the culture flask. Although these observations did not unequivocally prove that a single cell is transformed into a plant, Vasil and Hildebrandt (1965) showed that a single cell of a hybrid tobacco (*Nicotiana glutinosa* x *N. tabacum*) grown in isolation in a drop of defined liquid medium in a microculture chamber divides repeatedly to form a callus that is subsequently induced to form roots and shoots on a solid medium.

Demonstration of this new strategy to form a complete plant implied in the odyssey of single somatic cells subjected to different tissue culture manipulations was a remarkable developmental episode, as it did not fit comfortably into the conventional framework of plant development. Although the resemblance between certain cell aggregates in the carrot suspension culture and stages of zygotic embryogenesis in typical eudicots did not go unnoticed (Steward 1963b), the aforesaid investigations did not really show that single cells precisely duplicated the pathway of embryo-genesis normally followed by the zygote. This challenge was partially met by Reinert (1959) who first showed that a callus originating from a strain of carrot root, subsequent to a long period of culturing in a medium containing indoleacetic acid (IAA) and coconut milk, differentiates into bipolar embryo-like structures upon transfer to a synthetic medium enriched with an elaborate mixture of amino acids, amides, vitamins, a purine (hypoxanthin), and IAA. What was lacking in Reinert's work was proof that embryo-like units originate in single cells of somatic parentage. Proof of this was provided by Maheshwari and Baldev (1961) who cultured embryos of the common parasitic angiosperm, *Cuscuta reflexa*. This investigation showed that the cultured embryo differentiates numerous adventive embryos directly from its superficial cells or from a callus regenerated from its radicular end (Fig. 9.1). After establishing the single-celled origin of the adventive embryo from the radicle by conventional histological methods, it was found that the embryo passes through several stages typical of zygotic embryogenesis before appearing outside as a macroscopic bipolar structure. Besides dispelling the notion that the zygote is unique in its capacity to form the embryo, work definitely lends credence to the Haberlandt prophesy. Callus formation, followed by differentiation of adventive embryos from the callus was also noted on culturing globular embryos of another parasitic angiosperm, *Dendrophthoe falcata*, in a medium enriched with casein hydrolyzate and IAA, but the origin of adventive embryos from a specific cell of the callus was not traced in this work (Johri and Bajaj 1962).

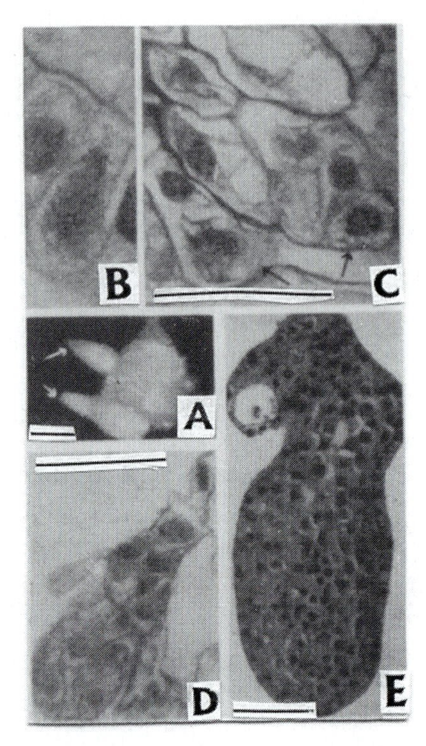

Fig. 9.1 Formation of adventive embryos from cultured embryos of *Cuscuta reflexa*. A, an embryo, cultured for 50 days showing adventive embryos at the radicular end (arrows). Scale bar = 1 mm. B, C, sections of an embryo, cultured for 40 days showing dividing cells and a two-celled proembryo (arrow). Scale bar, for both = 100 μm. D, E, two later stages in the development of adventive embryos. Scale bars = 100 μm (from Maheshwari and Baldev, 1961).

Universality of Somatic Embryogenesis

Follow-up investigations on carrot placed the phenomenon of somatic embryogenesis on a sound footing and unambiguously established the concept of totipotency of plant cells. This can be attributed to the efforts of Steward (1963a) and Wetherell and Halperin (1963) who reported that under certain experimental conditions, cells and cell clusters of carrot regenerate an enormous number of replicas of zygotic embryos. Steward (1963a) showed that free cells sloughed from the hypocotyl region of embryos of domestic or wild carrot germinated in a medium containing coconut milk are a prolific source of totipotent cells that exhibit typical embryogenic development. The effect is indeed dramatic when a cell suspension originating from an immature embryo of wild carrot is spread on a nutrient agar plate; here virtually every cell of the suspension yields an embryo-like unit (Steward et al. 1964). Wetherell and Halperin (1963) initially found that a callus obtained from the root tissue of wild carrot cultured on a solidified coconut milk-containing medium, following transfer to a liquid medium, produced numerous embryo-like structures which germinated into seedlings. Subsequently it was possible to dispense with coconut milk as an ingredient of the medium to promote somatic embryogenesis in carrot (Fig. 9.2); it was shown that various organs of the wild carrot including the root, peduncle, and petiole readily form a callus when cultured in a medium containing a moderately high level of the synthetic auxin 2,4-dichlorophenoxyacetic acid (2,4-D) and that embryogenic development of somatic cells is triggered by the simple expedient of transferring the callus to a medium containing a reduced level of the auxin (Halperin and Wetherell 1964; Halperin 1966a). This new idea introduced interpretations that led for a period of time to lively disagreements in the literature on somatic embryogenesis (Halperin 1966b). However, use of a defined medium and a single-step transfer of cells from a high auxin-containing medium to one containing a reduced amount of the hormone or none at all, was adopted as the standard protocol to study the physiology, biochemistry, and molecular biology of somatic embryogenesis in carrot in later years and became widely popular in inducing somatic embryogenesis in a number of other plants. In another work appearing at about this time, but largely overlooked, Kato and Takeuchi (1963) traced the fate of single cells separated from a carrot root callus formed on a solidified medium containing yeast extract and IAA and showed that they continued to divide in the embryogenic pathway and develop into plantlets. Another proof of the plasticity of carrot came with the simultaneous discovery made from two laboratories that somatic cells stripped of their cell walls (protoplasts) and cultured in an osmotically adjusted medium, promptly regenerated cell walls, divided, and formed

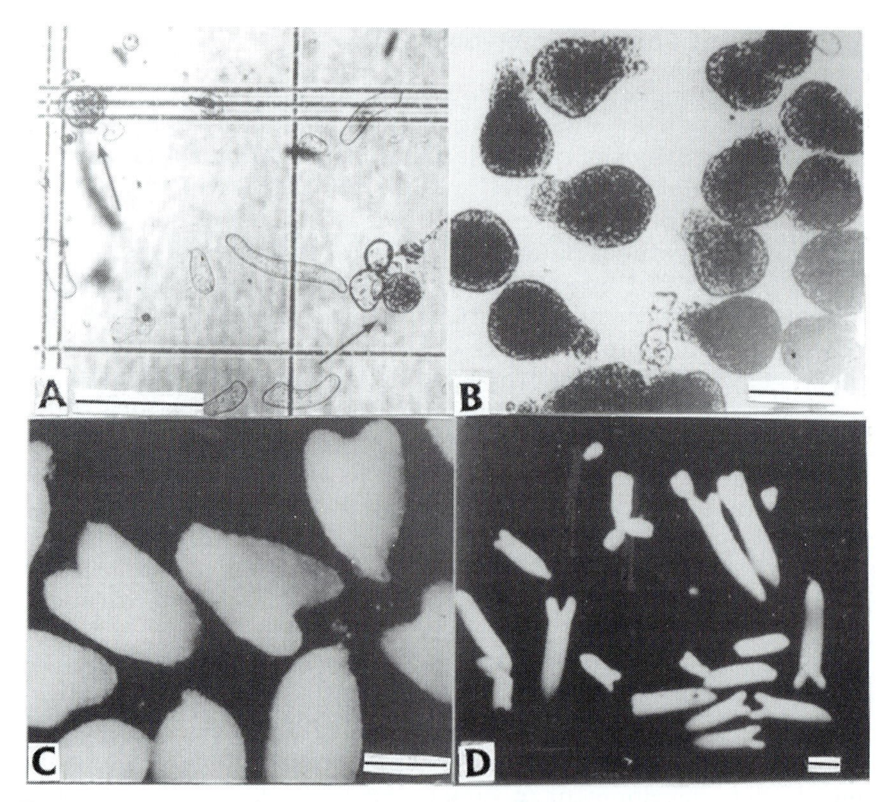

Fig. 9.2 Somatic embryogenesis in cultured petiole explants of *Daucus carota*. A-proembryogenic masses (arrows) in a suspension culture. B-globular embryos. C-heart-shaped embryos. D-mature embryos. Scale bars = 100 μm (from Halperin 1966a).

clusters of cells which regenerated embryo-like structures directly or indirectly from a callus (Kameya and Uchimiya 1972; Grambow et al. 1972). The convergence of experiments and observations from several laboratories was finally capped by the demonstration that an isolated single cell originating from a callus culture of carrot grown in a microculture chamber initially forms a mass of embryogenic and nonembryogenic cells from which a somatic embryo emerges, thus reinforcing the conclusion that embryo-like structures observed in suspension cultures have originated in single cells (Backs-Hüsemann and Reinert 1970). Collectively these studies led to the conclusion that the plantlet originates from a single cell that recapitulates stages of embryogenesis as it grows in the company of thousands of other cells and cell clusters in the same culture milieu.

The feasibility of somatic embryogenesis occurring directly on the cultured explant without an intervening callus phase was demonstrated by the culture of young flower buds of *Ranunculus sceleratus* in a medium supplemented with coconut milk and IAA; the dense callus that formed on the cut end of the explant subsequently differentiated numerous somatic embryos without further treatment (Konar and Nataraja 1965a, b). Some of the somatic embryos set free in the medium, instead of completing normal development, germinated precociously into seedlings that bore a second generation of somatic embryos all along their stem and hypocotyl by division of the epidermal cells (Fig. 9.3). In a further refinement of the direct somatic embryogenic system, embryos have been found to arise spontaneously without the intervention of a callus by dedifferentiation of superficial cells of cultured leaves, ovules, nucellar tissues, and embryos of various plants (Raghavan 1986). Leaves of *Dactylis glomerata* display an unusual type of embryogenic response: segments taken from the more basal parts of the leaf close to the intercalary meristem always form a callus which regenerates somatic embryos, whereas those excised from the distal portions away from the intercalary meristem show direct somatic embryogenesis (Conger et al. 1983). Apparently, actively dividing cells tend to produce a callus before they regenerate somatic embryos, whereas direct somatic embryogenesis is perpetuated by recrudescence of growth in fully differentiated cells.

Similarity in the appearance of zygotic embryos and somatic embryos and their undisputed single-celled origin do not necessarily mean that both follow the same sequence of divisions in their early development. Embryo development in carrot is of the Solanad type in which both the apical and basal cells generated by the first division of the zygote undergo two further rounds of divisions by transverse walls to form a filament of eight cells. The three terminal cells of the filamentous proembryo divide longitudinally and their descendants eventually form most of the embryo except the root tip. A limited number of divisions in the other five cells of the filament produce the root tip and a massive suspensor (Borthwick 1931; Lackie and Yeung 1996). In an investigation in which a single somatic cell of carrot was monitored in isolation during embryogenesis, the cell was found to yield by irregular divisions a mass of parenchymatous cells from which a bipolar embryo-like structure emerged (Backs-Hüsemann and Reinert 1970). However, embryo formation from a cell cluster in a suspension culture of carrot is accompanied by a series of orderly divisions. Here, following a transverse division of the embryogenic initial in a cell cluster, the terminal cell produces by further divisions the embryo proper, whereas the basal cell closest to the cellular aggregate divides transversely to form topographically an incipient suspensor-like structure (McWilliam et al. 1974). Based on these observations, the early division sequences of

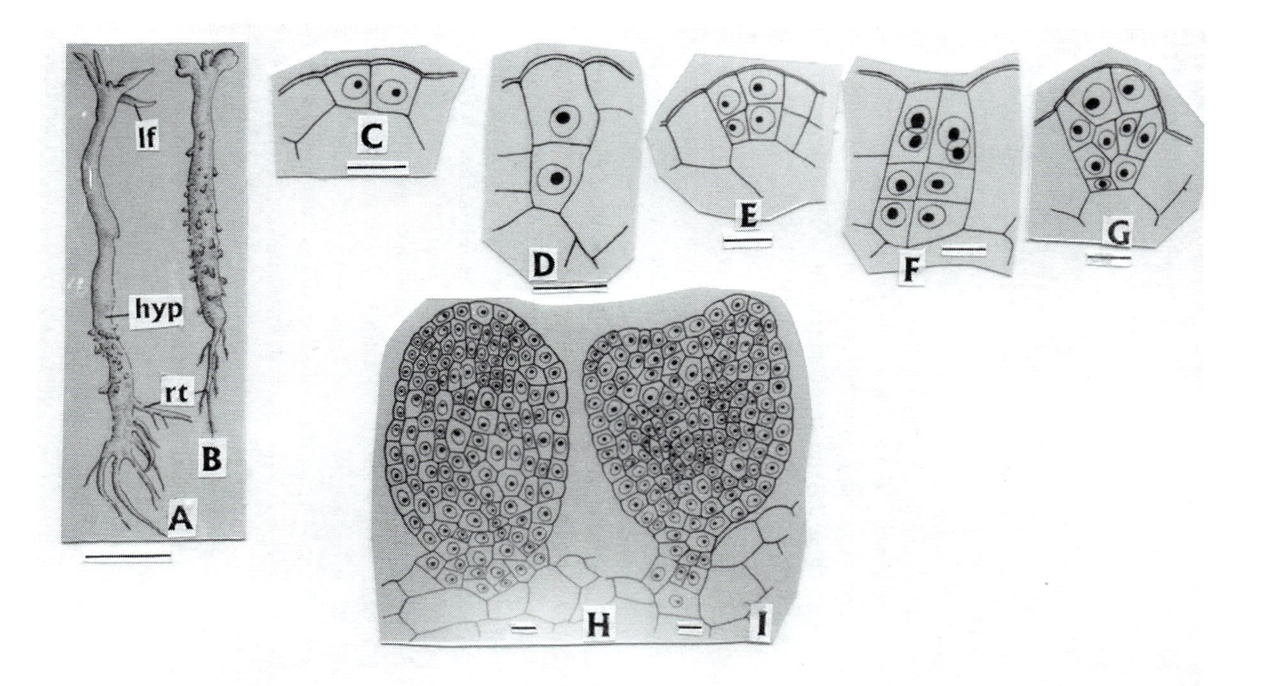

Fig. 9.3 Direct somatic embryogenesis in *Ranunculus sceleratus*. A, B-one-month-old seedlings bearing embryos on the basal part (A) and all over the stem (B); lf, leaf; hyp, hypocotyl; rt, root. Scale bar = 5 mm. C-two potentially embryogenic epidermal cells. D-G-further stages in the division of the epidermal cells in the embryogenic pathway. H-globular embryo. I-heart-shaped embryo. Scale bars = 10 μm (from Konar and Nataraja, 1965b).

somatic embryos of carrot seem to have more in common with the Crucifer type exemplified by *Arabidopsis thaliana* and *Capsella bursa-pastoris* than with the Solanad type.

Starting from the very first accounts of somatic embryogenesis in carrot, the number of reports of embryogenic induction from somatic cells of a variety of plants has steadily increased, initially confined to members of the carrot family (Umbelliferae) but later spreading to members of a number of angiosperm and gymnosperm families. No consolidated listing of these plants is currently available, but separate listings of herbaceous eudicots (Brown et al. 1995), herbaceous monocots (Krishnaraj and Vasil 1995), woody angiosperms and gymnosperms (Dunstan et al. 1995), and angiosperms in general (Thorpe and Stasolla 2001) have been published in recent years. The modes of somatic embryogenesis through a callus and directly on the explant have no generic barriers since both types of embryogenic development may be found in various genera of the same family. Among eudicots, Fabaceae stands out with the largest number of reports of somatic embryogenesis in about 82 species and hybrids, followed by Rutaceae with about 45 listings and Umbelliferae with 28 species. With about 96 listings, Poaceae has by far the largest number of species, hybrids, and transgenic plants reportedly showing somatic embryogenesis among monocots, followed distantly by Aracaceae with 16 listings (Thorpe and Stasolla 2001). However, optimized protocols for inducing somatic embryogenesis along with developmental and structural details are available for only a few systems, such as carrot, soybean, and alfalfa. Following the spectacular success achieved by Vasil and Vasil (1981) in inducing somatic embryogenesis in suspension cultures of *Pennisetum americanum* (Poaceae) by the judicial choice of explants, particularly with regard to their physiological age at culture, use of selected hormonal additives, and unorthodox manipulations of the medium at critical stages of culture, the same or modified protocols have been used to obtain prolific embryogenic cell suspension cultures of many cereals and grasses (Raghavan 1986). Embryogenic cell cultures of various cereals have served as the source of totipotent protoplasts for the generation of somatic hybrids and transgenic plants; transgenic cereals have also been recovered by somatic embryogenesis from cell suspension, callus, or scutellar tissue of immature embryos following microprojectile-mediated delivery of DNA (Krishnaraj and Vasil 1995).

In recent years, much genetic and molecular work on zygotic embryogenesis has become possible by drawing largely on the experimental advantages of *Arabidopsis thaliana* as a model system. Incorporating improvements in earlier published protocols for inducing somatic embryogenesis from immature zygotic embryos of *A. thaliana*, Ikeda-Iwai, et al. (2002) published a reproducible tissue culture system for large-scale production of somatic embryos by culture of immature zygotic embryos and maintenance of high embryogenic competence in tissues for a long period (Fig. 9.4). The protocol essentially involves the culture of embryos at the bent-cotyledon stage in a medium containing 2,4-D to induce formation of embryogenic cell clusters and transfer of two-week-old cultures of embryogenic cell clusters to an auxin-free medium to promote formation of somatic embryos. Mordhorst et al. (1998) have shown that seeds of *A. thaliana* monogenic mutants *primordial timing* (*pt*) and *clavata* (*clv*) and *pt clv* double mutants directly germinated in a liquid medium containing 2,4-D regenerate stable embryogenic cultures and somatic embryos from seedlings, thus circumventing the tedious dissection of immature embryos. Since both monogenic and digenic mutants are characterized by unusually enlarged embryonic shoot apical meristems, it has been suggested that noncommitted stem cells of the meristem facilitate somatic embryogenesis in *A. thaliana*. This does not necessarily mean that an active shoot apical meristem is required to produce embryogenic cells and somatic embryos, as embryos isolated from *A. thaliana* mutants defective in embryonic shoot apical meristem formation such as *shoot meristemless* (*stm*), *wuschel (WUS)*, and *zwill/pinhead* (*zll/pin*) also

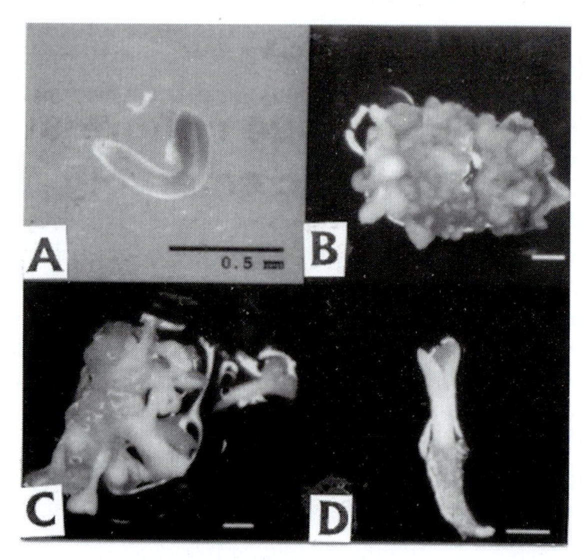

Fig. 9.4 Induction of somatic embryos from cultured zygotic embryos of *Arabidopsis thaliana*. A-bent-cotyledon stage embryo at the time of culture. B-embryogenic cell clusters formed on the embryo 10 days after culture in a hormone-containing medium. C-transformation of embryogenic cell clusters into somatic embryos in a hormone-free medium. D-magnified view of a somatic embryo from C. Scale bars, = 1 mm (from Ikeda-Iwai et al. 2002).

readily formed somatic embryos in the same medium that favors embryogenesis in the wild type (Mordhorst et al. 2002).

Using *A. thaliana* as the experimental system, molecular approaches involving manipulation of certain embryo-specific genes have led to the identification of genes that play a role in the transition of vegetative cells to embryogenic cells in the absence of hormones. One is the *LEAFY COTYLEDON1* (*LEC1*) gene which encodes subunit B of the heterotrimeric transcriptional activator complex *HAP3*; ectopic expression of this gene induces the formation of embryo-like structures from leaf cells of transgenic plants (Lotan et al. 1998). Another gene, *LEC2*, also encodes a transcription factor with similarity to the B3 domain characteristic of several plant transcription factors; like *LEC1*, overexpression of *LEC2* in transgenic plants leads to the formation of seedlings with a range of morphologies, including fleshy cotyledons and short hypocotyls and roots, all of which differentiate somatic embryos from their surface (Stone et al. 2001). In a recent investigation, Zuo et al. (2002) identified two alleles of a gene designated *PLANT GROWTH ACTIVATOR6* (*PGA6*) and showed that gain-of-function mutations at this locus cause somatic embryo formation in profuse numbers from various tissues and organs of *A. thaliana*. The homeodomain protein-encoding *WUS*, referred to in the previous paragraph, which plays a dominant role in specifying cell fate in the shoot apical meristem, is another gene peripherally functioning in maintaining embryogenic competence of somatic cells of *A. thaliana*. Supporting this statement is the observation that transgenic plants carrying a *WUS* construct phenocopy *pga* mutations and cause high-frequency somatic embryogenesis from vegetative tissues (Fig. 9.5). The use of innovative genetic and molecular techniques, combined with traditional tissue culture approaches promises to provide deeper insight into the mechanism underlying the transformation of somatic cells of *A. thaliana* to embryogenic cells.

SOMATIC EMBRYOGENESIS IN CARROT AS A MODEL SYSTEM

In the years following the excitement engendered by the discovery of somatic embryogenesis, remarkable strides have been made in developing protocols for obtaining somatic embryos reproducibly and in large numbers from carrot. A widely used method begins with the culture of root or hypocotyl segments excised from seedlings of domestic carrot aseptically germinated on the surface of solidified high-nitrogen containing Murashige-Skoog medium supplemented with sucrose, *myo*-inositol, a cytokinin, and 2,4-D. A piece of the callus regenerated on the explant is transferred to an agitated liquid medium of the same composition, but with a reduced level of 2,4-D to produce a suspension culture consisting

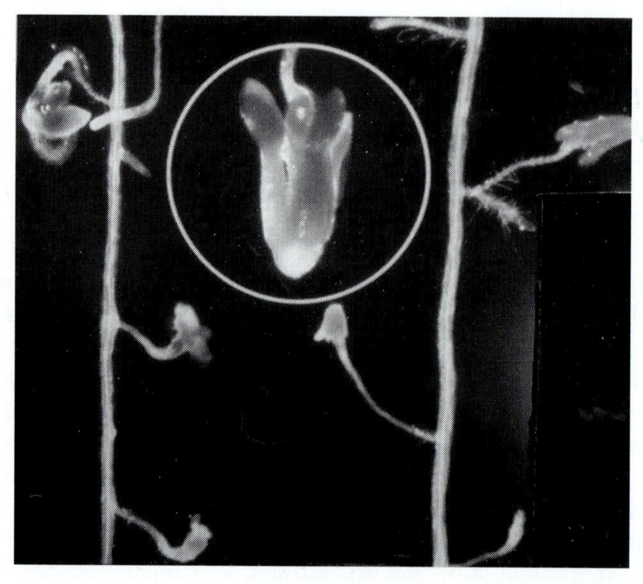

Fig. 9.5 Reprogramming of vegetative cells of transgenic *Arabidopsis thaliana* to somatic embryos by the *WUSCHEL* gene. Overexpression of the gene induces formation of somatic embryos around the root tips of transgenic plants. Circled insert shows an enlarged view of a cluster of somatic embryos (from Zuo et al. 2002).

of single cells and proliferating cell clusters known as proembryogenic masses. The cell population can be stably maintained in this medium for several months by repeated subcultures. Embryogenesis is induced by transferring an aliquot of the suspension to a medium totally deprived of 2,4-D (induction medium). In four to five days in the induction medium, the proembryogenic masses are transformed into globular embryos, followed in rapid succession by the initiation of cotyledons and establishment of bipolarity, formation of the shoot and root apices, and embryo maturation. By appropriate culture manipulations, conditions have been established for obtaining high yields of synchronously developing somatic embryos starting with a cell suspension or a population of embryogenic single cells potentially capable of becoming embryos (Fujimura and Komamine 1979; Giuliano et al. 1983; Nomura and Komamine 1985). A video cell-tracking system has identified both cytoplasmic single cells and vacuolated single cells in a carrot cell suspension that can develop into somatic embryos, thus reinforcing the view of their single-celled origin (Toonen et al. 1994). In hypocotyl segments of carrot exposed to 2,4-D, Guzzo et al. (1994, 1995) traced the origin of embryogenic single cells to the provascular cells, the only cells in the explant which divide in the presence of auxin. Continued divisions of these cells generate proembryogenic masses of cells. Cytohistological analysis of the cultured explant has shown that from the thousands of isodiametric cells produced by the activated provascular tissues, only a small number of elongate, oval to triangular cells actually become embryogenic. These cells are undoubtedly released from the proliferating explant and apart from their ability to produce proembryogenic masses, they are indistinguishable from the majority of isodiametric cells generated. Evidently, totipotency is not an intrinsic property of all cells of the cultured explant, but is acquired by certain meristematic cells subjected to hormonal treatment.

The problem of identifying by chemical or molecular markers the potentially embryogenic cells in a suspension culture consisting of a heterogeneous array of several different cell types and cell clusters with ill-defined morphology is quite vexing. Pennell et al. (1992) suggested use of the monoclonal antibody JIM8 which reacts with cell wall arabinogalactan proteins or arabinogalactans (glycoproteins with high levels of arabinosyl and galactosyl residues as well as high levels of alanine, serine, threonine, and hydroxyproline residues) as a molecular marker for an early stage in the developmental pathway of cells in a carrot cell suspension culture to somatic embryos. However, the usefulness of this method to reliably predict whether single cells will develop into somatic embryos was questioned when a later work using the improved video cell-tracking method to follow the developmental fate of the labeled cells showed poor correlation between cells reacting with the antibody and somatic embryo formation (Toonen et al. 1996). A breakthrough in identifying embryogenic cells from nonembryogenic cells came when it was shown that expression of the *SOMATIC EMBRYOGENESIS RECEPTOR KINASE* (*SERK*) gene, isolated from carrot cells, coincides with acquisition of embryogenic competence by cells. The protein product of this gene contains the signature of a leucine-rich transmembrane receptor-like kinase and probably functions in a signal transduction pathway. It was shown by tracking of single cells in a carrot cell suspension culture transformed with a *SERK* promoter-luciferase construct that the transgene-expressing cells develop into somatic embryos (Schmidt et al. 1997). Using a whole mount *in situ* hybridization method, the carrot *SERK* gene was also found to be a good molecular marker for cells competent to form somatic embryos directly on cultured leaf explants of *Dactylis glomerata* (Somleva et al. 2000). Thus, the closest we have come to identifying with some degree of precision a cell type endowed with embryogenic competence is through use of this marker. Constitutive overexpression of *AtSERK*, the *A. thaliana* homologue of the carrot *SERK* gene, is found to mark cells competent to form embryos in culture and to confer increased embryogenic competence of callus cells derived from cultured seedlings of *A. thaliana*. The demonstrated expression of this gene in germline cells and in early stage zygotic embryos of *A. thaliana* might indicate that the same signal transduction pathway is activated during embryogenic episodes from somatic cells and germ cells (Hecht et al. 2001).

Role of Auxin

As pointed out earlier, sequences of transfer of cells from a medium containing the auxin 2,4-D to a medium lacking auxin essentially define the protocol for inducing somatic embryogenesis in carrot. The fact that cells of a carrot hypocotyl segment acquire the capacity to form embryos only after a certain period of growth in the medium containing 2,4-D, has engendered the notion that embryogenic competence is attained by a specific gene expression program under the influence of auxin. Although the main focus of much of the physiological investigations on somatic embryogenesis in carrot during the past two decades has been on the role of auxin in making cells competent, the seemingly intractable problem of auxin action in the carrot system has been given a new lease on life by the publication of some reports of embryogenic development of cells in a completely auxin-free medium. Smith and Krikorian (1988, 1989, 1990) have shown that the mericarp (one-seeded half of the ovary) and mechanically wounded zygotic embryo of carrot produce somatic embryos in an auxin-free medium. Cell clusters generated from these explants are maintained as preglobular stage somatic embryos by continuous subculture in a buffered medium (pH 4.0) with 1-5 mM NH_4^+ as the sole source of nitrogen. Development of cell clusters into regular somatic embryos occurs when the pH of the medium is increased to 4.5 or above. Somatic embryogenesis in an auxin-free medium is induced without an intervening callus phase when meristem explants of carrot seedlings are cultured in media containing heavy metal ions (Kiyosue et al. 1990) or are subjected to salt stress (Kiyosue et al. 1989), osmotic stress (Kamada et al. 1993), or heat stress (Kamada et al. 1994). These results have led to the suggestion that cells express their innate embryogenic potential due to physical or chemical stresses. This view is also supported by a report that carrot seedlings cultured in a medium containing ABA—well-known for its role in stress signal transduction—as the sole growth hormone regenerate somatic embryos directly from epidermal cells of the hypocotyl (Fig. 9.6) (Nishiwaki et al. 2000). Here, ABA might be considered to act as a signal transducer in stress-induced somatic embryogenesis. These successful attempts to develop the carrot somatic embryogenic system in an auxin-free medium offer new opportunities for analyzing embryogenic induction from a new perspective.

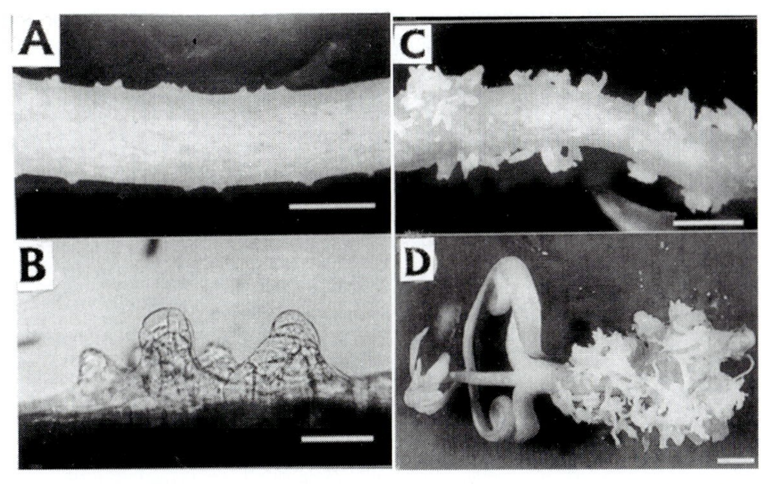

Fig. 9.6 Induction of somatic embryos on the hypocotyl of a seedling of *Daucus carota* by abscisic acid. A, B-protuberances seen on the hypocotyl treated with ABA for 10 days. C-somatic embryos 40 days after treatment with ABA. D-somatic embryos 50 days after treatment with ABA. Scale bars = A, C, D = 1 mm; B = 100 μm (from Nishiwaki et al. 2000).

Role of secreted molecules

Hari (1980) found that a quantitative reduction in embryogenesis occurs when carrot cell suspension is diluted to low cell density and that this is partially alleviated by the addition of a conditioned cell suspension medium. This observation initially led to the view that cells and cell clusters release into the medium components that favor cell proliferation and somatic embryogenesis, and later to the isolation of the secreted molecules and their characterization as glycosylated extracellular proteins. The potency of these proteins in inducing differentiation of somatic cells of carrot into embryos is reinforced by two observations: a conditioned medium from embryogenic cell clusters accelerates the acquisition of embryogenic competence by newly initiated cultures and addition of a purified protein mixture to cell lines blocked in the synthesis of these proteins and in somatic embryogenesis either by a glycosylation inhibitor or a mutation, restores embryogenesis (de Vries et al. 1988a, b; Lo Schiavo et al. 1990; Cordewener et al. 1991; de Jong et al. 1992). Four sets of proteins designated as EP1 (for extracellular protein 1), EP2, EP3, and EP4 have been further characterized from this mixture. EP1 and EP4 proteins are not associated with proembryogenic masses or somatic embryos, but are secreted by nonembryogenic cells of the suspension. EP1 proteins are localized primarily in the pectic cell wall material of nonembryogenic single cells. Unlike the unfractionated protein mixture, EP1 proteins by themselves are not sufficient to restore somatic embryogenesis in cell lines disturbed in the process. Nonembryogenic cells localizing EP4 proteins are mostly clustered rather than single cells, with the walls separating adjacent cells showing high levels of localization (van Engelen et al. 1991; van Engelen et al. 1995). An *EP2* protein secreted by embryogenic cells was identified as a lipid transfer protein. Expression abundance of transcripts of *EP2* gene in the epidermal cells of developing somatic and zygotic embryos of carrot has led to the view that the *EP2* protein might be involved in the transport of cutin monomers to the epidermal cells which are the traditional sites of cutin synthesis. Expression of the *EP2* gene transcripts in the proembryogenic mass of cells has suggested a role for the protein product of the gene as a marker of embryogenic potential of cells. In support of this view, it was found that the expression pattern of *ARABIDOPSIS THALIANA LIPID TRANSFER PROTEIN1* (*AtLTP1*) promoter-luciferase gene construct in transformed carrot proembryogenic masses of cells is identical to that of the endogenous carrot *EP2* gene. Moreover, by cell tracking it was also established that somatic embryos are invariably formed from cell clusters expressing the *AtLTP1*-luciferase gene construct (Sterk et al. 1991; Toonen et al. 1997). Another protein, EP3, purified from the wild-type culture medium was capable of overcoming the developmental arrest of globular embryos in a temperature-impaired cell line of carrot (*ts11*) at nonpermissive temperatures. The rescue protein was identified as a glycosylated acidic endochitinase Class IV, which is apparently secreted by mutant cells at a reduced level during a transient period of growth at the nonpermissive temperature (de Jong et al. 1992; Kragh et al. 1996). However, van Hengel et al. (1998) found no correlation between the presence of *EP3* gene transcripts and the potential of somatic cells of carrot to produce embryos. *In situ* hybridization studies showed that the *EP3* gene is expressed in cells of both embryogenic and nonembryogenic cultures, mostly in a subset of single cells morpho-logically identifiable by their elongated, strongly curved or vacuolated nature as well as in cells of the *ts11* line. Although the gene is not expressed in somatic or zygotic embryos of carrot, it is expressed in the integumentary cells of young fruits and in some endosperm cells of carrot seeds. From these results, the EP3 endochitinase is considered part of a nursing cell system active during zygotic embryogenesis; the nursing system is apparently functional during somatic embryogenesis by the presence of *EP3*-expressing cells in the suspension culture.

Arbinogalactan proteins have been identified as secreted molecules affecting cell fate in carrot suspension cell cultures in a dramatic way. For example, addition of a very small amount of this glycoprotein purified from embryogenic cell cultures of carrot and from carrot seeds not only enhances the embryogenic potential of established, embryogenic cell lines of carrot, but is also able to potentiate embryogenesis in an old cell line which has lost its embryogenic potential. That the activity of the crude arabinogalactans depends on the presence of both promotive and inhibitory molecules became apparent by the demonstration that when the glycoprotein from carrot or tomato (*Lycopersicon esculentum*) is fractionated, not all fractions are active to the same extent in enhancing the embryogenic potential of cultured cells (Kreuger and van Holst, 1993, 1995). By grouping discrete populations of single cells differing in morphology and embryogenic competence in a carrot cell suspension by their ability to recognize JIM8 in their cell wall and in conjunction with the use of secondary antibodies to label and sort pure populations of JIM8 (-) and JIM8 (+) cells, it was shown that JIM8 (-) cells which do not develop into somatic embryos do so when they are cultured in a medium conditioned by the culture of JIM8 (+) cells. This observation has suggested a role for cell-cell interaction involving the release of soluble signals of the arabinogalactan type by cells in the JIM8 (+) population in the control of a transient stage in the early developmental pathway to somatic embryogenesis in carrot (McCabe et al. 1997).

It appears from these investigations that the type of proteins secreted plays a role in determining the developmental state of cells in the carrot cell suspension culture, specifically whether they are diverted in the embryogenic pathway or whether they bide their time by repeated divisions. Localization of some of the proteins in the walls of both embryogenic and nonembryogenic cells has fueled the speculation that these proteins may somehow be involved in regulating the twin processes of division and elongation of cells, but further work is necessary to define their role in these cellular processes (van Engelen and de Vries 1992).

Embryonic Proteins and Regulation of Gene Expression

Carrot cell suspension has been widely used as a model system to study the biochemical and molecular changes associated with somatic embryogenesis. Much of the evidence favors the view that depletion of 2,4-D from the medium modulates embryogenesis by eliciting the synthesis of new mRNA and proteins. Yet it has not appeared unequivocally clear that gene activity for embryogenic induction in cells is initiated upon their transfer to a medium lacking auxin. The issue is whether proteins synthesized in cells nurtured in the auxin-free medium are coded on newly formed mRNA or on mRNA transcribed when cells are bathed in the auxin-containing medium. Wilde et al. (1988) found striking similarities between the *in vitro* translation products of mRNA populations of proembryogenic masses and torpedo-shaped somatic embryos of carrot, leading to the suggestion that a gene expression program for somatic embryogenesis is initiated when cells are bathed in the auxin-containing medium. The view that 2,4-D controls gene activity for embryogenic induction of somatic cells of carrot has been reinforced by a study of the translational profile of cells growing in the presence and absence of auxin in the medium (Sung and Okimoto 1981). Comparison of the spectrum of proteins synthesized by cells growing for 12 days in the presence or absence of 2,4-D showed no pronounced differences in the nearly 200 or so polypeptides spotted on gels, except that the embryogenic cells growing in the absence of auxin had two additional proteins, E1 and E2, designated as embryonic proteins. The surprising finding is that regardless of the presence or absence of 2,4-D in the medium, these two proteins were synthesized by cells as early as four hours of growth in the fresh media but, in the presence of the auxin, gradually diminished and eventually disappeared. Synthesis of embryonic proteins appears to be an early event of embryogenic induction triggered by 2,4-D, although by its very presence in the medium auxin also inhibits or prevents the continued synthesis of these proteins and execution of the embryogenic program of cells. As shown in a later investigation, high-density callus type growth of somatic cells of carrot in the presence of 2,4-D in the medium is marked by synthesis of two callus-specific proteins, designated C1 and C2. A variant cell line of carrot that is resistant to cycloheximide, an inhibitor of protein synthesis in both callus and embryogenic modes of growth, when nurtured as a callus in a medium containing 2,4-D, synthesizes the embryonic ones, but not the callus-specific ones. This finding implies that whereas callus growth can proceed even when the synthesis of callus-specific proteins is inhibited, synthesis of embryonic proteins alone does not necessarily evoke somatic embryogenesis (Sung and Okimoto 1983). Speculatively it might be said that callus growth and embryogenic induction in carrot somatic cells are coordinately expressed by a common mechanism, with the caveat that activation of one function is accompanied by elimination of the other.

In addition to the differential utilization of genetic information leading to the synthesis of new mRNA and proteins in the reprogramming of cells in the embryogenic pathway, progressive differentiation of single cells and cell clusters to form somatic embryos might be expected to require operation of a program of continued gene expression. However, given the extensive overlap between the *in vivo* synthesized proteins of embryogenic and nonembryogenic cells, it is possible that only minor changes accompany embryogenic induction and that understanding the molecular regulation of somatic embryogenesis in carrot may hinge on some rare class of genes. Of the many genes isolated by differential screening of cDNA libraries of undifferentiated callus cells and somatic embryos, a few appear to be structurally and functionally homologous to the late embryogenesis abundant (LEA) genes; others are expressed coordinately during embryogenesis beginning with formation of globular embryos; and still others, encoding proteins related to the cell wall, are expressed at the onset of embryogenesis (Zimmerman, 1993; Raghavan 1997, 2000). Unfortunately, the array of genes studied thus far does not include those rare genes whose expression is causal for embryogenic development of somatic cells.

The aforesaid results have led to an alternative interpretation, namely, that the transition of somatic cells to proembryogenic masses or embryos might not involve changes in the most abundant proteins or mRNAs, but rather be programmed by the down-regulation of some of the genes expressed in the somatic cells. In support of this view, characterization of the expression and regulation of a collection of 38 different genes isolated by a subtracted-probe strategy using mRNA from carrot seedlings to screen embryo-enhanced genes from carrot somatic embryos has shown that most of the genes are not only expressed in the callus, but some even expressed at higher levels in the callus than in somatic embryos (Lin et al. 1996). However, this work also has not identified any rare genes whose down-regulation modulates the transition of somatic cells to embryos.

CONCLUDING COMMENTS

It is the peculiar virtue of gymnosperms and angiosperms that they can regenerate facsimiles of embryos from somatic cells through a developmental sequence similar to that of the zygote. Whereas the inaccessibility of the zygote has prevented critical biochemical and physiological investigations on embryogenesis, embryogenic single somatic cells have provided a useful alternative system for basic research into angiosperm embryo development. On the applied side, knowledge of the factors controlling somatic embryo development in useful agronomic, horticultural, and forestry plants has led to the development of methods for their clonal propagation and the creation of artificial or synthetic seeds. Unlike zygotic embryos, somatic embryos do not develop within a protective seed coat and are not associated with nutritive tissues such as the endosperm or the megagametophyte; Unlike zygotic embryos, somatic embryos fail to lapse into a period of dormancy. These properties of somatic embryos have impeded efficient production of synthetic seeds in most species. Nonetheless, the similarity between a zygotic embryo and a somatic embryo is remarkable considering the fact that the zygote and the somatic cell as their respective precursors are subjected to different types of physical stresses and chemical stimuli during embryogenic episodes.

REFERENCES

Backs-Hüsemann D, Reinert J (1970) Embryobildung durch isolierte Einzelzellen aus Gewebekulturen von *Daucus carota.* Protoplasma 70:49–60.

Borthwick HA (1931) Development of the macrogametophyte and embryo of *Daucus carota*. Bot. Gaz. 92:23–44.

Brown DCW, Finstad KI, Watson EM (1995) Somatic embryogenesis in herbaceous dicots. *In:* TA Thorpe, (ed.). *In Vitro Embryogenesis in Plants*. Kluwer Acad. Publ. Dordrecht, Netherlands, pp. 345–415.

Conger BV, Hanning GE, Gray DJ, McDaniel JK (1983) Direct embryogenesis from mesophyll cells of orchard grass. Science 221:850–851.

Cordewener J, Booij H, van der Zandt H, van Engelen F, van Kammen A, de Vries S (1991) Tunicamycin-inhibited carrot somatic embryogenesis can be restored by secreted cationic peroxidase isoenzymes. Planta 184:478–486.

de Jong AJ, Cordewener J, Lo Schiavo F, Terzi M, et al. (1992) A carrot somatic embryo mutant is rescued by chitinase. Plant Cell 4:425–433.

de Jong AJ, Hendriks T, Meijer EA, Penning M, et al. (1995) Transient reduction in secreted 32 kDa chitinase prevents somatic embryogenesis in the carrot (*Daucus carota* L.) variant ts11. Devel. Genet. 16:332–343.

de Vries S C, Booij H, Janssens R, Vogels R, et al. (1988a) Carrot somatic embryogenesis depends on the phytohormone-controlled presence of correctly glycosylated extracellular proteins. Genes Devel. 2:462–476.

de Vries SC, Booij H, Meyerink P, Huisman G, et al. (1988b) Acquisition of embryogenic potential in carrot cell-suspension cultures. Planta 175:196–204.

Dodeman VL, Ducreux G, Kreis M (1997) Zygotic embryogenesis versus somatic embryogenesis. J. Exp. Bot. 48:1493–1509.

Dunstan DI, Tautorus TE, Thorpe TA (1995) Somatic embryogenesis in woody plants. *In:* TA Thorpe, (ed.). *In Vitro Embryogenesis in Plants*. Kluwer Acad. Publ. Dordrecht, Netherlands, pp. 471–538.

Fujimura T, Komamine A.(1979) Synchronization of somatic embryogenesis in a carrot cell suspension culture. Plant Physiol. 64:162–164.

Giuliano G, Rosellini D, Terzi M (1983) A new method for the purification of the different stages of carrot embryoids. Plant Cell Repts. 2:216–218.

Grambow HJ, Kao KN, Miller RA, Gamborg OL (1972) Cell division and plant development from protoplasts of carrot cell suspension cultures. Planta 103:348–355.

Guzzo F, Baldan B, Levi M, Sparvoli E, Lo Schiavo F, Terzi M, Mariani P (1995) Early cellular events during induction of carrot explants with 2,4-D. Protoplasma 185:28–36.

Guzzo F, Baldan B, Mariani P, Lo Schiavo F, Terzi M (1994) Studies on the origin of totipotent cells in explants of *Daucus carota* L. J. Exp. Bot. 45:1427–1432.

Halperin W (1966a) Alternative morphogenetic events in cell suspensions. Amer. J. Bot. 53:443–453.

Halperin W (1966b) Single cells, coconut milk, and embryogenesis in vitro. Science 153:1287–1288.

Halperin W, Wetherell DF (1964) Adventive embryony in tissue cultures of the wild carrot, *Daucus carota.* Amer. J. Bot. 51:274–283.

Hari V (1980) Effect of cell density changes and conditioned media on carrot cell embryogenesis. Z. Pflanzenphysiol. 96:227–231.

Hecht V, Vielle-Calzada JP, Hartog, MV, Schmidt EDL, et al (2001) The Arabidopsis *SOMATIC EMBRYOGENESIS RECEPTOR KINASE 1* gene is expressed in developing ovules and embryos and enhances embryogenic competence in culture. Plant Physiol. 127:803–816.

Ikeda-Iwai M, Satoh S, Kamada H (2002) Establishment of a reproducible tissue culture system for the induction of *Arabidopsis* somatic embryos. J. Exp. Bot. 53:1575–1580.

Johri BM, Bajaj YPS (1962) Behaviour of mature embryo of *Dendrophthoe falcata* (L. f) Ettingsh. *in vitro.* Nature 193:194–195.

Kamada H, Ishikawa K, Saga H, Harada H (1993) Induction of somatic embryogenesis in carrot by osmotic stress. *Plant Tissue Cult. Lett.* 10:38–44.

Kamada H, Tachikawa Y, Saitou T, Harada H (1994) Heat stress induction of carrot somatic embryogenesis. Plant Tiss. Cult. Lett. 11:229–232.

Kameya T, Uchimiya H (1972) Embryoids derived from isolated protoplasts of carrot. Planta 103:356–360.

Kato H, Takeuchi M (1963) Morphogenesis *in vitro* starting from single cells of carrot root. Plant Cell Physiol. 4:243–245.

Kiyosue T, Kamada H, Harada H (1989) Induction of somatic embryogenesis by salt stress in carrot. Plant Tiss. Cult. Lett. 6:162–164.

Kiyosue T, Takano K, Kamada H, Harada H (1990) Induction of somatic embryogenesis in carrot by heavy metal ions. Can. J. Bot. 68:2301–2303.

Konar RN, Nataraja K (1965a) Production of embryos on the stem of *Ranunculus sceleratus* L. Experientia 21:395.

Konar RN, Nataraja K (1965b) Experimental studies in *Ranunculus sceleratus* L. Development of embryos from the stem epidermis. Phytomorph. 15:132–137.

Kragh KM, Hendriks T, de Jong AJ, Lo Schiavo F, et al. (1996) Characterization of chitinases able to rescue somatic embryos of the temperature-sensitive carrot variant ts11. Plant Mol. Biol. 31:631–645.

Kreuger M, van Holst GJ (1993) Arabinogalactan proteins are essential in somatic embryogenesis of *Daucus carota* L. Planta 189:243–248.

Kreuger M, van Holst GJ (1995) Arabinogalactan-protein epitopes in somatic embryogenesis of *Daucus carota* L. Planta 197:135–141.

Krishnaraj S, Vasil IK (1995) Somatic embryogenesis in herbaceous monocots. *In:* TA Thorpe, (ed.). *In Vitro Embryogenesis in Plants*. Kluwer Acad. Publ. Dordrecht, Netherlands, pp. 417–470.

Lackie S, Yeung EC (1996) Zygotic embryo development in Daucus carota. Can. J. Bot. 74:990–998.

Lin X, Hwang GJH, Zimmerman JL (1996) Isolation and characterization of a diverse set of genes from carrot somatic embryos. Plant Physiol. 112:1365–1374.

Lo Schiavo F, Giuliano G, de Vries SC, Genga A, et al. (1990) A carrot cell variant temperature-sensitive for somatic embryogenesis reveals a defect in the glycosylation of extracellular proteins. Molec. Gen. Genet. 223:385–393.

Lotan T, Ohto M, Yee KM, West MAL, et al. (1998) *ARABIDOPSIS LEAFY COTYLEDON1* is sufficient to induce embryo development in vegetative cells. Cell 93:1195–1205.

Maheshwari P, Baldev B (1961) Artificial production of buds from the embryos of *Cuscuta reflexa.* Nature 191:197–198.

McCabe PF, Valentine TA, Forsberg LS, Pennell RI (1997) Soluble signals from cells identified at the cell wall establish a developmental pathway in carrot. Plant Cell 9:2225–2241.

McWilliam AA, Smith SM, Street HE (1974) The origin and development of embryoids in suspension cultures of carrot (*Daucus carota*). Ann. Bot. 38:243–250.

Mordhorst AP, Toonen MAJ, de Vries SC (1997) Plant embryogenesis. Crit. Rev. Plant Sci. 16:535–576.

Mordhorst AP, Hartog MV, el Tamer MK, Laux T, de Vries SC (2002) Somatic embryogenesis from *Arabidopsis* shoot apical meristem mutants. Planta 214:829–836.

Mordhorst AP, Voerman KJ, Hartog MV, Meijer EA, et al. (1998) Somatic embryogenesis in *Arabidopsis thaliana* is facilitated by mutations in genes repressing meristematic cell divisions. Genetics 149:549–563.

Nishiwaki M, Fujino K, Koda Y, Masuda K, Kikuta Y (2000) Somatic embryogenesis induced by the simple application of abscisic acid to carrot (*Dacus carota* L.) seedlings in culture. Planta 211:756–759.

Nomura K, Komamine A (1985) Identification and isolation of single cells that produce somatic embryos at high frequency in a carrot suspension culture. Plant Physiol. 79:988–991.

Pennell RI, Janniche L, Scofield GN, Booij H, de Vries SC, Roberts K (1992) Identification of a transitional cell state in the developmental pathway to carrot somatic embryogenesis J. Cell Biol. 119:1371–1380.

Raghavan V (1986) *Embryogenesis in Angiosperms. A Developmental and Experimental Study*. Cambridge Univ. Press, New York, NY.

Raghavan V (1997) *Molecular Embryology of Flowering Plants.* Cambridge Univ. Press, New York, NY.

Raghavan V (2000) *Developmental Biology of Flowering Plants.* Springer-Verlag, New York, NY.

Rao KS (1996) Embryogenesis in flowering plants: recent approaches and prospects. J. Biosci. 21:827–841.

Reinert J (1959) Uber die Kontrolle der Morphogenese und die Induktion von Adventivembryonen an Gewebekulturen aus Karotten. Planta 53:318–333.

Schmidt EDL, Guzzo F, Toonen MAJ, de Vries SC (1997) A leucine-rich repeat containing receptor-like kinase marks somatic plant cells competent to form embryos. Development 124:2049–2062.

Smith DL, Krikorian AD (1988) Production of somatic embryos from carrot tissues in hormone-free medium. Plant Sci. 58:103–110.

Smith DL, Krikorian AD (1989) Release of somatic embryogenic potential from excised zygotic embryos of carrot and maintenance of proembryonic cultures in hormone-free medium. Amer. J. Bot. 76:1832–1843.

Smith DL, Krikorian AD (1990) Somatic embryogenesis of carrot in hormone-free medium: external pH control over morphogenesis. Amer. J. Bot. 77:1634–1647.

Somleva MN, Schmidt EDL, de Vries SC (2000) Embryogenic cells in *Dactylis glomerata* L. (Poaceae) explants identified by cell tracking and by SERK expression. Plant Cell Rep. 19:718–726.

Sterk P, Booij H, Schellekens GA, van Kammen A, de Vries SC (1991) Cell-specific expression of the carrot EP2 lipid transfer protein gene. Plant Cell 3:907–921.

Steward FC (1963a) The control of growth in plant cells. Sci. Amer. 209(4):104–113.

Steward FC (1963b) Totipotency and variation in cultured cells: some metabolic and morphogenetic manifestations. *In:* P Maheshwari, NS Rangaswamy, (eds.). *Plant Tissue and Organ Culture. A Symposium*. Univ. Delhi, Delhi, pp. 1–25.

Steward FC, Mapes MO, Mears K (1958a) Growth and organized development of cultured cells. II. Organization in cultures grown from freely suspended cells. Amer. J. Bot. 45:705–708.

Steward FC, Mapes MO, Smith J (1958b) Growth and organized development of cultured cells. I. Growth and division of freely suspended cells. Amer. J. Bot. 45:693–703.

Steward FC, Mapes MO, Kent AE, Holsten RD (1964) Growth and development of cultured plant cells. Science 143:20–27.

Stone SL, Kwong LW, Yee KM, Pelletier J, Lepiniec L, Fischer RL, Goldberg RB, Harada JJ (2001) *LEAFY COTYLEDON2* encodes a B3 domain transcription factor that induces embryo development. Proc. Natl. Acad. Sci. USA 98:11806–11811.

Sung ZR, Okimoto R (1981) Embryonic proteins in somatic embryos of carrot. Proc. Natl. Acad. Sci. USA 78:3683–3687.

Sung ZR, Okimoto R (1983) Coordinate gene expression during somatic embryogenesis in carrot. Proc. Natl. Acad. Sci. USA 80:2661–2665.

Thorpe TA, Stasolla C (2001) Somatic embryogenesis. *In:* SS Bhojwani, WY Soh, (eds.). *Current Trends in the Embryology of Angiosperms*. Kluwer Acad. Publ., Dordrecht, Netherlands, pp. 279–336.

Toonen MAJ, Verhees JA, Schmidt EDL, van Kammen A, de Vries SC (1997) AtLTP1 luciferase expression during carrot somatic embryogenesis. Plant J. 12:1213–1221.

Toonen MAJ, Hendriks T, Schmidt EDL, Verhoeven HA, van Kammen A, de Vries SC (1994) Description of somatic-embryo-forming single cells in carrot suspension cultures employing video cell tracking. Planta 194:565–572.

Toonen MAJ, Schmidt EDL, Hendriks T, Verhoeven HA, van Kammen A, de Vries SC (1996) Expression of the JIM8 cell wall epitope in carrot somatic embryogenesis. Planta 200:167–173.

van Engelen FA, de Vries SC (1992) Extracellular proteins in plant embryogenesis. Trends Plant Sci. 8: 66–70.

van Engelen FA, Sterk P, Booij H, Cordewener JHG, et al. (1991) Heterogeneity and cell type-specific localization of a cell wall glycoprotein from carrot suspension cells. Plant Physiol. 96:705–712.

van Engelen FA, de Jong AJ, Meijer EA, Kuil CW, et al. (1995) Purification, immunological characterization and cDNA cloning of a 47 kDa glycoprotein secreted by carrot suspension cells. Plant Molec. Biol. 27:901–910.

van Hengel AJ, Guzzo F, van Kammen A, de Vries SC (1998) Expression pattern of the carrot *EP3* endochitinase genes in suspension cultures and in developing seeds. Plant Physiol. 117:43–53.

Vasil V, Hildebrandt AC (1965) Differentiation of tobacco plants from single, isolated cells in microcultures. Science 150:889–892.

Vasil V, Vasil IK (1981) Somatic embryogenesis and plant regeneration from suspension cultures of pearl millet (*Pennisetum americanum*). Ann. Bot. 47:669–678.

Wetherell DF, Halperin W (1963) Embryos derived from callus tissue cultures of the wild carrot. Nature 200:1336–1337.

Wilde HD, Nelson WS, Booij H, de Vries SC, Thomas TL (1988) Gene-expression programs in embryogenic and non-embryogenic carrot cultures. Planta 176:205–211.

Zimmerman JL (1993) Somatic embryogenesis: a model for early development in higher plants. Plant Cell 5:1411–1423.

Zuo J, Niu QW, Frugis G, Chua NH (2002) The *WUSCHEL* gene promotes vegetative-to-embryonic transition in *Arabidopsis.* Plant J. 30:349–359.

10

Somatic Embryogenesis in Carrot Suspension Cultures as a Model System for Expression of Totipotency

Koji Nomura[1] and Atsushi Komamine[2]

[1]*Graduate School of Life and Environmental Sciences, University of Tsukuba*
1-1-1, Tennodai, Tsukuba, Ibaraki 305-8572, Japan
[2]*Research Institute of Evolutionary Biology*
2-4-28, Kamiyoga, Setagaya-ku, Tokyo 158-0098, Japan
**corresponding author: e-mail: khf10654@nifty.ne.jp*

THE PIONEERS

The history of culture of isolated plant cells begins with the publication in 1902 of the work done by Haberlandt. The cell theory proposed by Schleiden and Schwann had already established the principal view for multicellular organisms on which modern biology has been constructed. Knowledge of plant nutrition also increased in the 19th century. Culture solutions for hydroponics were established by Sachs and Knop. It seems very appropriate that *in vitro* culture of plant cells at this circumstance should be essayed at the point. The report by Haberlandt (1902) is a story of trial and error. He tried to identify whether the behavior of single plant cells is free from organs or tissues. Although he did not succeed in obtaining *in vitro* proliferation of single plant cells, probably due to premature technologies and information, his experiments that tried to demonstrate the totipotency of cells were the origin of *in vitro* biology, leading to the studies of the potency of single cells.

We should wait more then 25 years to see the fruit of *in vitro* culture of plant cells. In 1934, White cultured root segments of tomato, and Gauthert succeeded continuous tissue culture in 1937. Adventitious shoot differentiation was induced in tobacco callus in 1939 by White. *In vitro* proliferation of plant somatic cells was established in the 1930s. Totipotency of plant somatic cells was demonstrated by somatic embryogenesis in suspension cultures. Somatic embryogenesis was reported by Reinert (1958) and Steward et al. (1958) independently. In both of their experiments, carrot (*Daucus carota*) cell suspension cultures were used. Since somatic embryos developed from unorganized proliferating cells in suspension cultures, and their morphology was quite similar to natural zygotic embryos, somatic embryogenesis was regarded as a direct evidence for totipotency of plant somatic cells.

SYNCHRONOUS SYSTEM: THE CORNERSTONE FOR MOLECULAR INVESTIGATION

Since the first reports by Steward et al. (1958) and Reinert (1958), progress in morphology and physiology on somatic embryogenesis was achieved to some extent, and the number of plant species reported in which somatic embryogenesis was induced *in vitro* were increased. However, it took almost 25 years to establish experimental systems for biochemistry and molecular biology. The keywords for the ideal experimental systems for biochemical and molecular biological approaches to elucidate mechanisms of somatic embryogenesis are just two synchronization and high frequency. If a low frequency system is used, specific events for somatic embryogenesis should be diluted with other cells that are not involved in somatic embryogenesis. Furthermore, biochemical or molecular biological parameters measured in asynchronous systems indicate only averages of cells at different stages of somatic embryogenesis. Therefore, high frequency and synchronous systems are required.

The idea of synchronization was first imported into somatic embryogenesis by Halperin (1966). To induce synchronous somatic embryogenesis, Halperin fractionated cell clusters in carrot cell suspension cultures according to their sizes using a series of sieves. He succeeded in reducing morphological variability of somatic embryos and in inducing a near-synchronous development of hundreds of embryos. Another approach was reported by Warren and Fowler (1977). They separated various stages of somatic embryos using columns filled with glass beads. Culture medium was also improved. Fujimura and Komamine (1975) found that zeatin strongly promoted somatic embryogenesis in carrot suspension cultures. One of the most refined culture system for the induction of somatic embryogenesis was established by Fujimura and Komamine (1979a). They achieved synchronization of somatic embryogenesis in carrot suspension cultures by fractionation of initial heterogeneous cell population by size and density. Suspension cells were sieved through a Nylon sieve of 47 μm aperture, and trapped in another Nylon sieve of 31 μm. The cells between 31 and 47 μm were further fractionated by a density gradient centrifugation in a 10 to 18% (w/v) Ficoll solution, and rinsed by repeated centrifugations at a low speed (50 x g) for a short time (5 seconds). The cell clusters obtained from >18% Ficoll were composed of 3 to 10 cells. When they were transferred to a medium containing zeatin (0.1 M) but no auxin, the frequency of embryo formation reached more than 90% (Fig. 10.1). Globular stage embryos appeared 4 days after inoculation, followed by rapid increase in the number of globular stage embryos. The number of embryos reached its maximum on the 7th day. Cell clusters differentiated to globular stage embryos synchronously for these 3 days. Heart-shaped embryos appeared on the 8th day and their number increased rapidly until the 13th day. Thus, synchrony of the process of embryogenesis was observed at least in the early stages of the process. Establishment of this system was the initial step for investigation of somatic embryogenesis at the molecular level.

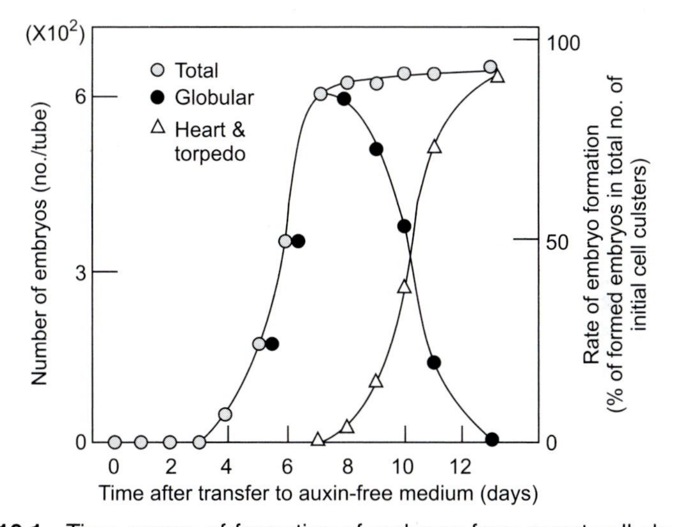

Fig. 10.1 Time course of formation of embryos from carrot cell clusters.

The first important result from this synchronous system was phase definition in somatic embryogenesis. Fujimura and Komamine (1980a) performed serial observations of the early stages of carrot somatic embryogenesis from a series of cell clusters fractionated by sieving, density gradient centrifugation in Ficoll solution, and low speed centrifugation. Three phases were recognized in each, of which the rate of cell division differed. The first phase was 0 to 3 days after transferring to an auxin-free medium. In this phase, cells in cell clusters divided slowly. The second phase was the 3rd to 4th day during which cells divided very rapidly in the determined locus of the cell cluster, leading to the formation of globular embryos. After formation of globular stage embryos, the 4th to 6th day was defined as the third phase. In the third phase cell division occurred at a slower rate than the second phase. Doubling times of the cells were 58 ± 22 h in the first phase, 6.3 ± 1.1 h in the second phase, and 20 ± 7.8 h in the third. Doubling time of actively dividing undifferentiated carrot cells in the medium containing auxin was 36 h. The rapid cell division in the second phase is considered to be a characteristic event in embryogenesis. Three regions were distinguishable in the cell clusters at the end of the second phase: a transparent region destined to develop into the shoot, an opaque region destined to develop into the root and another opaque region which had ceased to grow, suggesting that determination of differentiation had already occurred by the end of the second phase.

ACTION OF PLANT GROWTH REGULATORS

The initial stage of somatic embryogenesis is the induction of *in vitro* cell culture. In carrot, cell culture is usually initiated from the hypocotyl of a seedling. Exogenous auxin in the medium is necessary to induce cell division of cells in the explant. Changes in the signal transduction pathway were detected in this process (Ooura and Nomura, unpubl. data), but little is known about the role of exogenous auxin in the initiation of plant cell division, which is sometimes called dedifferentiation. Somatic cells undergo proliferation in a medium containing auxin. Cells growing in the presence of auxin show variability in morphology from single cells to cell clusters, which are composed of hundreds of cells.

Usually somatic embryogenesis is controlled by auxin in the culture medium. Transferring embryogenic cell clusters which have been induced and subcultured in a medium containing auxin, to an auxin-free medium is the common trigger for development of somatic embryogenesis. Addition of auxin, 2,4-D or IAA, inhibited somatic embryogenesis from embryogenic cell clusters at concentrations higher than 10^{-7} or 10^{-8} M respectively (Fujimura and Komamine 1979b). Antiauxins, 2,4,6-trichlorophenoxyacetic acid and p-chlorophenoxyisobutyric acid also inhibited development of somatic embryos from embryogenic cell clusters when they were added to the medium. The most sensitive phase to both auxin and antiauxin was the initial couple of days after transferring the embryogenic cell clusters to an auxin-free medium, but auxin (2,4-D) was less inhibitory when added after development of globular stage embryogenesis. Endogenous auxin, IAA, could be detected in embryos;

however, the level did not change significantly during embryogenesis. These results suggested that a polarity of endogenous auxin is essential for the induction of embryogenesis from embryogenic cell clusters. Exogenously supplied auxin could cancel the polarity by diffusion into cell clusters, resulting in inhibition of embryogenic development. Anti-auxin should also inhibit the action of endogenous auxin, resulting in the inhibition of embryogenic induction by auxin. Indoleacetic acid, β-naphthoxyacetic acid, indolebutyric acid, and NAA have reportedly induced roots at lower concentrations and somatic embryos at higher concentrations in hypocotyl segments of carrot. Short-term application of 2,4-D stimulated formation of somatic embryos and continuous application induced embryogenic calluses. Phenoxyacetic acid did not stimulate embryo formation, whereas o-chlorophenoxyacetic acid slightly stimulated root formation, and 2,4,5-trichlorophenoxyacetic acid, 2-methyl-4-chlorophenoxyacetic acid, and p-chlorophenoxyacetic acid stimulated somatic embryo formation (Kamada and Harada 1979). Effects of aryloxyalkanecarboxylic acids have been examined by Chandra et al. (1978). The yield of somatic embryos was enhanced when carrot cells cultured in the presence of 2,4-D were transferred into media containing α-(2-chlorophenoxy) isobutyric acid or 3,5-dichlorophenoxyacetic acid, but 2-chlorophenoxyacetic acid suppressed the number of embryos formed compared to cultures in auxin-free medium.

Zeatin promoted embryogenesis from embryogenic cell clusters in a narrow concentration range around 10^{-7} M, but other synthetic cytokinins such as 6-benzyladenine (BA) and kinetin showed no promotive effects (Fujimura and Komamine 1975).

Embryogenesis was shown to be suppressed by ethephon and ethylene. Depression of embryogenesis by 2,4-D was unrelated to ethylene evolution (Tisserat and Murashige 1977).

Gibberellic acid (GA_3) also inhibited somatic embryo formation (Fujimura and Komamine 1975; Kamada and Harada 1979). Endogenous gibberellin-like substances were also examined in suspension cultures of a hybrid grape during somatic embryogenesis. Free and highly soluble GA-like substances, expressed on a dry weight basis, decreased during development of embryos (Takeno et al. 1983).

Abscisic acid (ABA) concentration increased during seed development and promoted the onset and maintenance of dormancy. Addition of ABA during embryo growth and development in caraway cells suppressed abnormal embryo morphology. Growth in a medium with 10^{-7} M ABA in darkness produced normal embryos which closely resembled their zygotic embryo counterparts (Ammirato 1974). Ammirato (1977) also observed the interaction of ABA, zeatin, and gibberellic acid during the development of somatic embryos from cultured cells of caraway, and demonstrated that the balance between ABA on the one hand, and zeatin and gibberellic acid on the other could effectively control somatic embryo development. Exogenous ABA was found to be inhibitory to the development of embryos from embryogenic cell clusters in carrot (Fujimura and Komamine 1975). The amount of endogenous ABA in cultured cells and globular stage somatic embryos of carrot remained low but concentrations increased during further development. However, the content of ABA increased continuously from the 7th to the 13th day, and then decreased in a culture containing 2,4-D (Kamada and Harada 1981). The ABA content in developing embryos reached maximum on the 10th day, then decreased in a culture lacking 2,4-D. Addition of ABA to the medium suppressed formation of abnormal embryos but also decreased the total number of somatic embryos. The effect of ABA was especially marked when applied during embryo development (Kamada and Harada 1981).

THE COMPETENCE

Carrot cells require exogenous auxin for undifferentiated proliferation. Instead of this requirement, the cell clusters fractionated by Halperin (1966) or Fujimura and Komamine (1979) did not cease growth but developed to somatic embryos in an auxin-free medium. Those cell clusters should have some particular ability to give rise to somatic embryos in a medium in which undifferentiated proliferation is not sustained. Although carrot suspension cells in culture are totipotent, not all can develop into somatic embryos in a certain condition. Here, we use the term competent cells for those which have the ability to differentiate including embryogenesis if they receive inducers of differentiation. Halperin (1969) also proposed a similar concept of competence.

EMBRYOGENESIS FROM SINGLE CELLS

A system was established in which single cells differentiated into embryos at a high frequency in carrot cell suspension cultures (Nomura and Komamine 1985). Cells in a carrot suspension culture were sieved through a Nylon screen with 15 μm aperture, then trapped on a 10 μm screen. Cells separated by size were fractionated by a discontinuous density gradient centrifugation using Percoll solutions, and small single cells were obtained in the fraction from 20 to 25% Percoll. Some of these single cells divided and formed small cell clusters in a medium containing auxin. About 30% of those cell clusters developed to embryos when transferred into an auxin-free medium. Further selection was done by manually picking up small spherical single cells (Fig. 10.2). Eighty-five to

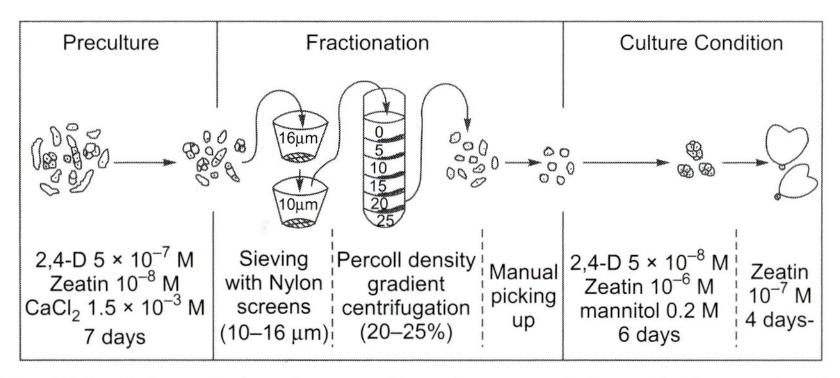

Fig. 10.2 Culture system for somatic embryogenesis from single carrot cells.

ninety percent of these small single cells differentiated into embryos when they were cultured in a medium containing 2,4-D (5×10^{-7} M), zeatin (10^{-6} M), and mannitol (0.2 M) for 7 days, followed by transfer to a medium containing zeatin (10^{-7} M) but no auxins (Fig. 10.3). Exogenous auxin was necessary to induce embryogenic cell clusters from single cells. The small spherical single cells could not differentiate into embryos nor divide in an auxin free medium. Thus, the single cells were totipotent but had not yet expressed the competence for embryogenesis in an auxin-free medium. Since cell clusters competent to embryogenesis may arise in several ways, single cells are likely not the sole origin of embryogenic cell clusters. For example, breakdown of a large cell cluster produced many small ones, but their competence was not predictable. Although the origin of those single cells is not specified, this work suggested that there was a phase of differentiation to embryogenic cell clusters from single cells before the development of embryos in an auxin-free medium. Thus, auxin, which inhibits development of embryos from cell clusters, is essential for differentiation to embryogenic cell clusters from single cells. Another system for the induction of somatic embryogenesis from single carrot cells was reported by Yasuda et al. (1998, 2000). They established an experimental system of embryogenesis of free cells isolated from hypocotyl segments of seedlings.

PHASES IN SOMATIC EMBRYOGENESIS

Morphological observation and physiological aspects show that early stages of somatic embryogenesis from single carrot cells can be classified into at least five phases:- 1, 0, 1, 2 and 3. We defined the state of the competent single cells (Nomura and Komamine 1985) as State 0. State 0 single cells are competent but have not yet expressed the potency to form

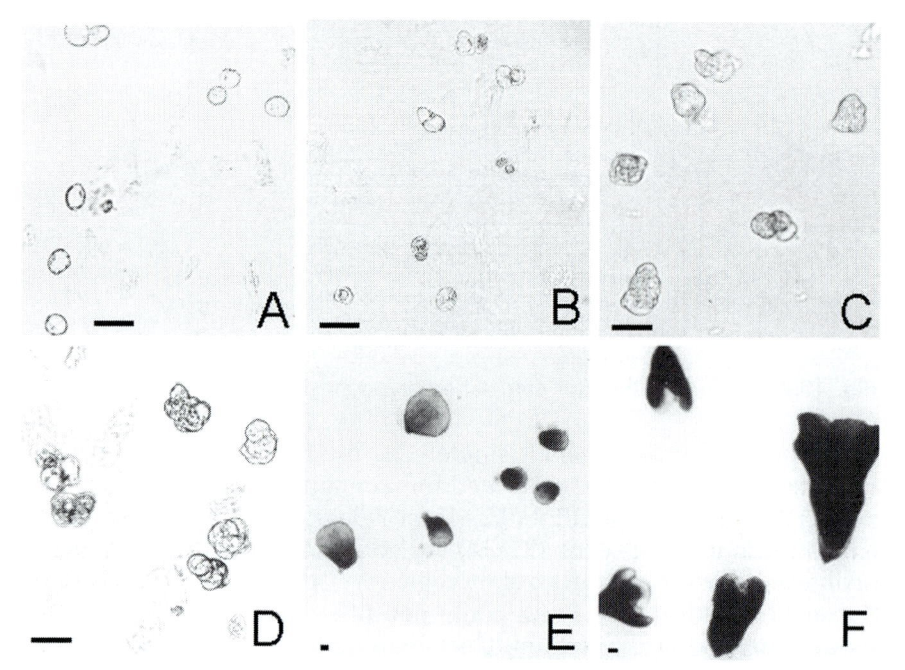

Fig. 10.3 Photographs of somatic embryogenesis from single cells. (A) Small single cells that had been picked up manually. (B) Cells on the 2nd day in medium that contained 5×10^{-8} M 2,4-D, 10^{-6} M zeatin, and 0.2 M mannitol showing the first division of single cells. (C) Culture on the 4th day. (D) Culture on the 7th day showing cell clusters. (E) Globular embryos formed by the 7th day after transfer of the cell clusters shown in D to embryo-inducing medium. (F) Torpedo-shaped embryos formed by the l4th day after transfer. Bars indicate 20 μm.

embryos. During Phase 0, competent single cells divide and gain the ability to form embryogenic cell clusters, State 1, in the presence of exogenous auxin. State 1 embryogenic cell clusters are ready to develop into globular embryos. The subsequent phase, Phase 1, is initiated by transfer of State 1 cell clusters to an auxin-free medium. If State 1 cell clusters are cultured in the presence of auxin, they do not give rise to somatic embryos but continue cell division to form larger cell clusters like a small callus. During Phase 1, cell clusters proliferate relatively slowly and apparently without any differentiation. After Phase 1, rapid cell division occurs in certain parts of cell clusters, leading to the formation of globular stage embryos. This phase is designated as Phase 2. In the next phase, Phase 3, globular stage embryos develop with relatively slow cell division into heart-shaped embryos, torpedo-shaped embryos, and then plantlets. When State 0 single cells are cultured in the absence of auxin, they do not proliferate and become elongated. In this process, designated as Phase-1, the totipotency of State 0 cells may be lost (see Fig. 10.4).

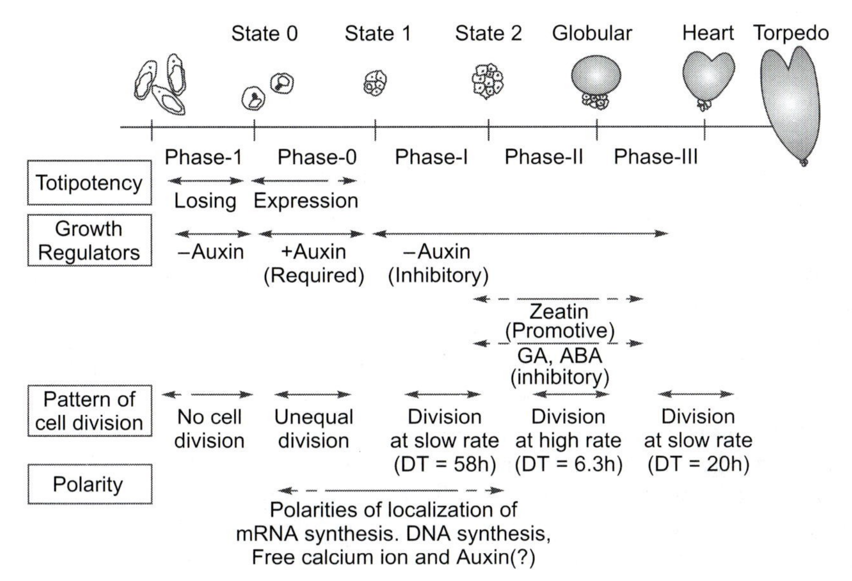

Fig. 10.4 Morphological and physiological aspects of somatic embryogenesis in carrot suspension cultures.

UNEQUAL CELL DIVISION

Unequal cell division of a zygote is observed in diverse species such as *Fucus*, a brown alga, and *Arabidopsis*. A small daughter cell and a large one are produced in the first division of a zygote; the small cell will subsequently give rise to a thallus in *Fucus* and a globular embryo in *Arabidopsis*, while the large cell will be rhizoid and the suspensor of an embryo respectively. The axis of the embryo is determined before the first division of the zygotes. Somatic embryogenesis from a single carrot cell also shows unequal cell division. Backs-Hüsemann and Reinert (1970) reported that an unequal cell division was necessary for somatic embryogenesis from a single cell in carrot suspension cultures. They observed that a relatively slender single cell divided into a small cell and a large cell. The small cell became the upper part of the somatic embryo leaving the large cell at the end of the suspensor part. Unequal cell division of a single cell was confirmed in a high frequency system of carrot suspension cultures by Nomura and Komamine (1986a). Spherical single cells (State 0 cells) divided on the 2nd day in a medium containing auxin. Even though small in size, they divided in an unequal manner to form a relatively small and a large cell. Observation by transmission electron microscopy revealed that the large cell retained its vacuole. On the 5th to the 7th day, the smaller cells began to divide, whereas the larger cells ceased further cell division. The cell clusters formed had a large cell in one part and small cells in the other part. Embryogenic cell clusters derived from single cells had morphological polarity within.

POLARITY OBSERVED IN THE DIFFERENTIATION OF SOMATIC EMBRYOS

Unequal cell division occurring during formation of an embryogenic cell cluster from a single cell is the initiation of expression of polarity in a multicellular tissue. Nomura and Komamine (1986a) showed a biochemical base of this polarity. Using an experimental system in which single cells differentiate to embryos at high frequency, [^{3}H]thymidine incorporation into cells and cell clusters during the initial stage of somatic embryogenesis was traced using autoradiography. Polarized incorporation of [^{3}H]thymidine was observed after the cell clusters passed the four-cell stage. Only the cell(s) at one position of cell clusters incorporated [^{3}H]thymidine in the nuclei. This polarity was preserved under conditions inducing embryogenesis when the cell clusters were transferred into a medium containing no auxin. On the other hand, the polarized incorporation of [^{3}H]thymidine was canceled when nonembryogenic growth was induced by transferring into a medium containing auxin. This indicated that the polarity of DNA synthesis activity may correlate with the competence for somatic embryogenesis.

Nomura and Komamine (1986b) detected another polarized event in somatic embryogenesis. Distribution of poly(A)$^+$RNA was polarized in one part of an embryogenic cell cluster. Prior to this observation, Raghavan (1981) reported uneven distribution of poly(A)$^+$RNA during embryogenesis of pollen in *Hyoscyamus niger*. These results suggested that cells in an embryogenic cell cluster are not equal in the expression of totipotency.

Calcium ions play an important role in the regulation of cellular events. Calcium ions mediate signal transduction in many cases. The role of Ca^{2+} ions and calmodulin in carrot somatic embryogenesis was examined by Overoorde and Grimes (1994). Formation of embryos was not affected until the concentration of Ca^{2+} ions was below 200 μM, and beyond this threshold the rate of embryo formation decreased with decreasing concentrations of Ca^{2+} ions. Treatment of developing embryos with Ca^{2+} channel blockers or a Ca^{2+} ionophore inhibited embryo formation. The results of Overoorde and Grimes (1994) indicated that exogenous Ca^{2+} ions or the maintenance of a gradient of Ca^{2+} ions was required for embryogenesis. These researchers also found that calmodulin-Ca^{2+} complexes were localized in regions that contained the developing meristem of both the cotyledon tips and rhizoid regions while the calmodulin protein appeared to be more uniformly distributed. Levels of mRNA for calmodulin increased slightly when cell clumps were induced to form embryos.

The distribution of cytosolic Ca^{2+} ions was traced using the fluorescent indicator fluo 3, antimonate precipitation, and proton-induced X-ray emission analysis. Embryogenesis was found to coincide with an increase in the level of free cytosolic Ca^{2+} ion. The highest level of Ca^{2+} ions was found in the protoderm of embryos from the late globular to the torpedo stage. The gradient of Ca^{2+} ions was observed along the longitudinal axis of the embryo. The nucleus gave the most conspicuous signals.

Signal transduction systems mediated by calmodulin have been investigated by Ooura and Nomura (unpubl. data). Levels of some calmodulin-binding proteins were found to decrease before the formation of globular embryos. Proteins appeared when developing embryos were transferred to a medium that contained 2,4-D. These results suggested that some signal transduction system(s) might play a role in embryogenesis and the proliferation of undifferentiated somatic cells.

Endogenous electrical currents across somatic embryos of wild carrot were first observed by Brawley et al. (1984). Development of a vibrating probe allowed them to measure currents around small cell clusters and embryos. In fully developed globular embryos, the efflux of ionic current was observed in the region near the suspensor, with an influx at the apical pole. Current also entered the exposed surfaces of early globular embryos that were developing from parts of large clusters of cells. In addition to the morphological asymmetry, this electrical current is the first detectable evidence of polarity in developing somatic embryos. A localized current can be observed at both ends of the embryo at subsequent stages of embryogenesis. At later stages, an inward current is found at the cotyledon and an outward current is found at the root in heart- and torpedo-shaped embryos and in plantlets. This current is reversibly inhibited by exogenous 1AA at 3×10^{-6} M.

Electrical polarity was also recognized in cell cultures during apparently unorganized proliferation in the presence of 2,4-D (Gorst et al. 1987). This electrical polarity is similar to that found in developing somatic embryos in an auxin-free medium. This observation suggested that embryogenesis was suppressed in embryogenic cell clusters that were proliferating in the presence of 2,4-D. Gorst et al. (1987) confirmed this suppression of embryogenesis by scanning electron microscopy. From their observations, they concluded that the potential for embryogenesis exists even in the presence of 2,4-D, but exogenous auxins inhibit its expression.

The inward current at the cotyledon is composed largely of K^+ ions, and the outward current at the radicle is mainly the result of the active extrusion of protons (Rathore et al. 1988). In the heart-shaped embryos, an inward current of $1.2 \pm 0.1\ \mu Acm^{-2}$ was detected at the cotyledon and an outward current of $1.0 \pm 0.1\ \mu Acm^{-2}$ was found at the radicle at pH 5.5. When the pH was raised to 5.75, the currents increased by 0.2 to 0.3 μAcm^{-2}. The sites of entry and exit of the current were more acidic than the rest of the medium. Removal of K+ ions from the medium reversibly reduced the currents to about 25% of their original value at both the cotyledon and radicle; removal of Cl^- ions decreased the currents slightly; and removal of Ca^{2+} ions resulted in a rapid doubling of currents. Addition of N, N′-dicyclohexylcarbodiimide, an inhibitor of plasma-membrane ATPase, or tetraethyl ammonium chloride, a K^+-channel blocking agent, substantially reduced the overall currents, and their removal resulted in partial recovery of the currents. These observations suggest that such currents are due mainly to K^+ and H^+ ions.

Changes in electrical and ionic currents might lead to alterations in the pattern of development. Application of a low-voltage electrical field enhanced development of somatic embryos from protoplasts of alfalfa (Dijak et al. 1986). An electrical field might alter the distribution of proteins or channels on the cellular membrane.

REMARKS ON MORPHOLOGY AND PHYSIOLOGY OF SOMATIC EMBRYOGENESIS FROM SINGLE CELLS

The key morphological and physiological events during somatic embryogenesis from single cells in carrot cell cultures are summarized in Figure 10.4. The state of cells and the phases in differentiation of somatic embryos are classified according to the observations and physiological investigations already described. Developmental stages of animal embryos have been defined according to the number of cells and morphology of embryos; on the other hand, no identities have been given in plant embryos before the globular stage. The concept of states and phases for the early stages of somatic embryogenesis makes it possible to clearly denote and precisely discuss the events of somatic embryogenesis.

Genes Involved in Somatic Embryogenesis

Many efforts have been made to identify molecular markers that are specific for somatic embryos (Sung and Okimoto 1981, 1983; Chibbar et al. 1989). Several molecular markers were also found in our carrot embryogenesis system described above. Two-dimensional gel electrophoresis showed the existence of three proteins, a, b and c, which could be detected throughout the process of expression of totipotency (Phases 0–3) but disappeared during the process of loss of totipotency. Two mRNAs, 1 and 2, also showed the same pattern of appearance as proteins a, b and c. In addition, protein d appeared during Phase 0. In Phases 1 and 2, more than 99% of polypeptides produced by translation *in vitro* gave the same pattern when mRNAs from embryogenic and nonembryogenic cultures were translated. Two mRNAs that encoded polypeptides appeared exclusively in embryogenic cultures, while two others appeared in nonembryogenic cultures (Nomura and Komamine, unpubl. data). These results indicate that only a few proteins may play important roles during embryogenesis and that changes in protein pattern are regulated at the transcriptional level. Recently, Zou et al. (2002) found a gene which promoted vegetative-to-embryonic transition in *Arabidopsis* using a chemical-inducible activation tagging system. This gene was identical to WUSCHEL, a homeodomain protein shown to specify stem cell fate in shoot and floral meristems. Genes involved in early phases of somatic embryogenesis may reveal the mechanisms of expression of totipotency.

Smith et al. (1988) described a monoclonal antibody designated 21D7 that reacted with a nuclear protein associated with cell division. We applied 21D7 to our system and, using Western blot and immunocytochemical methods, examined whether antigen 21D7 (21D7 protein) might be a candidate for a molecular marker of totipotency. The 21D7 protein was detected throughout the process of expression of totipotency, while it disappeared within 48 h during the process of losing totipotency, that is, when State 0 single cells were cultured in the absence of auxin. Furthermore, when State 0 single cells were microinjected with the 21D7 monoclonal antibody, they elongated and failed to divide or differentiate even when cultured in the presence of auxin. These results indicate that the expression of the 21D7 protein may be essential for the expression of totipotency. However, since expression of totipotency involves initiation of cell division, it seemed possible that 21D7 protein was involved in cell division. In fact, 21D7 protein was subsequently shown to be associated with cell division (Smith et al. 1993).

One of the most attractive approaches to the elucidation of mechanisms of somatic embryogenesis involves the isolation of genes expressed specifically during embryogenesis and characterization of their functions. Many attempts have been made to pursue these approaches (Choi et al. 1987; Borkird et al. 1988; Wilde et al. 1988; Dure III et al. 1989; Aleith and Richter 1990). Choi et al. (1987) isolated several cDNA clones from mRNA that were preferentially expressed during somatic embryogenesis in carrot by a combined immunoadsorption and epitope-selection method. Developmental regulation was analyzed in detail for two clones and the expression of the genes corresponding to them was found to be associated with the heart-stage embryos (Borkird et al. 1988). Sequence analysis showed that one of the clones encodes an analogue of LEA (late embryogenesis abundant) proteins (Dure III et al. 1989). Aleith and Richter (1990) described cDNA clones that had been isolated by differential screening from cDNA libraries established using mRNA extracted from somatic embryos of carrot. The expression of genes corresponding to some of the clones was roughly associated with the first morphogenetic or globular stage. Thus, many genes involved in embryogenesis have been isolated but the functions of some still remain unclear.

Organ specific genes

Lambda-gt11 cDNA libraries were constructed from poly(A)$^+$ RNA isolated from the hypocotyls and roots of carrot seedlings, and were differentially screened to isolate the hypocotyl- or root-specific cDNAs. Two cDNAs, CAR3 and CAR4, were isolated, which specifically expressed in hypocotyls, and two other cDNAs, CAR5 and CAR6, were found to specifically express in roots.

Expression of these genes was investigated during Phases 1-3 by Northern hybridization. The level of expression of the genes that corresponded to CAR4 and CAR5 increased after globular embryos were formed, and that of the gene that corresponded to CAR6 increased a little sooner (after heart-shaped embryos had formed), while CAR3 mRNA was expressed earlier (before globular embryos had formed) i. e. during Phase 2. Expression of these mRNAs was very limited in cells cultured in medium with 2,4-D, and it was strongly suppressed when 2,4-D was added to cultured heart-shaped embryos. *In situ* hybridization showed that CAR4 mRNA was expressed in the epidermis and in the regions around tracheary element in torpedo-shaped embryos. The predicted amino acid sequence of the protein encoded by CAR4 was rich in proline (N-terminal region) and leucine (C-terminal region) residues (Matsumoto and Komamine, unpubl. data). A characteristic repeated motif was found in the proline-rich region, resembling repeated sequences and in proline-rich cell wall proteins, such as p33 (carrot) (Chen and Varner 1985) or PRP (soybean) (Hong et al. 1987).

CEM1 gene

Isolation of genes specific for early stages of embryogenesis was attempted. Five cDNAs were cloned by differential screening between State 1 cell clusters, cultured in the absence of auxin for 5 days (preglobular embryos), and State 1 cell clusters cultured in the presence of auxin for 5 days. These clones were designated CEM1, 2, 3, 4 and 5. One of them, CEM1, was revealed by Northern hybridization to be expressed preferentially prior to and after the globular stage of embryogenesis. The nucleotide sequence of CEM1 was determined. The predicted amino acid sequence of the protein encoded by CEM1 was found to exhibit high homology to elongation factor (EF-1α) of eukaryotic cells. The extent of homology was 76.4% for human EF-1α, 76.8% *Xenopus*, 73.1 % yeast, 81.0% *Euglena*, and 94.2% for *Arabidopsis*. EF-1αis essential for elongation of the peptide chain during protein synthesis on the ribosome. The distribution of CEM1 mRNA was investigated during somatic embryogenesis in carrot cells by *in-situ* hybridization. Accumulation of the specific mRNA was observed in the spherical regions of the globular embryos and in the meristematic regions of heart- and torpedo-shaped embryos (Kawahara et al. 1994). It seemed that CEM1 mRNA was expressed in close association with cell division activity during embryogenesis, because active protein synthesis is required in actively dividing tissues.

21D7 gene

As described previously, 21D7 protein reported by Smith (Smith et al. 1988) is expressed from State 0 to 3, but is never expressed in Phase-1, that is, the process of losing the totipotency. To confirm whether or not 21D7 is associated with cell division, which is essentially involved in the expression of totipotency, a monoclonal antibody against 21D7 protein (mAB 21D7) was used to screen a carrot cDNA library. The cDNA clone isolated using this antibody was expressed in bacteria and the recombinant fusion protein was used to raise a new polyclonal antibody, FP13. Both mAB21D7 and FP13 recognized the same carrot protein, indicating that the cloned cDNA encodes the 21D7 protein. RNA blot analysis using carrot and *Catharanthus roseus* cDNAs showed that the levels of 21D7 mRNA correlated with the activity of cell proliferation and changed significantly during the cell cycle of synchronized *C. roseus* cell cultures (Smith et al. 1993). 21D7 protein is similar to the deduced sequence encoded by the *Saccharomyces cerevisiae* gene *SUN2* (Kawamura et al. 1996). In this study, the expression of plant 21D7 cDNA rescued the yeast *sun2* mutant. Fractionation of carrot and spinach crude extracts showed that the 21D7 protein sedimented with the active 26S proteasomes. Cessation of cell proliferation in carrot suspensions at the stationary phase caused 26S proteasome dissociation and correspondingly the 21D7 protein sedimented together with the free regulatory complexes of the 26S proteasomes. Large-scale purification of carrot 26S proteasomes resulted in co-isolation of the 21D7 protein. Polyacrylamide gel electrophoresis under nondenaturing conditions showed that the 21D7 protein had the same mobility as the 26S proteasome and that proteasome dissociation changed the mobility of the 21D7 protein accordingly. It is concluded that the 21D7 protein is a subunit of the plant 26S proteasome and that it probably belongs to a proteasome regulator. Because proteasome may play important roles in the progression of the cell cycle which is critically involved in somatic embryogenesis, 21D7 is an essential factor for progression of somatic embryogenesis.

Homeobox genes

Homeodomains (HDs) are DNA binding domains that have been well characterized in animals, and HD proteins are thought to be regulators of transcription. To investigate the regulation of gene expression during somatic embryogenesis in carrot, an attempt was made to isolate cDNA clones that encode HD proteins, and independent clones (CHB1 through CHB6) were isolated. Transcripts corresponding to CHB1 through CHB6 were expressed at different times during somatic embryogenesis. In particular, transcripts corresponding to CHB2 were expressed in close association with the early development of embryos, indicating that CHB2 may be an embryo specific gene. CHB4 and CHB5 were expressed after torpedo-shaped embryos, and CHB4 in hypocotyls and CHB5 in hypocotyls and roots, indicating CHB4 and CHB5 are involved in the differentiation of hypocotyls and/or roots during somatic embryogenesis (Kawahara et al. 1995).

CEM6 gene

To isolate a truly embryo specific gene, subtractive differential screening was performed. The cDNA library was constructed from proglobular embryos. For use as a probe in screening, the same cDNA used for library construction was enriched for specific sequences using subtractive hybridization. The cDNA used for subtraction was prepared from suspension cultures 5 days after subculturing in auxin-supplemented medium. Nine independent differentially expressed cDNA clones were obtained from a screen of 150,000 recombinant phages. Northern analyses indicated one of them, CEM6, to be expressed specifically during somatic embryogenesis, i.e. only from globular to torpedo-shaped embryos. The amino acid sequence deduced from the nucleotide sequence of the CEM6 cDNA indicates that it encodes a glycine-rich protein containing a hydrophobic signal-sequence-like domain. Its early embryo specific expression and sequence characteristics suggest that it has an important role of being a protein that surrounds the cell wall in embryogenesis (Sato et al. 1995). *In-situ* hybridization showed that transcripts of CEM6 were localized in the peripheral area of cells in globular to torpedo-shaped embryos, but signals of CEM6 began to disappear from root parts in late stage torpedo-shaped embryos and no signal could be observed in early seedlings. Antisense CEM6 was introduced to carrot cells in suspension cultures. Transformant cells were selected by kanamycin to obtain a stable transformant cell population, which was checked by GUS assay, PCR, and RT-PCR. From six transformant lines obtained, embryogenesis was induced according to Fujimura and Komamine's method (1979a). Frequency of embryogenesis was much lower (10 to 40%) than in control (<90%). Even if embryogenesis was induced, the first appearance of globular stage embryos was much delayed and most formed embryos were polyembryos. Because antisense cDNA cannot completely suppress the expression, these findings indicate that antisense CEM6 suppressed embryogenesis, suggesting that CEM6 plays an essential (or at least important) role in somatic embryogenesis (Asami et al. unpubl. data).

Summary of Gene Expression During Somatic Embryogenesis

Several genes expressed during somatic embryogenesis were isolated. They are classified into three categories as described below.

Genes involved in cell division: 21D7 and CEM1 belong to this category. In somatic embryogenesis, initiation and maintenance of cell division are essential. It was found that 21D7 encoded a regulatory subunit of 26S proteasome, functioning to regulate the progress of the cell cycle. Thus expression of 21D7 gene is essential for initiation of cell division and the expression of totipotency. CEM1 encodes Elongation Factor I, an essential protein for protein synthesis and required for active cell division prior to formation of globular embryos.

Genes ivolved in organ formation: In the late stages of embryogenesis, the torpedo stage in particular, formation of organs, hypocotyls, and roots is initiated. Thus, CAR3, CAR4, and CHB4 involved in hypocotyl formation, while CAR5, CAR6, and CHB5 involved in hypocotyl and/or root formation were expressed from globular and torpedo-stage embryos. Interestingly the expression of all these genes was suppressed by exogenously supplied auxin, which also inhibits embryogenesis.

Embryo specific genes: Two genes, CHB2 and CEM6, which were expressed specifically only in globular to torpedo-shaped stage of embryogenesis, were isolated. CHB2 is one of the homeobox genes. CEM6 encodes a glycine-rich protein containing a hydrophobic signal sequence, and may be a glycoprotein. CEM6 transcripts were localized in peripheral cells of torpedo-shaped embryos in particular. Introduction of antisense CEM6 suppressed somatic embryogenesis, suggesting that CEM6 plays an important role in somatic embryogenesis. Functions of these genes are under investigation.

Figure 10.5 summarizes the pattern of gene expression during somatic embryogenesis in carrot.

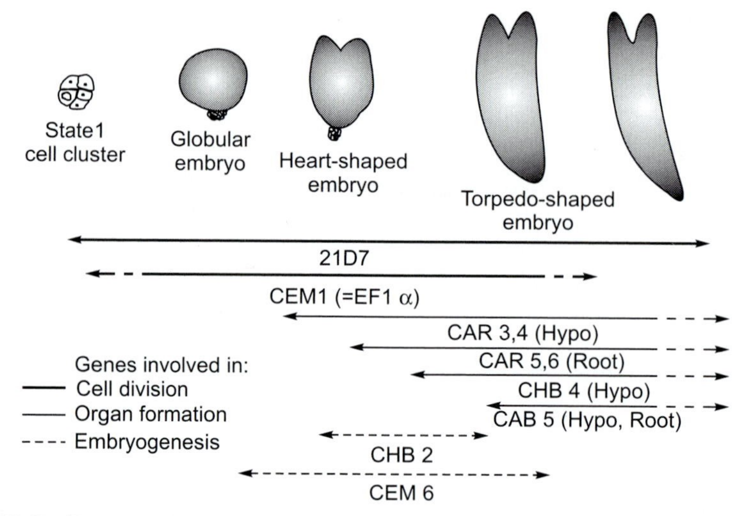

Fig. 10.5 Pattern of gene expression during somatic embryogenesis in carrot suspension cultures.

CONCLUDING REMARKS

Somatic embryogenesis demonstrate the totipotency of somatic cells in higher plants. These cells have been used as an alternative system of zygotic embryos for the investigation of plant embryogenesis. Advances in molecular biological methods have made it possible to apply molecular probes obtained from somatic embryos to zygotic embryos. This kind of comparative study reveals the whole process of expression of totipotency in both zygotic and somatic embryos. On the other hand, we find other values in somatic embryos: The practical or economical value in plant biotechnology and the biological value to reveal the whole life of plants.

At present, most plant scientists are conducting genetical analysis using *Arabidopsis thaliana* for elucidation of mechanisms of cell functions and phenomena. It is truly a more powerful tool than differential screening or an other method using culture systems. Using the *Arabidopsis thaliana* system, numerous genes involved in embryogenesis were isolated and their functions actively investigated. However, each gene is still responsible for one point in the whole process of embryogenesis in most cases. The next step of research in the whole process of embryogenesis is to elucidate networks of genes involved in embryogenesis and to reconstruct the assembly of all genes isolated. At that time the high-frequency and synchronous culture systems described above will again become useful systems for achieving the final goals of research on embryogenesis.

REFERENCES

Aleith F, Richter G. (1990) Gene expression during induction of somatic embryogenesis in carrot cell suspensions. Planta 183:17-24.

Ammirato PV (1974) The effects of abscisic acid on the development of somatic embryos from cells of caraway (*Carum carvi* L.). Bot. Gaz. 135:328-337.

Ammirato PV (1977) Hormonal control of somatic embryo development from cultured cells of caraway. Plant Physiol. 59:579-586.

Backs-Hüsemann D, Reinert J (1970) Embryobildung durch isolierte Einzelzellen aus Gewebekulturen von *Daucus carota*. Protoplasma 70:49-60.

Borkird C, Choi JH, Jin Z, Franz G, et al. (1988) Developmental regulation of embryonic genes in plant. Proc. Natl. Acad. Sci. USA 85:6399-6403.

Brawley SH, Wetherell DF, Robinson KR (1984) Electrical polarity in embryos of wild carrot precedes cotyledon differentiation. Proc. Natl. Acad. Sci. USA 81:6064-6067.

Chandra H, Lam TH, Street HE (1978) The effects of selected aryloxyalkanecarboxylic acids on the growth and embryogenesis of a suspension culture of carrot (*Daucus carota* L.). Z. Pflanzenphysiol. 86:55-60.

Chen J, Varner JE (1985) Isolation and characterization of cDNA clones for carrot extension and a proline-rich 33-kDa protein. Proc. Natl. Acad. Sci. USA 82:4399-4403.

Chibbar RN, Polowick PL, Newsted WJ, Shyluk J, Georges F (1989) Identification and isolation of a unique esterase from the mediun of non-embryogenic cell line of cultured carrot cells. Plant Cell, Tiss. Org. Cult. 18:47-53.

Choi JH, Liu LS, Borkird C, Sung ZR (1987) Isolation of cDNA clones for rare embryo specific antigens of cultured carrot cells. Proc. Natl. Acad. Sci. USA 84:1906-1910.

Dijak M, Smith DL, Wilson TJ, Brown D (1986) Stimulation of direct embryogenesis from mesophyll protoplasts of *Nechiego sativa*. Plant Cell Repts. 5:468-470.

Dure III L, Crouch M, Harada J, Ho TD, et al. (1989) Common sequence domains among the LEA proteins of higher plants. Pl. Molec. Biol. 12:475-489.

Fujimura T, Komamine A (1975) Effects of various growth regulators on the embryogenesis in a carrot cell suspension culture. Plant Sci. Lett. 5:359-364.

Fujimura T, Komamine A (1979a) Synchronization of somatic embryogenesis in a carrot cell suspension culture. Plant Physiol. 64:162-164.

Fujimura T, Komamine A (1979b) Involvment of endogenous auxin in somatic embryogenesis in a carrot cell suspension culture. Z. Pflanzenphysiol. 95:13-19.

Fujimura T, Komamine A (1980a) The serial observation of embryogenesis in a carrot cell suspension culture. New Phytol. 86:213-218.

Fujimura T, Komamine A (1982) Molecular aspects of somatic embryogenesis in a synchronous system. *In:* A Fujiwara (ed.). Plant Tissue Culture. Maruzen, Tokyo, p. 105.

Gautheret RJ (1937) Nouvelles recherches sur la culture du tissu cambial. C. R. Acad. Sci., Paris 205:572-574.

Gorst J, Overall RL, Wernicke W (1987) Ionic currents traversing cell clusters from carrot suspension cultures reveal perpetuation of morphogenetic potential as distinct from induction of embryogenesis. Cell Diff. 21:101-109.

Haberlandt G (1902) Culturversuche mit isolierten Pflanzenzellen. Sitsungsberchte der mathematixh-naturwissenschaften Classe der Kaiserlichen Akademie der Wissenshaften 111:69-95.

Halperin W (1964) Morphogenic studies with partially synchronized cultures of carrot embryos. Science 146:408-410.

Halperin W (1966a) Alternative morphologic events in cell suspensions. Amer. J. Bot. 53:443-453.

Halperin W (1969) Morphogenesis in cell cultures. Ann. Rev. Plant Physiol. 20:395-418.

Hong JG, Nagao RT, Key JL (1987) Characterization and sequence analysis of a developmentally regulated putative cell wall protein gene isolated from soybean. J. Biol. Chem. 262:8367-8376.

Kamada H, Harada H (1979) Studies on the ontogenesis in carrot tissue cultures I. Effects of growth regulators on somatic embryogenesis and root formation. Z. Pflanzenphysiol. 91:255-266.

Kamada H, Harada H (1981) Changes in the endogenous level and effects of abscisic acid during somatic embryogenesis of *Daucus carota* L. Plant Cell Physiol. 22:1423-1429.

Kawahara R, Komamine A, Fukuda H (1995) Isolation and characterization of homeobox-containing genes of carrot. Pl. Molec. Biol. 27:155-164.

Kawahara R, Sunabori S, Fukuda H, Komamine A (1992) A gene expressed preferentially in the globular stage of somatic embryogenesis encodes elongation-factor 1α in carrot. Eur. J. Biochem. 209:157-162.

Kawahara R, Sunabori S, Fukuda H, Komamine A (1994) Analysis by in situ hybridization of the expression of elongation factor 1α in the carrot cells during somatic embryogenesis. J. Pl. Res. 107:361-364.

Kawamura M, Kominani K, Takeuchi J, Toh-e A (1996) A multicopy suppressor of sin 1-1 of the yeast *Saccharomyces cerevisiae* in a counter part of the *Drosophila melamogaster* diphenol oxidase A2 gene, Dox-A2. Molec. Gen. Genet. 251:146-152.

Nomura K, Komamine A (1985) Identification and isolation of single cells that produce somatic embryos at a high frequency in a carrot suspension culture. Plant Physiol. 79:988-991.

Nomura K, Komamine A (1986a) Polarized DNA synthesis and cell division in cell clusters during somatic embryogenesis from single carrot cells. New Phytol. 104:25-32.

Nomura K, Komamine A (1986b) In situ hybridization on tissue sections. Pl. Tiss. Cult. Lett. 3:92-94.

Overoorde PJ, Grimes HD (1994) The role of calcium and calmodulin in carrot somatic embryogenesis. Plant Cell Physiol. 35:135-144.

Raghavan V (1981) Distribution of poly (A) - containing RNA during normal pollen development and during induced pollen embryogenesis in *Hyoscyamus niger* J. Cell Boil. 89: 593-606.

Rathore KS, Hodges TK, Robinson KR (1988) Ionic basis of currents in somatic embryos of *Daucus carota*. Planta 175:280-289.

Reinert J (1958) Untersuchngen uber die Morphogenese an Gewebekulturen. Ber. dt. bot. Ges. 71:15.

Sato S, Toya T, Kawahara R, Whittier RF, Fukuda H, Komamine A (1995) Isolation of a carrot gene expressed specifically during early-stage somatic embryogenesis. Pl. Molec. Biol.28:39-46.

Smith JA, Kraus MR, Borkird C, Sung ZR (1988) A nuclear protein associated with cell division in plants. Planta 174:462-472.

Smith MW, Ito M, Yamada T, Suzuki T, Komamine A (1993) Isolation and characterization of a cDNA clone for plant nuclear antigen 21D7 associated with cell division. Plant Physiol. 101:809-817.

Steward FC, Mapes MO, Mears K (1958) Growth and organized development of cultured cells II. Organization in cultures grown from freely suspended cells. Amer. J. Bot. 45:705-708.

Sung ZR, Okimoto R (1981) Embryogenic proteins in somatic embryos of carrot. Proc. Natl. Acad. Sci. USA 78:3683-3687.

Sung ZR, Okimoto R (1983) Coordinate gene expression during somatic embryogenesis in carrots. Proc. Natl. Acad. Sci. USA 80:2661-2665.

Takeno K, Koshiba M, Pharis RP, Rajasekaran K, Mullins MG (1983) Endogenous gibberellin-like substances in somatic embryos of grape (*Vitis vinifera* × *Vitis rupestris*) in relation of embryogenesis and chilling requirement for subsequent development of mature embryo. Plant Physiol. 73:803-808.

Tisserat B, Murashige T (1977) Effects of ethephon, ethylene, and 2,4-dichlorophenoxyacetic acid on asexual embryogenesis *in vitro*. Plant Physiol. 60:437-739.

Warren GS, Fowler MW(1977) A physical method for the separation of various stages in the embryogenesis of carrot cell culture. Plant Sci. Lett. 9:71-76.

White PR (1934) Potentially unlimited growth of excised tomato root tips in a liquid medium. Plant Physiol. 9:585-600.

White PR (1939) Potentially unlimited growth of plant callus in an artificial nutrient. Amer. J. Bot. 26:59-64.

Wilde DH, Nelson WS, Booij H, De Vries SC, Thomas TL (1988) Gene-expression program in embryogenic and non-embryogenic carrot culture. Planta 176:205-211.

Yasuda H , Nakajima M, Masuda H, Ohwada T (2000) Direct formation of heart-shaped embryos from differentiated single carrot cells in culture. Pl. Sci. 152:1-6.

Yasuda H, Satoh T, Masuda H (1998) Rapid and frequent somatic embryogenesis from single cells of regenerated carrot plantlets. Biosci. Biotechnol. Biochem. 62: 1273-1278.

Zuo JR, Niu QW, Frugis G, Chua NH (2002) The WUSCHEL gene promotes vegetative-to-embryonic transition in *Arabidopsis*. Plant Journal 30:349-359.

11

Zygotic Embryogenesis

V. Raghavan
Department of Plant Cellular and Molecular Biology, The Ohio State University, 318 West 12th Avenue, Columbus, OH 43210, USA
e-mail: raghavan.1@osu.edu

INTRODUCTION

Our present understanding of the complex processes involved in the formation of embryos in angiosperms unifies the first description of embryo development in certain eudicotyledons (dicots or eudicots) and monocotyledons (monocots) by Hanstein in 1870 with discoveries of syngamy in plants by Strasburgher in 1877 and of double fertilization independently by Nawaschin and Guignard in 1898 and 1899 respectively. The traditional setting for embryogenesis in angiosperms is the sanctum sanctorum of the female gametophyte—popularly known as the embryo sac—which itself is wrapped in several layers of cells of the nucellus and integuments constituting the ovule. Two groups of four haploid nuclei are initially embedded in a typical embryo sac, one at the micropylar end, the other at the opposite, chalazal end. The demarcation of groups of four nuclei, each nucleus surrounded by its own cytoplasm as a distinct compartmentalized, membrane-bound cell, is the primary determinant of form of the mature embryo sac. Three nuclei at the micropylar pole become organized as the egg apparatus, consisting of a large egg cell flanked on either side by a cellular synergid. Three nuclei at the opposite pole become the antipodal cells. The main body of the embryo sac remaining after the egg apparatus and antipodals are cut off is the central cell consisting of the two remaining nuclei which fuse prior to fertilization to form the diploid polar fusion nucleus. The mature embryo sac is thus a seven-celled, eight-nucleate supercell, in which fertilization and embryogenesis follow in quick succession. An important feature of angiosperm reproductive biology is the occurrence of double fertilization, that results in two developmentally different cell lineages. During double fertilization, one sperm fuses with the egg cell to produce the zygote which initiates embryogenesis; the other sperm fuses with the polar fusion nucleus to form the endosperm nucleus that gives rise to the nutritive tissue of the endosperm. Following these fusion events, the ovule containing the nascent embryo and endosperm developing inside the embryo sac is transformed into a seed.

This Chapter presents an overview of the establishment of the structural and functional body plan of embryos of representative eudicot and monocot species and the factors that control the growth and differentiation of embryos as studied by embryo culture techniques. The representatives have been chosen as models for comprehending in an uncomplicated manner the successive divisions of the zygote to form the embryo and in recognition of their relatively widespread use in experimental investigations. In keeping with the theme of this book, both old and new contributions which have laid the foundation for the current explosion of knowledge in the genetic and molecular aspects of embryogenesis are considered, but understandably are not exhaustive.

GLOBAL VIEW OF ZYGOTIC EMBRYOGENESIS

After years of focus on the division patterns of the zygote, other aspects of this cell, such as polarity and organelle disposition, have begun to unravel their secrets. Whereas the zygote possesses every structural and functional quality of a typical plant cell, it is also highly differentiated as a storehouse of developmental information in anticipation of a complex division program. A comparison of the ultrastructural profile of the zygote with that of the egg indicates that fertilization results in considerable metabolic turmoil. Although egg cells of angiosperms have a similar overall organization characterized by inherent polarity due to a tapered, basal micropylar end attached to the embryo sac wall and a broad, free, terminal chalazal end, polarization of the cytoplasm resulting in ultrastructural distinctions between the terminal and basal parts of the egg is a hallmark of many plants. For example, superimposed upon the predetermined polarity of egg cells of cotton (*Gossypium hirsutum*) (Jensen 1963, 1965), *Capsella bursa-pastoris* (Schulz and Jensen 1968), turnip (*Brassica campestris*) (Sumner and van Caeseele 1989), *Arabidopsis thaliana* (Mansfield et al. 1991), and *Pelargonium zonale* (Kuroiwa et al. 2002) is a large vacuole toward the micropylar end, with the cytoplasmic organelles including the nucleus displaced toward the chalazal end. The total amount of cytoplasm present in the egg cells of these plants is sparse and spread in a thin strip surrounding the vacuole except near the nucleus. Plastids, mitochondria, and dictyosomes are randomly and parsimoniously distributed in the egg cytoplasm of cotton, turnip, and *C. bursa-pastoris*. Although the egg cytoplasm of turnip and *A. thaliana* also has a low chloroplast count, a large number of undifferentiated starch-containing plastids surround the nucleus at the chalazal end, leading to the suggestion that the egg cells may serve as a sink for carbohydrates prior to fertilization. Strands of endoplasmic reticulum (ER) are relatively abundant in the egg of cotton, where they seem to partially surround the plastids, mitochondria, and dictyosomes. Occasional strands of ER also appear unique in having an internal network of tubes probably formed by the invagination of the inner membrane. By contrast, egg cells of turnip and *C. bursa-pastoris* have very little ER, which occurs in the form of short, randomly oriented strands. Eggs of both cotton and *C. bursa-pastoris* also contain liberal supplies of ribosomes that exist predominantly as monosomes. Stacks of cup-shaped giant mitochondria filled with large amounts of DNA flood the egg cytoplasm of *P. zonale*. Polarity is conferred in the egg of maize (*Zea mays*) by the presence of vacuoles of various sizes at the chalazal end, whereas the cytoplasm along with the nucleus is confined to the basal end. A close structural relationship of the egg to its maturation stage is underlined by the observation that whereas the immature egg is small and nonvacuolate, the mature egg is large with a proportionately conspicuous chalazal vacuole (van Lammeren 1986; Mól et al. 2000). From the functional point of view, the ultrastructural simplicity of the mature egg cells of the species considered here, in particular the comparative poverty of their cytoplasm, tends to suggest that the angiosperm egg is an inactive cell whose metabolism is on the decline. Since synergids are intimately bound with the egg in the egg apparatus, metabolic quiescence of the egg is often compensated by the presence of metabolically active synergids. A different situation is observed in the egg of *Plumbago zeylanica*, whose embryo sac lacks synergids. The major ultrastructural features of the egg cell are the elaboration of wall ingrowths at the micropylar end corresponding to the filiform apparatus normally found in synergids and the presence of a metabolically active cytoplasm with large numbers of relatively well-developed mitochondria and dictyosomes, and ER studded with polysomes (Cass and Karas 1974). Here the egg not only performs the function of a synergid, but relies in large measure on its own synthetic capacity as a gamete.

The chalazal end of egg cells of many angiosperms examined under the electron microscope has been found to be attenuated, a feature achieved in large measure by a decreasing amount of organized cell wall material. In most cases, this is manifest by the presence of wall material around the micropylar half of the cell, the chalazal half being covered by just the plasma membrane, or by the deposition of patches of wall material dotting the chalazal part of the egg, or by the wall disappearing from the chalazal part with maturity (Jensen 1965; Folsom and Peterson 1984; Yan et al. 1991). Possibly the naked or partially naked chalazal part of the egg facilitates the entry of the sperm for fertilization and the absorption of nutrients from the central cell. Whether signals from the stigma, style, or central cell at the time of pollination are involved in the differential accumulation of wall material on the egg is not known.

Initial responses of the egg to fertilization do not follow well-orchestrated patterns, but in various species investigated, they involved changes in size, laying down of wall material at the chalazal end, and overhauling of the cytoplasmic contents. Cotton provides a good model showing that before the zygote nucleus divides, there is a dramatic decrease in cell size to nearly half the volume of the egg. This is accompanied by a decrease in size of the vacuole soon after fertilization, presumably due to loss of water into the central cell, continuing even after the cell size ceases to decrease (Jensen 1963, 1968). In contrast, within a few hours after fertilization, the egg cell of *A. thaliana* executes a nearly threefold elongation along the apicobasal axis. The prime mover of this event is also the vacuole, whose reorganization involves, replacement of the large micropylar vacuole of the egg by numerous small vacuoles, which finally give way to a large vacuole filling most of the volume of the zygote (Mansfield and Briarty 1991; Jürgens and Mayer 1994). As shown diagrammatically in Fig. 11.1, zygote growth in *A. thaliana* coincides with gradual change in the configuration of the array of microtubules from a perpendicular to

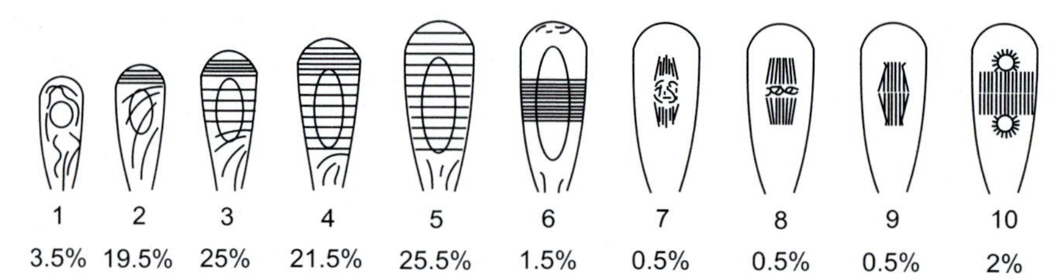

Fig. 11.1 Diagrams showing the changing patterns of distribution of microtubules during development and division of the zygote of *Arabidopsis thaliana*. Stages 1–2 show the distribution of microtubules in the young zygote. Stages 3-5, the distribution of microtubules during the elongation phase of the zygote; stage 5 represents a mature zygote. Stage 6, formation of the preprophase band preparatory to the first division of the cell. Stages 7-10, the distribution of microtubules during the first zygote mitosis. The frequency of each pattern observed in the 200 zygotes examined is shown as percentage (from Webb and Gunning 1991).

transverse cortical alignment, predominantly in a subapical band, during the elongation phase of the cell (Webb and Gunning 1991). This implies that cortical microtubules provide the force for elongation of the zygote and that this activity is largely restricted to the apical region of the cell. In some plants it has also been shown that laying down of wall materials at the chalazal end of the egg as an early post-fertilization event involves increased activity of the cortical microtubules as well as of dictyosome vesicles in the cytoplasm; interesting questions regarding the regulatory mechanism of wall formation and the biological role of the newly formed wall are raised by these observations (Jensen 1968; Schulz and Jensen 1968; Yan et al. 1991; Sumner 1992). Perhaps in the microcosm of the embryo sac a complete wall promotes subsequent divisions of the zygote insulated from the influence of the endosperm nucleus of a different genotype.

Other fertilization-related events contribute to an accentuation of the inherent polarity of the egg and to an increase in the metabolic state of the zygote. Implicated in the further development of polarity following fertilization of the egg in cotton are ER, ribosomes, plastids, and mitochondria which gather around the nucleus at the chalazal end. A consistent feature of this change is that the organelles complete their migration from the periphery of the cell to their new positions within 24 hours after fertilization (Jensen 1968). Underpinning zygote establishment, a complete overhaul of the cytoplasm occurs in *Papaver nudicaule* and *Zea mays*, resulting in the migration of the nucleus and cytoplasmic contents from their prefertilization micropylar locations to the chalazal pole (Olson and Cass 1981; van Lammeren 1986). Aggregation of ribosomes into polysomes, formation of new populations of ribosomes, increase in number of lipid bodies, mitochondria, and dictyosomes, fragmentation of mitochondria and decrease in their DNA content, change in ER from rough to smooth or tubular type, and increase in RNA and protein contents observed in zygotes of various plants are consistent with the scenario associated with increased metabolic activity of the zygote (Jensen 1968; Schulz and Jensen 1968; Cocucci and Jensen 1969; Mogensen 1972; Mansfield and Briarty 1991; Sumner 1992; Kuroiwa et al. 2002).

There is now compelling evidence that fertilization activates a cascade of changes in the egg. Formation of a highly polarized zygote not only ensures fidelity in its subsequent divisions in the embryogenic pathway, but also allows for the formation of phenotypically and functionally different cells. Despite the limited accessibility of the angiosperm egg to experimental manipulations, sorting out the molecular interactions that operate after fertilization is of importance in deciphering the genetic switches that are turned on and off to augur the changes just described.

Morphogenetic Pathways of the Zygote

Within the zygote lies the potential to form an entire plant, a feat which is accomplished by extensive changes in form in defined and drammatic ways and by the progressive change of an undifferentiated cell to a mass of differentiated cells. Common cellular processes involved in the transformation of the zygote into the embryo are cell division, expansion, maturation, and differentiation; these are terminated by the formation of meristems and embryonic organs. In a very young embryo, all cells divide faithfully to produce a new generation of daughter cells. However, during progressive embryogenesis, cell divisions are restricted to certain parts of the embryo predictable to some extent by their position in the cell lineage to produce specialized cells, tissues, and organs. Unfortunately, in the still-unfolding molecular biology of embryogenesis, not much is known about the mechanisms that restrict functional activities of cells in the developing embryo.

The zygotic genome is activated and poised to divide within a few hours after fertilization, but the division does not occur until after the endosperm nucleus has generated a syncytium of free nuclei. The first division of the zygote is almost invariably asymmetric and transverse to its long axis, cutting off a large, vacuolate, basal cell toward the micropylar end and a small, densely cytoplasmic, terminal cell toward the chalazal end. The embryo proper is derived from the terminal cell with varying degrees of contribution from the basal cell; however, a common scenario is one in which the basal cell wholly or partially forms a suspensor. The suspensor is believed to anchor the embryo to the embryo sac wall and facilitate nutrient absorption from the surrounding tissues. Obviously, the fates of the terminal and basal cells are reflected by the polarity of the zygote described earlier. The plane of the subsequent division of the terminal cell and the extent of contribution of the basal cell to the formation of the embryo proper have been tied together to provide the framework for a widely used classification of embryo development types in angiosperms; the basis of this classification has changed little since it was introduced more than 50 years ago (Johansen 1950; Maheshwari 1950). In this classification, embryo development types are separated into two major groups: in one group, the first division of the terminal cell of the two-celled proembryo is longitudinal; in the second group, the division is transverse. A small number of plants in which the first division of the zygote itself is oblique or vertical have been lumped into a third group. Within the first two groups, different embryo segmentation types are recognized and identified to customize the division sequence for specific families and designated by the name of the family in which many examples of the type are found or in which the type was first described.

These include the Crucifer (or Onagrad) and Asterad types in the first group, Chenopodiad, Solanad, and Caryophyllad types in the second group and Piperad type in the third group. However, some families show great diversity in their embryogeny and more than one type of embryo development represented by one or two genera in each case is not uncommon in them. A case in point is Fabaceae, in which as many as four (Asterad, Caryophyllad, Crucifer, and Solanad) of the six possible types and many varieties of the basic types of embryo development are known to exist in the subfamily Papilionoideae alone (Prakash 1987). Cell lineage studies in the widely investigated species such as *Capsella bursa-pastoris* and *Arabidopsis thaliana* have validated the essential basis of the Crucifer type of embryo development in these species, including the contribution of the terminal and basal cells of the two-celled proembryo to the formation of the mature embryo. However, lack of critical cell lineage studies in representatives of other types has infused the field of descriptive embryogenesis with unanswered questions about the precise contribution of the descendants of the basal cell in the genesis of the mature embryo. The basic features of this embryo classification are reviewed by Natesh and Rau (1984), to which reference is made for additional details. Embryogenesis in representative plants of each group is described in the books of Maheshwari (1950) and Johri, Ambegaokar, and Srivastava (1992).

Models of Embryogenesis in Eudicots and Monocots

The above classification encompasses embryos of both eudicots and monocots and indeed that embryos of both initially exhibit largely identical and orderly series of cell divisions is a generally accepted fact. Our understanding of the challenging problem of how the fertilized egg gives rise to a diverse array of organs and tissues constituting the embryo is based on the careful documentation of division sequences of the zygote in a broad spectrum of both eudicots and monocots. Although *Capsella bursa-pastoris* has served for many years as a text book example of embryogenesis in eudicots, recent elucidation of the precise embryo division patterns in the related *Arabidopsis thaliana* has had considerable symbolic significance in opening up powerful genetic and molecular approaches to the study of embryogenesis. As an introduction to embryogenesis in a representative eudicot, it is therefore appropriate to describe the cellular details during progressive divisions of the zygote of *A. thaliana*; there are common threads connecting the embryogenic division sequences in *C. bursa-pastoris* and *A. thaliana*, no doubt anticipated given their membership in the same family (Brassicaceae).

The first division of the zygote of *A. thaliana* is unequal and gives rise to a small terminal cell and a large basal one, constituting a two-celled proembryo. This division is marked by the concentration of microtubules in a discrete band girdling the nucleus as a preprophase band; the appearance of a preprophase band of microtubules marking the future cell plate bisecting the zygote is surprising since a similar alignment of these cytoskeletal elements is suppressed in the several divisions that occur during megasporogenesis and megagametogenesis (see Fig. 11.1; Webb and Gunning 1991). Mutations in *A. thaliana* designated as *tonneau* (*ton1* and *ton2*) which are unable to form the interphase and preprophase bands of microtubules in dividing cells predictably cause irregular cell expansion and inability to align the division planes in cells, yet the regenerated phenotypes produce tissues and organs in their correct spatial positions. This observation is of particular significance: by linking the mutant phenotype with abnormalities in the cytoskeleton, it negates the notion that genes affecting polarized cell expansion and division plane alignment are necessary for spatial positioning of tissues and organs during embryogenesis (Traas et al. 1995). Indeed, this message would not have been discernible or even imaginable without the experimental analysis of mutants.

A major developmental decision appears to be made during the first division of the zygote as descendants of the terminal cell become the organogenetic part of the embryo, whereas cells derived from the basal cell form the suspensor. The basic differences between these cells have been put into ultrastructural and cytological frameworks by descriptions of subtle variations in the distribution of cytoplasmic organelles and in the concentrations of macromolecules in them (Jensen 1963). To underscore the difference in cell fates at the molecular level, transcripts of the *ARABIDOPSIS THALIANA MERISTEM L1 LAYER* (*ATML1*) gene, encoding a transcription factor belonging to the homeodomain class of proteins, are found to accumulate in the terminal, but not in the basal cell born from division of the zygote of *A. thaliana* (Lu et al. 1996). Cytologically, the divergence in fates of the two cells of the proembryo is highlighted by a longitudinal division in the terminal cell and a transverse division in the basal cell. Another variable that contributes to the divergent cell fates is the unlimited divisions in the former and the limited number of cells generated in the latter. Much research will be required to determine the mechanism that underlies the predictability of cell fates of the two-celled proembryo, although it might involve polarization of the zygote alluded to earlier. The subsequent divisions of the basal cell to form the suspensor and of the terminal cell to form the embryo proper are now fairly well under-stood. The basal cell divides first and does so once or occasionally twice transversely. The cell closest to the terminal cell, designated as the suspensor cell, undergoes additional transverse divisions to form a filament of seven to nine cells connected to one another by end-wall plasmodesmata. These cells are terminated at the micropylar end of the embryo sac by the enlarged basal cell. The entire filamentous structure, including the basal cell is known as the suspensor. The basal cell and cells produced by the suspensor cell display specific ultrastructural features to facilitate the absorption and transport of solutes from the surrounding endosperm (Raghavan 2001). Investigations on some embryo-defective mutants in *A. thaliana* have led to a new level of understanding of the basis for suspensor and embryo cell identities. Two mutations designated as raspberry and suspensor (sus), stymie the growth of the embryo proper at the same time that the suspensor begins to grow into a many-celled structure in place of the normal one of a limited number of cells (Schwartz et al. 1994; Yadegari et al. 1994). An important theme emerging here is that products of raspberry and sus genes are required for the growth of the embryo and that the full developmental potential of the suspensor is not attained when the embryo is in an active mode of growth.

The terminal cell divides longitudinally when the zygote has produced three to four cells. An additional longitudinal division in each of the two daughter cells of the terminal cell (quadrant stage) followed by a transverse division produces an octant embryo, comprising an upper and lower tier of four cells each. At this stage, the basal cell may have produced four or five cells (Mansfield and Briarty 1991; Jürgens and Mayer 1994). Histological techniques combined with clonal analysis have shown that fates of the different cell groups are already fixed in the octant embryo, with the caveat that derivatives of more than one group may be integrated into specific organs of the mature embryo (Fig. 11.2). Thus, the upper tier of cells is destined to form exclusively the shoot apex and most of the cotyledons; the lower tier, in addition to providing derivatives to part of the cotyledons, generates the hypocotyl and radicle, including most of the root apical meristem; the central region of the root cap known as the columella and the remainder of the root apical meristem comprising the quiescent center are derived from the terminal cell of the suspensor closest to the embryo known as the hypophysis (Dolan et al. 1993; Scheres et al. 1994; Jürgens 2001). The apicobasal pattern of the future plant, to be built up by the reiterative action of the meristems or the stem-cell systems in the shoot and root apices, is thus established in the octant embryo. A series of tangential divisions, separating eight peripheral cells from a core of eight inner cells herald the next phase of development of the embryo. In the 16-celled embryo, the eight external cells form the protoderm or the precursor cells of the epidermis and the eight internal cells differentiate into the procambium and ground meristem (precursors of the vascular tissue and ground tissue of the cortex, respectively). These divisions initiate formation of the radial pattern elements made up of concentric tissue layers in the basal part of the embryo. Identification of certain gene markers

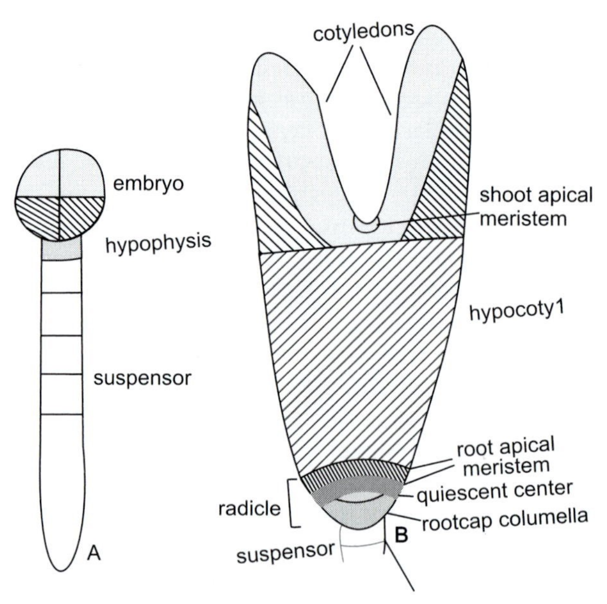

Fig. 11.2 Schematic diagrams showing cell fate determination in the octant-stage embryo leading to formation of the apico-basal axis in the torpedo-shaped embryo of *Arabidopsis thaliana.* A-delimitation of the upper (light shade) and lower (hatched) tiers of cells of the octant-stage embryo. The dark-shaded cell of the suspensor is the hypophysis. B-a torpedo-shaped embryo showing the formation of shoot apical meristem and part of the cotyledons from the upper tier of cells and the rest of the cotyledons, hypocotyl, and radicle, including most of the root apical meristem from the lower tier. The hypophysis contributes to formation of the quiescent center and the root cap columella. Part of the root apical meristem incorporating the quiescent center is also known to be derived from the hypophysis, whereas the bulk of the root apical meristem is derived from the lower tier of cells of the octant embryo. Clonal boundaries of cells derived from the octant-stage embryo and suspensor hypophysis are indicated by lines.

activated during pattern formation in developing embryos has served importantly to gain an understanding of the mechanism underlying the allocation of cell fates along the pattern-forming pathways. Evidence that the Kunitz trypsin inhibitor (*Kti*) mRNA is a marker of the root pole of early division stage embryos of soybean (*Glycine max*) has been obtained by *in situ* hybridization with transcripts of a cloned *Kti3* gene. The transcripts are found to persist in the axial cells of the ground meristem of developing soybean embryos to signal the progressive establishment of the apicobasal axis (Perez-Grau and Goldberg 1989). Apicobasal and radial pattern formation in *A. thaliana* embryos is reflected in the expression of transcripts of the *ARABIDOPSIS THALIANA LIPID TRANSFER*

PROTEIN (*AtLTP1*) gene, encoding a lipid transfer protein and the *ATML1* gene. That pattern formation is mediated in part by the position-specific expression of the *AtLTP1* gene is inferred from its initial expression in the protoderm of early division stage embryos and later expression in the cotyledons and upper end of the hypocotyl of later stage embryos (Vroemen et al. 1996). Although the *ATML1* mRNA is expressed uniformly in all cells of the early stage embryo, specificity of the gene as a marker for radial-pattern-forming elements is vividly seen in the disappearance of transcripts from the inner cells of the 16-celled embryo and restriction of transcripts to the protoderm cells (Lu et al. 1996). Probably, different combinations of marker genes mediate in region- and cell layer-specific interpretation of some basic positional information during pattern formation.

At the 16-celled stage of the embryo of *A. thaliana*, the hypophysis is formed by a transverse division of the uppermost suspensor cell. At the same time, another round of divisions in the derivatives of the terminal cell of the two-celled proembryo produces a globular embryo consisting of an epidermis and a central core, each of 16 cells. The first division of the hypophysis yields a small lens-shaped upper cell which abuts the lower end of the globular embryo and a large lower cell which contacts laterally with the embryo epidermis and at its basal end with the uppermost suspensor cell. The suspensor has now attained its genetically permissible number of cells and apparently, having fulfilled their function, these cells gradually begin to lose connection with one another and the embryo and disintegrate. The globular stage of the embryo is complete by approximately three additional rounds of divisions, mostly of the inner core of cells. The end of the globular stage also signifies a change from radial to bilateral symmetry of the embryo which initially flattens and attains a transient triangular stage. Emerging from the triangular stage, the embryo expands laterally by cell divisions to forecast the imminent formation of a pair of cotyledons and assumes the heart-shaped stage at the same time as the two hypophyseal cells divide twice vertically to form two layers of four cells each. It has been suggested that polar auxin transport may be involved in directing localized cell divisions in the globular embryo preparatory to the outgrowth of cotyledons. The basis for this suggestion is the observation that treatment of cultured globular embryos of *Brassica juncea* with auxin transport inhibitors leads to the formation of a ringlike structure around the shoot apex, akin to fused cotyledons, instead of two separate cotyledons. Since the effect is specific to the globular embryo, it appears that auxin signaling is an integral part of the mechanism that directs localized cell divisions in the embryo to form cotyledons (Liu, Xu, and Chua 1993b). In the context of embryogenesis in *A. thaliana*, it is noteworthy that a small percentage of embryos of the flower mutant, *pin-formed*, which has a defect in polar auxin transport, strikingly resembles embryos of *B. juncea* treated with auxin transport inhibitors (Okada et al. 1991).

Further growth of cotyledons and elongation of the embryo axis which occur during the heart-shaped stage are accompanied by the appearance of meristems from which the future seedling organs are derived. For example, the shoot apical meristem is organized in the depression between the cotyledons and appears as a mound of rapidly dividing cells. At the opposite end of the embryo, cells of the lower hypophyseal layer divide horizontally to produce four superimposed layers of four cells each (Fig. 11.3). By further divisions, these cells become the root cap columella; cells of the lateral root cap and root epidermis generated by accompanying periclinal divisions of cells adjacent to the hypophyseal cells contribute additional derivatives to the generation of the root apex from the lower tier of cells of the octant embryo and hypophysis. Formation of the root apex is complete by incorporation into the root apical meristem

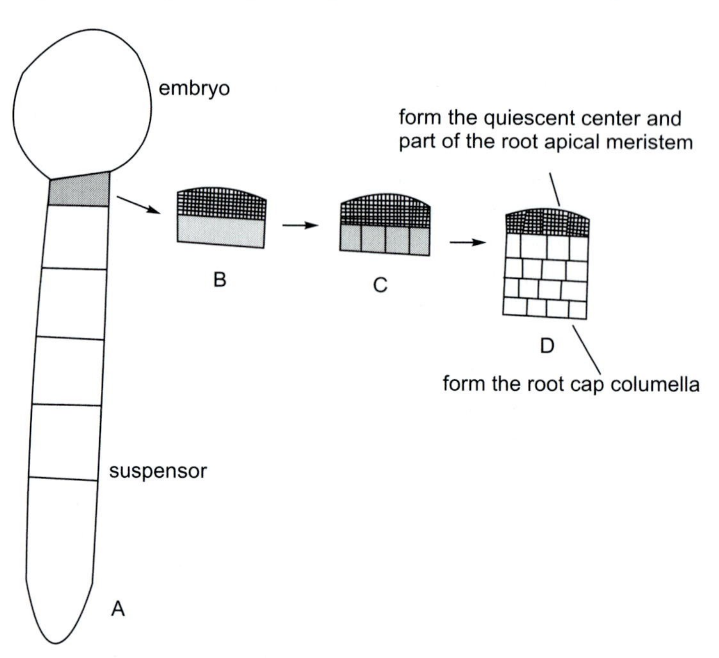

Fig. 11.3 Schematic diagrams showing the divisions of the hypophysis beginning at about the 16-celled stage of the embryo of *Arabidopsis thaliana*. A-embryo with the suspensor: hypophysis shaded. B-division of the hypophysis to form a lens-shaped upper cell and a lower cell. C-vertical divisions of both cells to form two layers of four cells each. The four upper cells form the quiescent center and a few root meristem initials. D-horizontal divisions of cells of the lower layer to form four superimposed layers of four cells each; the root cap columella is generated by these cells.

of the four upper hypophyseal layer of cells as the quiescent center. Cells of the quiescent center may also provide a few initials to the root apical meristem, most of which is formed by the activity of the lower tier of cells of the octant-stage embryo. The heart-shaped stage of the embryo is followed by the torpedo-shaped stage when further elongation of cotyledons and the hypocotyl as well as extension of the vascular tissues carved out from the inner core of cells, occur. Although the embryo continues to increase in size and to exhibit further changes in shape and organizational complexity as it goes through the bent-cotyledon and mature stages, the basic body plan of a shoot-root axis becomes unmistakably clear in the torpedo-shaped embryo. The striking morphological feature of the bent-cotyledon stage embryo is the curvature of cotyledons toward the hypocotyl; at the mature stage, because of space restrictions within the ovule, tips of cotyledons come to lie opposite the root pole. It has been estimated that a mature embryo of *A. thaliana* has about 15,000 to 20,000 cells and under favorable conditions of growth, about nine days from the time of fertilization are required to reach the mature embryo stage. The main tissues formed in the embryo are the protoderm, cortex, endodermis (the innermost layer of cells of the cortex), pericycle (the outermost layer of cells of the procambium), xylem, and phloem (formed from the procambium) (Mansfield and Briarty 1991; Jürgens and Mayer 1994). Most of the stages of embryogenesis just described are easily seen in sections of *A. thaliana* ovules of different ages (Fig. 11.4).

Based on the foregoing account, one can conceptually divide embryogenesis in *A. thaliana* into an early stage when all cells of the embryo engage in divisions to generate a population of new cells and a later phase when divisions are restricted to cells of certain embryonic regions to produce tissues and organs. The apicobasal and radial patterns of the embryo are established during the first phase; as reviewed elsewhere (Vroemen and de Vries 1999; Raghavan 2000a, b; Perez-Grau 2002), sensitive genetic screens have begun to unravel the molecular components of the pattern forming system and the morphogenetic control mechanisms during embryogenesis. In brief, characterization of mutants impaired in embryogenesis has suggested that patterning along the apicobasal and radial axes of the embryo is controlled by specific genes. These include genes such as *SHOOT-MERISTEMLESS* (*STM*) for the shoot apical meristem, *MONOPTEROS* (*MP*) for the hypocotyl and root apical meristem, *FACKEL* (*FK*) for the central region, *GURKE* (*GK*) for the apical region along the apicobasal axis, *KNOLLE* (*KN*) specifying the protoderm, and *SHORT ROOT* (*SHR*) specifying the endodermis along the radial axis. Some of these genes have been cloned, their protein products identified, and expression patterns characterized.

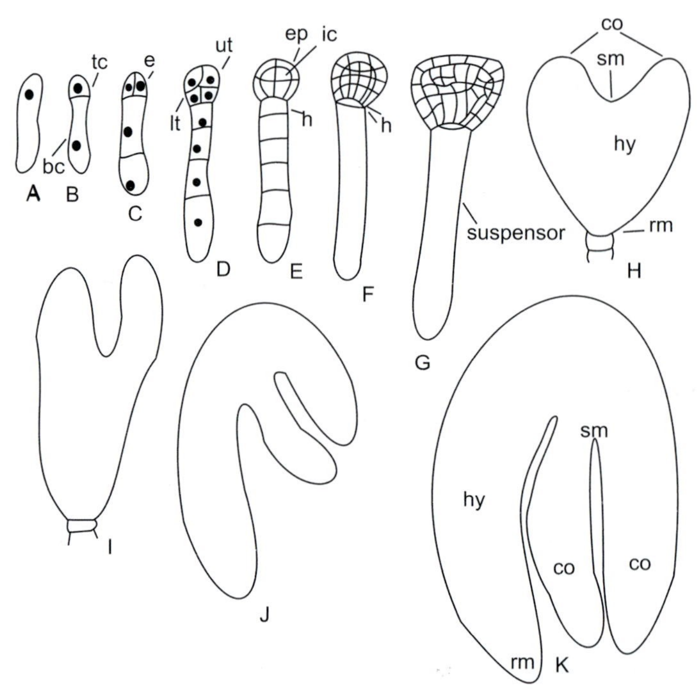

Fig. 11.4 Figures showing stages in development of the embryo of *Arabidopsis thaliana*. A-fully elongated zygote. B, first asymmetric division forming a small terminal cell (tc) and a large basal cell (bc). C-quadrant-stage embryo (e) attached to a file of two cells derived from the basal cell. D-octant-stage embryo attached to a file of four cells of the suspensor. The embryo proper now has eight cells constituting an upper tier (ut) and lower tier (lt) of four cells each. E-16-celled embryo consisting of 8 epidermal precursor cells (ep) and an inner core (ic) of 8 cells. The suspensor has a file of 6 cells including the basal cell; the uppermost cell of the file is the hypophysis (h). F-globular embryo of about 32 cells. The hypophysis has divided to produce two cells (h). The suspensor is fully developed. G-triangular stage embryo marking the transition from radial to bilateral symmetry. H-heart-shaped embryo showing the cotyledon primordia (co), hypocotyl (hy), incipient root apical (rm) and shoot apical (sm) meristems. I-torpedo-shaped embryo. J-bent-cotyledon shaped embryo. K-mature embryo. Figures based on cleared whole mount preparations and sections.

Because of the presence of a single cotyledon, embryos of monocots present a picture strikingly different from those of eudicots. There is now general agreement that the development of the embryo up to the octant stage is almost identical in monocots and eudicots and that in the former both the shoot apex and the cotyledon share a common origin from the terminal cell of a three-celled proembryo (Raghavan and Sharma 1995). However, in their ontogeny and mature structure, embryos of the large monocot family of grasses (Poaceae) do not have much in common with other monocots. This is exemplified by an account of embryogenesis in *Zea mays*. As in *A. thaliana*, an asymmetric division producing a small terminal cell and large basal cell is the hallmark of the zygote in *Z. mays*, but subsequent divisions are variable. One or two longitudinal divisions in the terminal cell are further followed by irregular divisions in the daughter cells as well as in the basal cell to produce a club-shaped embryosuspensor complex in about five days after pollination. Consistent with the polarity of the zygote, the upper part formed by descendants of the terminal cell generates the embryo proper, whereas descendants of the basal cell form the suspensor. Coincidentally, the embryo proper which is radially symmetrical at this stage contains small and dense cells compared to the large and vacuolated cells of the suspensor. Differentiation of the protoderm as a distinct layer of homogeneous cells, followed by the for-mation of new cells in the embryo part and elongation of the suspensor, moves the embryo into the transitional stage. This stage is considered to correspond to the globular stage of embryos of eudicots (Randolph 1936; Clark 1996; Elster et al. 2000). Based on localization of transcripts of the transcription-factor-encoding homeobox gene *ZEA MAYS OUTER CELL LAYER* (*ZmOCL1*) in the protoderm layer throughout maize embryogenesis, this gene has been identified with a commitment to epidermis specification. Since the gene is also expressed in the emerging shoot and root apical meristems, it might additionally be considered to function in the apicobasal pattern formation in the embryo (Ingram et al. 1999).

In about seven days after pollination, the maize embryo assumes bilateral symmetry and undergoes morphogenesis as it initiates organs and tissues of the adult plant. Although the bulk of the flattened embryo contributes to the formation of the scutellum (considered equivalent to the single cotyledon), differentiation of two groups of cells within the embryonic mass foreshadows the future shoot and root apices. A small elevation on the anterior face of the embryo, seen as early as the transitional stage, demarcates the shoot apex, whereas the root apex arises endogenously as a dark-staining region. The first sign of the coleoptile, the sheathing structure around the shoot apical meristem and the embryonic leaves, is formation of a bulging ring of cells on the face of the scutellum encircling the shoot apex (coleoptilar stage). This is followed by the appearance of the first leaf primordium on the surface of the shoot apical meristem. Considerable expansion and growth of embryonic organs occurs during this period and extends into the maturation period. It is during the latter period that the root apex becomes ensheathed by the coleorhiza which originates during the transitional stage by the division of cells in the lower part of the embryo axis. The final morphogenetic event before embryo maturation is formation of the mesocotyl as an internode lying between the nodes of the coleoptile and scutellum. The mature embryo is formed in about 45 days after pollination and reaches a length of 7 to 10 mm; it would have also generated five or six leaf primordia by this time (Fig. 11.5). These primordia are thus considered products of embryogenic events rather than of post-germinative episodes as in eudicots (Randolph 1936; Clark 1996; Elster et al. 2000). Along with the presence of organs such as the coleoptile, coleorhiza, mesocotyl, and epiblast (a flap on one side of the coleorhiza, absent in maize embryo, but present in embryos of many other members of Poaceae), the grass embryo represents one of those rare examples in which no counterparts to these organs are found in embryos of other angiosperms. However, these organs have functional similarities in the different genera and species, suggesting that mechanisms regulating their development have been evolutionarily conserved.

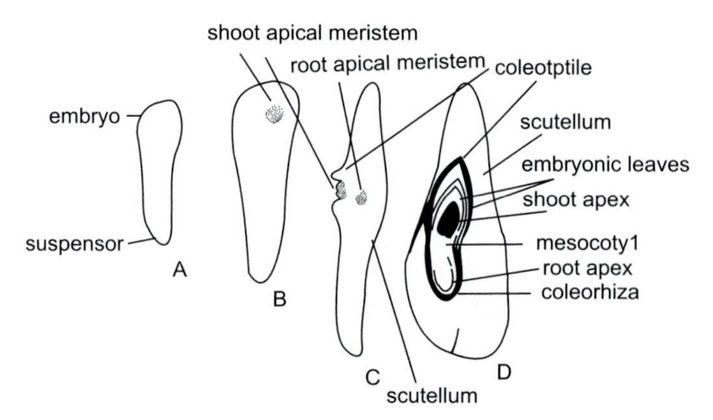

Fig. 11.5 Figures showing stages in the development of the embryo of *Zea mays*. A-club-shaped embryo-suspensor complex. B-transition stage embryo; shaded region is the shoot apical meristem. C-late coleoptilar stage showing the scutellum, coleoptile, and shoot apical meristem. D-section through a nearly mature embryo showing the individual parts. Figures based on whole mount preparations and hand sections.

ARE EMBRYONIC ORGANS AND TISSUES LINEAGE-RESTRICTED COMPARTMENTS?

The developmental patterns of embryos of most eudicots and many monocot species seem to involve an orderly pattern of transverse and longitudinal divisions to make it possible to assign by histological observations the ancestry of a tissue or organ of the mature embryo to a particular cell or groups of cells of the early division phase embryo. This view of cell lineage relationship inferred from the stereotyped segmentation of the zygote formed the basis of classification of embryo developmental types considered earlier and even led to the formulation of laws of embryogenesis to define each species based on the fundamental organization of its embryo. As an adjunct to conventional histological studies, clonal analysis has been used to construct fate maps of embryos of *A. thaliana* and maize; these investigations in which the distribution of genetically marked cells of early division phase embryos is followed in seedlings or adult plants have revealed that the acquisition of cell fate is less lineage-dependent than is predicted by histological analysis. In the strategy used for *A. thaliana*, excision of a maize transposable element *activator* (*ac*) from a transgenic marker made with a reporter gene construct [cauliflower mosaic virus (CaMV) 35S promoter-*GLUCURONIDASE* (*Gus*) gene] linked through the transposon was used to mark cell clones. As the reporter gene is expressed in sectors of the seedling constituted of cell progenies in which the *ac* excision occurs, sectors marking gene expression can be detected histologically by a characteristic blue stain. This method has beautifully confirmed predictions from cell lineage studies that the hypocotyl, an intermediate zone near the upper boundary of the root, the root, and the root meristem have their origin from the lower tier of cells of the octant embryo. This is not the entire story, however: strong evidence that there are no restricted lineages that result in the progressive allocation of cells to specify the root and hypocotyl came from the observation that the sector boundaries spanning these two adjacent organs of the seedling, unlike the cellular boundaries set up in the embryo, intrude into each other (Scheres et al. 1994). Analysis of *A. thaliana* shape mutant *fass* (*fs*) has also raised doubts about the notion of lineage-restricted cell compartments during embryogenesis. Embryos of this mutant, whose cells are disorganized and irregularly enlarged with misaligned division planes beginning with the first zygotic division, nonetheless produce misshapen seedlings with root, stem, and leaves in the correct places (Torres-Ruiz and Jürgens 1994). This is consistent with the thesis that seedling morphogenesis is not coupled to the production of lineage-derived cells.

Experiments in which fates of cells of maize proembryos were followed after X-irradiation of developing ears heterozygous for cell marker mutations that affect pigmentation of mature embryos and seedlings, also provide good evidence that the cell lineage of the embryo is variable and somewhat indeterminate. Using a maize stock which produces sectors in both the scutellum and seedling, it was found that marking cells after the first longitudinal division of the terminal cell of the two-celled proembryo by irradiation produced a few kernels in which the sectors occupied more than half of the scutellum, overlapping with the embryo axis and others in which the sectors occupied less than half of the scutellum. Obviously, this would not have been the case had the developmental potential within a lineage had been restricted before the first longitudinal division of the terminal cell; it also appears unlikely from the results that derivatives of this division contribute equally to the growth of the embryo (Poethig, et al. 1986).

While these studies have undoubtedly raised questions about the importance of lineage relationships in specifying embryonic organs, they do not show whether any far-reaching positional information is necessary for the allocation of prospective cell fates for organ formation. If long-range interactions are involved, they might be mediated by hormones such as auxins. Several investigations, summarized by Souter and Lindsey (2000) and Jürgens (2001), have assigned a major role to auxin in different aspects of embryogenesis, especially in providing positional cues for embryo pattern formation. It would be interesting to see whether cell fate determination also falls within the signaling network initiated by auxin during embryogenesis.

HOW ARE DEVELOPING EMBRYOS NOURISHED?

There is considerable indirect evidence to support the view that continued growth of the zygote through progressive embryogenesis depends upon an uninterrupted supply of nutrients from the milieu of the embryo sac, the endosperm, cells of the ovule surrounding the embryo sac, and from other parts of the mother plant. Although a mass of free nuclei or cells produced by the division of the primary endosperm nucleus is usually in place in the central cell of the embryo sac before the zygote begins to divide in the embryogenic pathway, there is scant evidence to support the dependency of the zygote on the initial products of division of the primary endosperm nucleus as a nutrient source. Rather, structural modifications such as haustorial outgrowths of the embryo sac and plasmodesmata and the legendary transfer-cell-type wall projections in the central cell described in numerous plants seem to implicate the female gametophyte in the large-scale absorption of nutrients from the neighboring cells. It is likely that these modifications of the female gametophyte are carried forward unchanged after fertilization and have some primary consequences for nutrition of the zygote (Raghavan 1997). Wall projections presumably involved in nutrient absorption have also been shown to originate from the inner wall of the embryo sac at the micropylar end close to the zygote in cotton (Schulz and Jensen 1977), soybean (Tilton et al. 1984; Chamberlin et al.1993), and *A. thaliana* (Mansfield and Briarty 1991). Consistent with the role of wall ingrowths in the absorption of nutrients, autoradiography of the fate of ^{14}C-labeled photosynthates in the ovule of soybean has shown that at the zygote stage the label is concentrated at the wall projections of the embryo sac and in the hypostase—a group of nucellar cells at the chalazal end of the embryo sac. These observations have led to the suggestion that the zygote obtains its nutrients through two major transport pathways, one from the outer integument to the base of the zygote and the central cell, and the other through the hypostase to the central cell (Chamberlin et al. 1993).

The suspensor has figured in a number of investigations related to the transfer of nutrients from the endosperm to the embryo. Most of these studies have focused on the plasma membrane-lined invaginations from the outer wall of suspensor cells that project into the endosperm or from the inner wall of suspensor cells whose surface area is considerably extended by formation of rows of wall projections. Ultrastructural modifications of suspensor cell walls, akin to transfer cells, have been conjured up to support the role of the suspensor in nutrient exchange between the endosperm and the growing embryo (Raghavan 2001). That the suspensor acts as a conduit for metabolites to the developing embryo has also become evident from studies showing that the administration of radioactively labeled sucrose or putrescine, a polyamine, to pods or isolated embryos of *Phaseolus coccineus* and *P. vulgaris* leads to the uptake, translocation, and accumulation of much of the radioactivity in the suspensor (Yeung 1980; Nagl 1990). That the embryo sac itself is actively involved in the transfer of cytoplasmic nutrients from the endosperm is supported by the presence of invaginations of the inner wall of the embryo interfacing with the endosperm and other sites of nutrient flux in ovules of several plants, including pea (*Pisum sativum*), *Stellaria media, Vigna sinensis, Medicago sativa,* sunflower (*Helianthus annuus*), cotton, *Haemanthus katherinae,* soybean, and rice (*Oryza sativa*) (Raghavan and Sharma 1995).

The role of the endosperm in embryo nutrition has been strengthened by two additional observations. First, chemical analysis has shown that the liquid endosperm of coconut (*Cocos nucifera*), widely used as a growth adjuvant in plant tissue culture media as coconut milk or coconut water, contains low concentrations of a variety of amino acids, vitamins, plant growth hormones, sugars, and sugar alcohols. Although data on the chemical composition of endosperms of other plants are scant, extracts or infusions made from the endosperm tissues of maize, horse chestnut (*Aesculus woerlitzensis*), and *Datura stramonium,* among others, have been used in media to promote growth of cultured plant organs and tissues (Raghavan 1976). Despite the lack of direct evidence of the utilization of the endosperm by the growing embryo, it is difficult to ignore a role for the nutritive and hormonal substances of the endosperm in fostering embryo growth *in vivo*. Second, in many interspecific and intergeneric crosses, embryo abortion is usually accompanied by simultaneous disintegration of the endosperm, although the extent to which disturbances in endosperm development in a nonviable cross make this tissue incapable of supporting the growth of the embryo remains uncertain (Raghavan 1977b). A comparative ultrastructural analysis of endosperm development in normal diploid maize and in an unsuccessful cross between the diploid and an autotetraploid line has shown that cells of the endosperm formed in the placentochalazal region of the hybrid grain lack wall ingrowths typical of the normal endosperm. Necrosis of the endosperm in the unsuccessful cross has been attributed to interference with the flow of nutrients from the endosperm to the embryo (Charlton et al. 1995). If hybrid embryos abort due to the inability of the endosperm to provide the exacting nutrients required for their continued growth, it should be possible to rescue embryos by culturing them in media supplied with nutrient substances normally present in the endosperm. In different wide hybrids, this assumption has been borne out by *in vitro* techniques such as embryo culture, embryo implantation, ovary and ovule culture, and by organogenesis and somatic embryogenesis (Raghavan 1985 1986a).

Morphological and anatomical studies of developing embryos of certain plants have provided evidence for the possible utilization of materials from cells of the ovule surrounding the embryo sac. For example, development of elaborate haustorial structures from the suspensor which come into contact with cells of the integument, nucellus, and placenta are common in plants of Rubiaceae, Fabaceae, Orchidaceae, and Trapaceae, whose seeds also lack a well-developed endosperm (Raghavan 1976). It is difficult to find structural evidence for the supply of nutrients from the vegetative parts of the plant to the growing embryo, although it is known that seeds that accumulate large quantities of storage reserves in the embryo or endosperm act as powerful sinks for metabolites from other parts of the plant. It is a matter of common observation that in developing seeds of pea and other legumes, most of the endosperm is consumed by the time storage protein synthesis is initiated in the embryonic cotyledons. However, storage protein synthesis coincides with a marked increase in the amount of vascular tissues in the ovule and funiculus, and in the phloem transfer cells of the latter, although it is uncertain whether nutrients transported through the vascular tissues from the vegetative parts of the plant contribute to the nutrition of the embryo or are converted into storage products (Hardham 1976). An investigation that traced the fate of labeled photosynthates in developing soybean seeds has suggested that materials transported through the vascular system into the micropylar, chalazal, and lateral poles of the embryo sac are used for the nurture of globular to heart-shaped embryos (Chamberlin et al. 1993).

In summary, it is clear that developing embryos employ different strategies involving whole plant physiology, structural modifications, and ultrastructure, for their nurture. Admittedly, much work remains to be done to demonstrate in a straightforward way the specific nutrient resources utilized by embryos of different ages.

IN VITRO TECHNIQUES TO STUDY NUTRITION, GROWTH, AND DIFFERENTIATION OF EMBRYOS

In further attempts to understand how the growing embryo is nourished and the nature of the nutrients required for its continued growth, considerable attention has been bestowed on the artificial culture of embryos and ovules. By growing embryos outside the environment of the ovule under aseptic conditions in media of known chemical composition, it is possible to identify the nutrient and hormonal requirements essential for their continued growth and differentiation, a feat which would be difficult to accomplish while the embryo is enclosed in the ovule or seed. Although the goal of monitoring progressive embryogenesis *in vitro* in a defined medium starting with the single-celled zygote has not been attained in a true sense, considerable information has accumulated on the culture of mature and differentiated embryos excised from seeds and on the changing nutritional requirements of progressively younger embryos excised from developing ovules.

Protocols for the culture of embryos involve isolating embryos from the ovule without injury, formulating a suitable nutrient medium, and inducing continued embryogenic growth and seedling formation. Since embryos are confined to the sterile environment of the ovule or seed, the latter are surface-sterilized and embryos removed aseptically. Seeds with hard seed coats are soaked in water for a few hours to a few days and then surface-sterilized before embryos are removed. Splitting open seeds and transferring embryos to the nutrient medium is perhaps the simplest technique that can be used with seeds, whereas relatively small embryos are best isolated from ovules under a dissecting microscope. The most important aspect of embryo culture is the selection of medium necessary to sustain continued growth of embryos. Included in the media used for embryo culture are the same elements found essential for plant growth in general, such as K, Ca, Mg, N, P, and S as major salts, and Fe, B, Mn, Cu, Zn, and Mo as trace elements. Vitamins are another group of compounds used in embryo culture media, although there is no hard evidence to justify their essentiality. Besides major salts, trace elements, and vitamins, it is the general experience that growth and survival of cultured embryos are enhanced by fortifying the medium with a carbon energy source such as sucrose. Compositions of commonly used embryo culture media are given in articles by Raghavan (1977a) and Williams, et al. (1987). Medium requirements for culture of embryos and physiological responses of cultured embryos to additives in the medium are found to depend upon their age at excision, the rule-of-thumb being that whereas seed embryos require a relatively simple medium, progres-sively younger embryos require more complex media.

Culture of Mature and Differentiated Embryos

With a few exceptions, the embryo enclosed within the seed is a fully differentiated organ, consisting of shoot and root apical meristems and one or two cotyledons. Since the cells of meristems are poised to undergo division, elongation, and differentiation, cultured seed embryos grow into plantlets when supplied with a limited diet consisting of inorganic salts and a carbon energy source such as sucrose. Plantlets are generally transferred to sterilized soil or vermiculite and grown to maturity in the greenhouse.

The large size of the embryo enclosed in seeds of many plants and the ease with which embryos can be isolated from seeds have contributed in a large measure to their use in metabolic and physiological studies, especially those relating to carbohydrate and nitrogen nutrition and the effects of plant hormones. It is the general experience that growth and survival of cultured embryos are markedly enhanced by supplementation of the medium with sucrose as a carbon energy source and rarely has any other carbohydrate been as successful. At the morphogenetic level, the effect of sucrose is manifest in the enhanced growth of the root and shoot primordia of cultured embryos (Raghavan 1980, 1993). With regard to the nitrogen nutrition of embryos, a series of investigations have led to three significant generalizations. One is that embryos are able to grow moderately well in a medium utilizing nitrates or ammonium salts as the sole source of nitrogen. The second is that amide glutamine is an efficient source of nitrogen for the growth of embryos of a number of plants (Paris et al. 1953; Rijven 1956; Matsubara 1964). Thirdly, mutual antagonism and synergism exist among different amino acids in the growth of embryos. This was first reported by Sanders and Burkholder (1948) who found that a mixture of 20 amino acids in the proportion in which they occur in casein hydrolyzate is as effective as the latter in promoting the growth of pre- and early heart-shaped embryos of *Datura stramonium* and *D. innoxia*. The reality of an interaction among amino acids became evident when it was found that the favorable effect of the complete mixture containing both beneficial and inhibitory amino acids on the growth of *D. stramonium* embryos is not reproduced by the beneficial compounds alone. Competitive interactions between individual amino acids also affected morphogenesis of embryos of *D. innoxia*. Here, embryos grown in a medium containing the beneficial amino acids alone were abnormal with long cotyledons, compared to those grown in the complete amino acid mixture or in casein hydrolyzate, which formed short cotyledons in proportion to the hypocotyl. In later investigations, negative interactions between individual amino acids and their alleviation by other amino acids in the biosynthetic pathways were noted in the growth of cultured embryos of oat (*Avena sativa*) (Harris 1956), barley (*Hordeum vulgare*) (Miflin 1969), wheat (*Triticum aestivum*) (Bright et al. 1978), and maize (Green and Donovan 1980), for possible selection of feedback-sensitive mutants.

The effects of the three major groups of plant hormones, namely, auxins, gibberellins, and cytokinins on the growth of seed embryos have been extensively studied. As reviewed elsewhere (Raghavan 1980, 1986b; Raghavan and Srivastava 1982), it can be stated as a secure generalization that the principal organs of cultured seed embryos respond to auxins, gibberellins, and cytokinins in a way quite similar to the corresponding organs of seedling plants. For example, culture of mature embryos of *Capsella bursa-pastoris* in a range of concentrations of indoleacetic acid (IAA) led to promotion of growth of the radicle at low concentrations (0.0001 to 0.001 mg L^{-1}), inhibition of growth of the root, hypocotyl, and shoot at intermediate concentrations (0.01 to 1.0 mg L^{-1}), and callus growth at high concentrations (2.0 to 10.0 mg L^{-1}). A range of concentrations of gibberellic acid (GA) was found to promote hypocotyl and root elongation, while kinetin generally suppressed root growth, but promoted leaf expansion and callus growth (Raghavan and Torrey 1964).

Nyman et al. (1986) found that seed embryos of taro (*Colocasia esculenta*) develop into plantlets only when cultured in the presence of the endosperm, including an intact aleurone layer. In a later work, they showed that naphthaleneacetic acid (NAA) and 6-dimethylaminopurine can substitute for the endosperm effect; this suggests a role for the endosperm in supplying hormonal substances for embryo growth (Nyman et al. 1987). Among hormonal effects on the morphogenesis of cultured embryos, perhaps the most important is the involvement of auxin in the production of callus. With the demonstration that the callus formed on embryos of the stem parasite *Cuscuta reflexa* grown in a medium containing casein hydrolyzate and IAA formed numerous adventive embryos (Maheshwari and Baldev 1961), the stage was set for the production of somatic embryos from various plants with the seed embryo as the starting material. Somatic embryogenesis is considered in a companion chapter (Chapter 9) in this book (Raghavan 2004).

Culture of Immature Embryos

In contrast to cultured mature embryos which develop into normal seedlings, cultured immature embryos skip the later stages of embryogenesis and the usual period of quiescence or dormancy and evolve into weak seedlings by a process known as precocious germination. This observation made during the early years of embryo culture investigations by Dieterich (1924) has been followed by other studies which showed that it is possible to control precocious germination by manipulation of the medium composition by provision of a high osmotic pressure (Ziebur et al. 1950) or the addition of abscisic acid (ABA) (Norstog 1972), and by changes in the environmental conditions of the culture such as provision for high light intensities, moderately high temperatures, and reduced oxygen tension in the culture milieu (Norstog and Klein 1972). The effect of ABA in curtailing precocious germination of embryos has been confirmed in many plants, raising the possibility that this hormone is a natural factor that suppresses precocious germination during normal embryogenesis (Fig. 11.6). At the biochemical level, Ihle and Dure (1970, 1972) showed that culture of immature embryos of cotton triggers precocious germination as well as the development of activity of certain proteases involved in the mobilization of food reserves of cotyledons by premature translation of their relevant mRNAs; addition of an aqueous extract of the ovule or ABA to the medium halts precocious germination of immature embryos and the development of enzyme activity in them. In cotton and other plants, the synthesis of a variety of proteins including late-embryogenesis abundant (LEA) proteins, storage proteins, and defense-related proteins

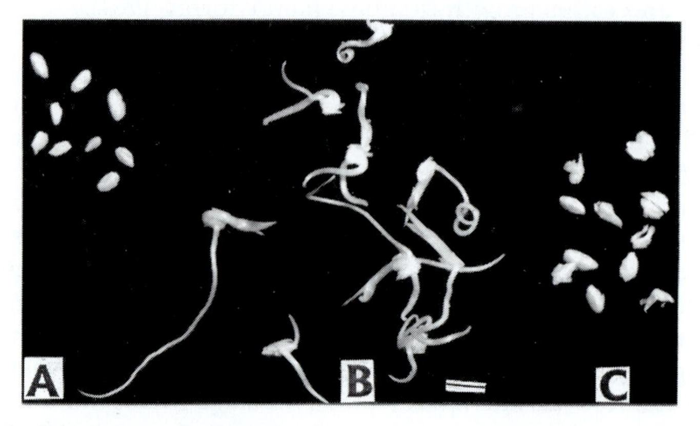

Fig. 11.6 Precocious germination of embryos of rice (*Oryza sativa*) and its reversal by ABA. A-embryos at the time of excision and culture. B-precocious germination in the basal medium. C-reversal of precocious germination in a medium containing 7.5 mg L^{-1} ABA. Scale bar = 5 mm.

is slowed down or inhibited in precociously germinating embryos, whereas in the presence of ABA in the medium which suppresses precocious germination, cultured embryos accumulate these proteins to the same extent as those growing *in situ* (Raghavan 1997).

Vivipary or germination of immature embryos in seeds within fruit still attached to the plant has been described as an infrequent occurrence in cultivated plants such as tomato (*Lycopersicon esculentum*) and maize and a natural way of life in mangroves, and trees and shrubs of estuarine habitats. Characteristically, embryos of viviparous plants skip the usual period of quiescence or dormancy typical of those of normal plants and display an uninterrupted transition from embryo development to germination. In maize, where the genetics of vivipary has been extensively studied, at least nine recessive genes are known to control vivipary. Genetic and physiological studies have shown that besides causing premature germination, mutations in some viviparous genes interfere with other facets of grain maturation (McCarty 1995). Since vivipary is analogous to precocious germination *in planta*, it is in some way related to a decreased ABA content of the seed or to a decreased ABA-sensitivity of the embryo. In embryos of viviparous mutants of maize, the genetic lesion apparently confers a greater resistance to growth inhibition by exogenous ABA than in embryos of the wild type (Robichaud et al. 1980). Involvement of a gene in the induction of vivipary in *Arabidopsis thaliana* was established by the isolation and characterization of ABA-deficient (*aba*) mutants whose seeds and fruits (siliques) have lower levels of endogenous ABA than those of the wild type. The relationship between a low level of ABA in the seed and vivipary was strengthened by the demonstration that whereas mature seeds of mutant lines germinate in siliques attached to plants held in an atmosphere of high humidity, with little or no evidence of arrested embryo growth, immature seeds germinate in siliques incubated on wet filter paper (Karssen et al. 1983). Three other genes implicated in embryo development, seed germination, and vivipary in *A. thaliana* are *ABSCISIC ACID INSENSITIVE* (*ABI*), *FUSCA* (*FUS*), and *LEAFY COTYLEDON* (*LEC*); this view is based on the observation that vivipary occurs in detached siliques of some recombinants between *lec/fus* and *aba/abi* stocks (Raz et al. 2001). A new twist has been added to the story by a recent demonstration that vivipary can be induced in siliques of different ages of *A. thaliana* by the simple expedient of culturing them in a mineral salt medium with no hormonal supplements. Implicating ABA in the induction of vivipary is the observation that addition of ABA to the medium inhibits vivipary in cultured siliques without affecting the growth of the embryo (Raghavan 2002).

In developmental terms, during precocious germination of embryos in culture, two control points of embryogenesis—the processes that lead to embryo maturation and a subsequent period of arrested embryo growth in the seed resulting in quiescence or dormancy—are replaced by an alternative program of germinative growth resulting in weak seedlings. In contrast, during germination manifest in vivipary, embryos skip the latter control point and evolve into robust seedlings.

Culture of Proembryos and Zygotes

It was pointed out in the previous section that the various structural modifications of the postfertilization embryo sac along with the presence of a nutrient-rich endosperm and a short-lived suspensor have a primary function in the nutrition of developing proembryos and even of the zygote. This has led to the suggestion, borne out by experiments described below, that the nutritional requirements for growth in culture of proembryos are more exacting than those found necessary for growth of immature and mature embryos. With different species, successful culture of proembryos has been possible by the use of nutrients of endospermic origin, modifications of the physical conditions of culture, application of hormonal and organic additives, and manipulations of the suspensor; with a few model systems, zygotes enclosed in cultured ovules, created by *in vitro* fertilization of single gametes, isolated after fertilization *in planta*, and harbored in the embryo sac, have also been reared into plants.

Use of nutritionally rich substances of endospermic origin, such as coconut milk, for the culture of proembryos arose from the work of van Overbeek, Conklin, and Blakeslee (1942). It was found that although it was possible to grow to plantlet stage torpedo-shaped and heart-shaped embryos of *Datura stramonium* in an inorganic nutrient medium enriched with a mixture of vitamins (nicotinic acid, thiamine, pyridoxine, ascorbic acid, and pantothenic acid) and assorted organic substances (adenine, glycine, and succinic acid), smaller embryos failed to grow in this medium or grew feebly before callusing. The clue to the recrudescence of growth in cultured proembryos lay in supplementation of the medium with non-autoclaved coconut milk (Fig. 11.7). In a later work, a hormonal factor, designated "embryo factor", which promoted growth of proembryos in a very low dilution was also isolated from coconut milk (van Overbeek et al. 1944). These studies spawned several successful attempts to culture proembryos of other plants by the use of plant extracts of endospermic or nonendospermic origin (Kent and Brink 1947; Matsubara 1962; Nakajima 1962), or by using the endosperm as a nurse tissue (Stefaniak 1987). Also noteworthy in this context is the successful attempt by Norstog and Smith (1963) to substitute requirement for coconut milk in the promotion of growth of undifferentiated barley embryos with a phosphate-enriched medium at pH 4.9 fortified with glutamine and alanine (400 mg L^{-1} each)

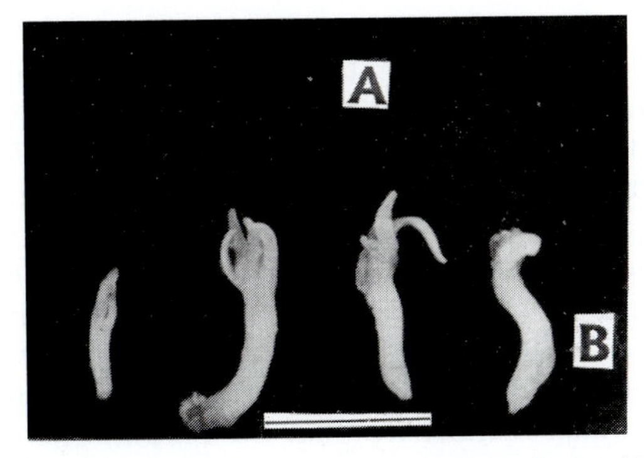

Fig. 11.7 Effect of coconut milk on growth of proembryos (140 μm diameter) of *Datura stramonium.* A-original size of the embryo. B-embryos of the same size cultured on an agar medium containing mineral salts, sucrose, vitamins, and nonautoclaved coconut milk. Scale bar = 1 mm (from van Overbeek, et al. 1942).

as major nitrogen sources, and lesser amounts of leucine (20 mg L^{-1}), and tyrosine, phenylalanine, cysteine, and tryptophan (all at 10 mg L^{-1}). Growth promotion of cultured proembryos by endosperm extracts thus appears to be mediated by specific chemical components.

It has been known for some time that the amorphous liquid endosperm in which proembryos are constantly bathed has a very low (more negative) osmotic potential that substantially decreases (becomes more negative) as the embryo matures (Ryczkowski 1960; Smith 1973; Yeung and Brown 1982). This has led to the view that the osmotic pressure of the medium might play a role in promoting growth of proembryos both *in vitro* and *in vivo*. Following the successful culture of proembryos of *D. stramonium* in a medium containing coconut milk, it was found that a mineral salt medium fortified with 8% to 12% sucrose or with 2% sucrose plus enough mannitol to be isotonic with 8% to 12% sucrose was optimal for continued growth and differentiation of globular and preheart-shaped embryos (Rietsema et al. 1953). Artificially increasing the osmotic concentration of the culture medium by the addition of high concentrations of sucrose or mannitol also led to the successful culture of proembryos of other plants which had previously not survived even in the most complex media tested; these include *Capsella bursa-pastoris* (Rijven 1952; Veen 1963), *Datura tatula* (Matsubara 1964), *Linum usitatissimum* (Pretová 1974), and *Triticum aestivum* (Fischer and Neuhaus 1995). From this it appears that sucrose functions more as an osmotic stabilizer in the medium than as a carbon energy source, although the latter role cannot be wholly discounted.

Two technical modifications of the culture system have eliminated the need to gradually reduce the osmolarity of the medium during growth of embryos without their sequential transfer from one medium to another. For the culture of embryos of *C. bursa-pastoris* as small as 50 µm in length this is achieved by using two media of different osmolarities solidified in juxtaposition in a petri dish. During the initial growth of embryos, the high osmolarity of the medium is gradually reduced by diffusion of water from the medium of low osmolarity (Monnier 1976). Continued growth and differentiation of 8- to 36-celled proembryos of *Brassica juncea* was obtained in a culture system composed of two agar layers, with the top layer having a higher osmolarity than the bottom one. Embryos were embedded in the top layer whose osmolarity decreased during culture (Liu et al. 1993a).

Promotion of growth of proembryos by hormonal additives in the medium is illustrated in a brief review of the requirements for growth in culture of progressively small embryos of *C. bursa-pastoris* (Raghavan 1965). Although heart-shaped and older embryos of *C. bursa-pastoris* have been routinely cultured in an inorganic liquid medium of high osmolarity secured by the addition of 12 to 18% sucrose (Rijven 1952; Veen 1963), later work opened up the feasibility of culturing heart-shaped embryos in an agar-solidified mineral salt medium supplemented with 2% sucrose. Growth of still smaller embryos (up to about 55 µm long) was secured by fortifying this medium with a balanced mixture of IAA, kinetin, and adenine sulfate (Raghavan and Torrey 1963). A requirement for kinetin in inducing growth of proembryos of *Linum usitatissimum* (Pretová 1986), for IAA and benzylaminopurine for growth of heart-shaped embryos of *Medicago scutellata* and *M. sativa* (Bauchan 1987), and for zeatin or benzylaminopurine for proembryos of maize (Matthys-Rochon et al. 1998) has also been reported.

A role for the suspensor in embryo nutrition implied from morphological and cytological studies has been strengthened by investigations in which the growth of the embryo severed of its connection with the suspensor was followed. From experiments using this approach, it was found that continued growth of proembryos of *Eruca sativa* (Corsi 1972), *Phaseolus coccineus* (Nagl 1974; Yeung and Sussex 1979), and *C. bursa-pastoris* (Monnier 1984) is more enhanced in the presence of an attached suspensor than in its absence. Indeed, the growth of proembryos of *P. coccineus* is promoted even by the presence of a detached suspensor kept in close proximity in the medium (Yeung and Sussex 1979). Other experiments on *P. coccineus* involving supplementation of the medium with growth hormones and determination of the growth hormone levels in the embryo and suspensor cells at specific stages of development have provided indirect evidence to show that the presumed suspensor function is due to the production of gibberellins and cytokinins (Alpi et al. 1975; Cionini et al. 1976; Lorenzi et al. 1978; Bennici and Cionini 1979). These results provide strong argument for the existence of hormonal gradients from the suspensor in regulating growth of the proembryo and might serve as baseline information for formulating a medium for culturing still smaller embryos and even the zygote of *P. coccineus*.

For many plants extracting proembryos from the confines of the ovule has remained a stumbling block in their successful culture. Now, some insights into the growth requirements of proembryos and even of zygotes of certain plants that hitherto defied attempts at excision and culture have been obtained by an alternative approach to ovule and ovary culture. Although growth in culture of isolated ovules of *Papaver somniferum* containing the zygote or two-celled proembryo in a mineral salt medium containing 5% sucrose was sporadic, growth of the nascent sporophyte was obtained when the medium was supplemented with casein hydrolyzate, yeast extract, or kinetin (Maheshwari 1958; Maheshwari and Lal 1961). Following this success, growth of the enclosed zygote or proembryo was induced in cultured ovules of *Zephyranthes* (Kapoor 1959), cotton (Stewart and Hsu 1977), and *Capsella bursa-pastoris* (Lagriffol and Monnier 1985), and ovaries of barley (Töpfer and Steinbiss 1985; Holm et al. 1995) and wheat (Zenkteler and Nitzsche 1985; Comeau et al. 1992). Ovules of cotton enclosing the zygote were successfully cultured to the mature embryo stage by supplementing a mineral salt medium with low concentrations of IAA, kinetin, GA, and 15 mM NH_4^+. Whereas the addition of hormones enabled the ovules to attain normal size, NH_4^+ promoted growth and differentiation of the zygote in the embryogenic pathway (Stewart and Hsu 1977).

Various strategies have been undertaken to induce growth *in vitro* of the zygote of maize. One, ovary culture (Schel and Kieft 1986). Another, is culture of zygote-containing embryo sac surrounded by the nucellus with or without a block of the endosperm (van Lammeren 1988; Campenot et al. 1992; Leduc et al. 1995). Initial culture of embryo sacs enclosing the zygote combined with a subsequent transfer of embryos to a medium of different composition proved also suitable for regenerating plants *in vitro* by division of the zygote (Mól et al. 1993, 1995). Using enzyme digestion and microdissection, zygotes isolated from embryo sacs have been induced to form fertile plants through stimulation of typical stages of *in vivo* embryogenesis. The unexpected finding was that a nurse tissue of embryogenic microspores of barley was necessary for inducing continued growth and division of explanted maize zygotes (Leduc et al. 1996). The same nurse tissue culture system was used to induce the growth of zygotes extruded from embryo sacs of barley (Holm et al. 1994) and wheat (Kumlehn et al. 1998) into normal plants.

Maize has also been used for the development of the first *in vitro* fertilization system and the regeneration of plants from the product of *in vitro* fertilization. The general approach followed in this research consists of isolation of viable embryo sacs by enzymatic maceration of pieces of ovular tissues. Sperm cells are isolated from osmotically shocked pollen grains. The gametes are selected by micropipettes and fertilization is accomplished on a cover slip containing a microdrop of a fusion medium (0.55 M mannitol) by applying an electric pulse, or chemically in a medium containing a high concentration of $CaCl_2$. The product of fusion is a zygote that promptly goes through embryogenic divisions (Kranz and Lörz 1993; Faure et al. 1994). Electrofusion between isolated male and female gametes of wheat resulting in multicellular structures during subsequent growth has also been reported (Kovács et al. 1995).

In vitro methods for the study of the nutrition, growth, and differentiation of embryos have led to a better understanding of the physiology of seed development in angiosperms. Compared with some other *in vitro* methods, embryo culture is a time-consuming operation involving a great deal of manipulative skills. Nonetheless, in various agricultural stations around the world, embryo culture has become the method of choice to obtain hybrids from wide crosses. Although methods for the culture of zygotes and proembryos for a wide range of plants have yet to be formulated, availability of this information will permit conclusions to be drawn about the biosynthetic pathways activated during progressive embryogenesis in angiosperms. The potential of zygote culture to transfer foreign genes in a single-step method affords hope for significant advances in the genetic engineering of plants.

CONCLUDING COMMENTS

The first cell of the angiosperm sporophyte is the zygote which undergoes repeated divisions, growth, and differentiation to form the embryo enclosed in the ovule. Embryogenesis culminates in generation of the seed as the unit of dispersal of the plant. During the past quarter century, research in plant embryogenesis has been defined by ongoing advances in molecular biology, recombinant DNA technology, and gene cloning. These studies have been undertaken with model systems like *Arabidopsis thaliana* and maize and have afforded insight into the genetic and molecular mechanisms involved in the development of the embryo from its single-celled progenitor. Since these later investigations have not been reviewed here, this Chapter constitutes only an overview of the structural aspects of development of the embryo from the zygote, intended solely to provide an appreciation of the discoveries that laid the foundation for the ever-ongoing research into genetic and molecular aspects of embryogenesis in flowering plants.

REFERENCES

Alpi A, Tognoni F, D'Amato F (1975) Growth regulator levels in embryo and suspensor of *Phaseolus coccineus* at two stages of development. Planta 127:153–162.

Bauchan GR (1987) Embryo culture of *Medicago scutellata* and *M. sativa*. Plant Cell, Tiss. Org. Cult. 10:21–29.

Bennici A, Cionini PG (1979) Cytokinins and *in vitro* development of *Phaseolus coccineus* embryos. Planta 147:27–29.

Bright SWJ, Wood EA, Miflin BJ (1978) The effect of aspartate-derived amino acids (lysine, threonine, methionine) on the growth of excised embryos of wheat and barley. Planta 139:113–117.

Campenot MK, Zhang G, Cutler AJ, Cass DD (1992) *Zea mays* embryo sacs in culture. I. Plant regeneration from 1 day after pollination embryos. Amer. J. Bot. 79:1368–1373.

Cass DD, Karas I (1974) Ultrastructural organization of the egg of *Plumbago zeylanica*. Protoplasma 81:49–62.

Chamberlin MA, Horner HT, Palmer RG (1993) Nutrition of ovule, embryo sac, and young embryo in soybean: an anatomical and autoradiographic study. Can J. Bot. 71:1153–1168.

Charlton WL, Keen CL, Merriman C, Lynch P, Greenland AJ, Dickinson HG (1995) Endosperm development in *Zea mays*; implications of gametic imprinting and paternal excess in regulation of transfer layer development. Development 121:3089–3097.

Cionini PG, Bennici A, Alpi A, D'Amato F (1976) Suspensor, gibberellin and *in vitro* development of *Phaseolus coccineus* embryos. Planta 131:115–117.

Clark JK (1996) Maize embryogenesis mutants. *In:* TL Wang, A Cuming, (eds.). *Embryogenesis. The Generation of a Plant.* Bios Scientific Publ., London, pp. 89–112.

Cocucci A, Jensen WA (1969) Orchid embryology: megagametophyte of *Epidendrum scutella* following fertilization. Amer. J. Bot. 56:629–640.

Comeau A, Nadeau P, Plourde A, Simard R, et al. (1992) Media for the *in ovulo* culture of proembryos of wheat and wheat-derived interspecific hybrids or haploids. Plant Sci. 81:117–125.

Corsi G (1972) The suspensor *Eruca sativa* Miller (Cruciferae) during embryogenesis *in vitro*. Giorn. Bot. Ital. 106:41–54.

Dieterich K (1924) Über Kultur von Embryonen ausserhalb des Samens. Flora 117:379–417.

Dolan L, Janmaat K, Willemsen V, Linstead P, Poethig S, Roberts K, Scheres B (1993) Cellular organization of the *Arabidopsis thaliana* root. Development 119:71–84.

Elster R, Bommert P, Sheridan WF, Werr W (2000) Analysis of four embryo-specific mutants in *Zea mays* reveals that incomplete radial organization of the proembryo interferes with subsequent development. Dev. Genes Evol. 210:300–310.

Faure JE, Digonnet C, Dumas C (1994) An *in vitro* system for adhesion and fusion of maize gametes. Science 263:1598–1600.

Fischer C, Neuhaus G (1995) *In vitro* development of globular zygotic wheat embryos. Plant Cell Repts. 15:186–191.

Folsom MW, Peterson CM (1984) Ultrastructural aspects of the mature embryo sac of soybean, *Glycine max* (L.) Merr. Bot. Gaz. 145:1–10.

Green CE, Donovan CM (1980) Effect of aspartate-derived amino acids and aminoethyl cysteine on growth of excised mature embryos of maize. Crop Sci. 20:358–362.

Hardham AR (1976) Structural aspects of the pathways of nutrient flow to the developing embryo and cotyledons of *Pisum sativum* L. Aust. J. Bot. 24:711–721.

Harris GP (1956) Amino acids as sources of nitrogen for the growth of isolated oat embryos. New Phytol. 55:253–268.

Holm PB, Knudsen S, Mouritzen P, Negri D, Olsen FL, Roué C (1994) Regeneration of fertile barley plants from mechanically isolated protoplasts of the fertilized egg cell. Plant Cell 6:531–543.

Holm PB, Knudsen S, Mouritzen P, Negri D, Olsen FL, Roué C (1995) Regeneration of the barley zygote in ovule culture. Sex. Plant Reprod. 8:49–59.

Ihle JN, Dure III LS (1970) Hormonal regulation of translation inhibition requiring RNA synthesis. Biochem. Biophys. Res. Commun. 38:995–1001.

Ihle JN, Dure III LS (1972) The developmental biochemistry of cottonseed embryogenesis and germination. III. Regulation of the biosynthesis of enzymes utilized in germination. *J. Biol. Chem.* 247:5048–5055.

Ingram GC, Magnard JL, Vergne P, Dumas C, Rogowsky PM (1999) *ZmOCL1*, an HDGL2 family homeobox gene, is expressed in the outer cell layer throughout maize embryogenesis. Plant Molec. Biol. 40:343–354.

Jensen WA (1963) Cell development during plant embryogenesis. *In:* Meristems and Differentiation. Brookhaven Symp. Biol. 16:179–202.

Jensen WA (1965) The ultrastructure and composition of the egg and central cell of cotton. Amer. J. Bot. 52:781–797.

Jensen WA (1968) Cotton embryogenesis: the zygote. Planta 79:346–366.

Johansen DA (1950) *Plant Embryology. Embryogeny of the Spermatophyta.* Chronica Botanica Co., Waltham, MA.

Johri BM, Ambegaokar KB, Srivastava PS (1992) *Comparative Embryology of Angiosperms,* Vols. 1, 2. Springer-Verlag, Heidelberg.

Jürgens G (2001) Apico-basal pattern formation in *Arabidopsis* embryogenesis. *EMBO J.* 20:3609–3616.

Jürgens G, Mayer U (1994) *Arabidopsis. In:* J Bard, (ed.). Embryos. Color Atlas of Development. Wolfe Publ., London, pp. 7–21.

Kapoor M (1959) Influence of growth substances on the ovules of *Zephyranthes. Phytomorphology.* 9:313–315.

Karssen CM, Brinkhorst-van der Swan DLC, Breekland AE, Koornneef M (1983) Induction of dormancy during seed development by endogenous abscisic acid: studies on abscisic acid deficient genotypes of *Arabidopsis thaliana* (L.) Heynh. Planta 157:158–165.

Kent NF, Brink RA (1947) Growth *in vitro* of immature *Hordeum* embryos. Science 106:547–548.

Kovács M, Barnabás B, Kranz E (1995) Electrofused isolated wheat (*Triticum aestivum* L.) gametes develop into multicellular structures. Plant Cell Repts. 15:178–180.

Kranz E, Lörz H (1993) *In vitro* fertilization with isolated, single gametes results in zygotic embryogenesis and fertile maize plants. Plant Cell 5:739–746.

Kumlehn J, Lörz H, Kranz E (1998) Differentiation of isolated wheat zygotes into embryos and normal plants. Planta 205:327–333.

Kuroiwa H, Nishimura Y, Higashiyama T, Kuroiwa T (2002) *Pelargonium* embryogenesis: cytological investigations of organelles in early embryogenesis from the egg to the two-celled embryo. Sex. Plant Reprod. 15:1–12.

Lagriffol J, Monnier M (1985) Effects of endosperm and placenta on development of *Capsella* embryos in ovules cultivated *in vitro.* J. Plant Physiol. 118:127–137.

Leduc N, Matthys-Rochon E, Dumas C (1995) Deleterious effect of minimal enzymatic treatments on the development of isolated maize embryo sacs in culture. Sex. Plant Reprod. 8:313–317.

Leduc N, Matthys-Rochon E, Rougier M, Mogensen L, Holm P, Magnard JL, Dumas C (1996) Isolated maize zygotes mimic *in vivo* embryonic development and express microinjected genes when cultured *in vitro*. Dev. Biol. 177:190–203.

Liu CM, Xu ZH, Chua NH (1993a) Proembryo culture: *in vitro* development of early globular-stage zygotic embryos from *Brassica juncea.* Plant J. 3:291–300.

Liu CM, Xu ZH, Chua NH (1993b) Auxin polar transport is essential for the establishment of bilateral symmetry during early plant embryogenesis. Plant Cell 5:621–630.

Lorenzi R, Bennici A, Cionini PG, Alpi A, D'Amato F (1978) Embryo-suspensor relations in *Phaseolus coccineus*: cytokinins during seed development. Planta 143:59–62.

Lu P, Porat R, Nadeau JA, O'Neill SD (1996) Identification of a meristem L1 layer-specific gene in *Arabidopsis* that is expressed during embryonic pattern formation and defines a new class of homeobox genes. Plant Cell 8:2155–2168.

Maheshwari N (1958) *In vitro* culture of excised ovules of *Papaver somniferum.* Science 127:342.

Maheshwari N, Lal M (1961) *In vitro* culture of excised ovules of *Papaver somniferum* L. Phytomorphology 11:307–314.

Maheshwari P (1950) *An Introduction to the Embryology of Angiosperms*. McGraw-Hill Book Co., New York, NY.

Maheshwari P, Baldev B (1961) Artificial production of buds from the embryos of *Cuscuta reflexa*. Nature 191:197–198.

Mansfield SG, Briarty LG (1991) Early embryogenesis in *Arabidopsis thaliana.* II. The developing embryo. Can. J. Bot. 69:461–476.

Mansfield SG, Briarty LG, Erni S (1991) Early embryogenesis in *Arabidopsis thaliana.* I. The mature embryo sac. Can. J. Bot. 69:447–460.

Matsubara S (1962) Studies on a growth promoting substance, "embryo factor", necessary for the culture of young embryos of *Datura tatula in vitro.* Bot. Mag., Tokyo 75:10–18.

Matsubara S (1964) Effect of nitrogen compounds on the growth of isolated young embryos of *Datura*. Bot. Mag., Tokyo 77:253–259.

Matthys-Rochon E, Piola F, le Deunff E, Mòl R, Dumas C (1998) *In vitro* development of maize immature embryos: a tool for embryogenesis analysis. J. Expt. Bot. 49:839–845.
McCarty DR (1995) Genetic control and integration of maturation and germination pathways in seed development. Annu. Rev. Plant Physiol. Plant Molec. Biol. 46:71–93.
Miflin BJ (1969) The inhibitory effects of various amino acids on the growth of barley seedings. J. Expt. Bot. 20:810–819.
Mogensen HL (1972) Fine structure and composition of the egg apparatus before and after fertilization in *Quercus gambelii*: the functional ovule. Amer. J. Bot. 59:931–941.
Mól R, Matthys-Rochon E, Dumas C (1993) *In vitro* culture of fertilized embryo sacs of zygotes and two-celled proembryos can develop into plants. Planta 189:213–217.
Mól R, Matthys-Rochon E, Dumas C (1995) Embryogenesis and plant regeneration from maize zygotes by *in vitro* culture of fertilized embryo sacs. Plant Cell Repts. 14:743–747.
Mól R, Idzikowska K, Dumas C, Matthys-Rochon E (2000) Late steps of egg cell differentiation are accelerated by pollination in *Zea mays* L. Planta 210:749–757.
Monnier M (1976) Action de la pression partielle d'oxygène sur le développement de l'embryon de *Capsella bursa-pastoris* cultivé *in vitro*. Compt. Rend. Acad. Sci. Paris 282D:1009–1012.
Monnier M (1984) Survival of young immature *Capsella* embryos cultured *in vitro*. J. Plant Physiol. 115:105–113.
Nagl W (1974) The *Phaseolus* suspensor and its polytene chromosomes. Z. Pflanzenphysiol. 73:–44.
Nagl W (1990) Translocation of putrescine in the ovule, suspensor and embryo of *Phaseolus coccineus*. J. Plant Physiol. 136:587–591.
Nakajima T (1962) Physiological studies of seed development, especially embryonic growth and endosperm development. Bull. Univ. Osaka Pref. Ser. B 13:13–48.
Natesh S, Rao MA (1984) The embryo. *In:* BM Johri, (ed.). *Embryology of Angiosperms*. Springer-Verlag, Berlin, pp. 377–443.
Norstog K (1972) Factors relating to precocious germination in cultured barley embryos. Phytomorphology 22:134–139.
Norstog K, Klein RM (1972) Development of cultured barley embryos. II. Precocious germination and dormancy. Can. J. Bot. 50:1887–1894.
Norstog K, Smith JE (1963) Culture of small barley embryos on defined media. Science 142:1655–1656.
Nyman LP, Webb EL, Gu Z, Arditti J (1986) Structure and *in vitro* growth of zygotic embryos of taro (*Colocasia esculenta* var. antiquorum). Ann. Bot. 57:623–630.
Nyman LP, Webb EL, Gu Z, Arditti J (1987) Effects of growth regulators and glutamine on *in vitro* development of zygotic embryos of taro (*Colocasia esculenta* var. *antiquorum*). Ann. Bot. 59:517–523.
Okada K, Ueda J, Komaki MK, Bell CJ, Shimura Y (1991) Requirement of the auxin polar transport system in early stages of *Arabidopsis* floral bud formation. Plant Cell 3:677–684.

Olson AR, Cass DD (1981) Changes in megagametophyte structure in *Papaver nudicaule* L. (Papaveraceae) following *in vitro* placental pollination. Amer. J. Bot. 68:1333–1341.

Paris D, Rietsema J, Satina S, Blakeslee AF (1953) Effect of amino acids, especially aspartic and glutamic acid and their amides, on the growth of *Datura stramonium* embryos *in vitro*. Proc. Natl. Acad. Sci. USA 39:1205–1212.

Perez-Grau L (2002) Plant embryogenesis—the cellular design of a plant. *In:* SD O'Neill, JA Roberts, (eds.). Plant Reproduction. Annual Plant Reviews, Sheffield Acad. Press, Sheffield, Vol. 6, pp. 154–192.

Perez-Grau, L Goldberg RB (1989) Soybean seed protein genes are regulated spatially during embryogenesis. Plant Cell 1:1095–1109.

Poethig RS, Coe Jr, EH, Johri MM (1986) Cell lineage patterns in maize embryogenesis: a clonal analysis. Dev. Biol. 117:392–404.

Prakash N (1987) Embryology of the Leguminosae. *In:* CH Stirton, (ed.). *Advances in Legume Systematics*, Part 3. Royal Botanic Gardens, Kew, pp. 241–278.

Pretová A (1974) The influence of the osmotic potential of the cultivation medium on the development of excised flax embryos. Biol. Plant. 16:14–20.

Pretová A (1986) Growth of zygotic flax embryos *in vitro* and influence of kinetin. Plant Cell Repts. 3:210–211.

Raghavan V (1965) Hormonal control of embryogenesis in *Capsella*. *In:* PR White, AR Grove, (eds.). *Proceedings of an International Conference on Plant Tissue Culture*. McCutchan Publishing Corp., Berkeley, CA. pp. 357–369.

Raghavan V (1976) *Experimental Embryogenesis in Vascular Plants*. Academic Press, London.

Raghavan V (1977a) Diets and culture media for plant embryos. *In:* M Rechcigl, Jr, (ed.). *CRC Handbook Series in Nutrition and Food*. Section G: Diets, Culture Media, Food Supplements. Vol 4. Culture media for Cells, Organs and Embryos. CRC Press, Cleveland, pp. 361–413.

Raghavan V (1977b) Applied aspects of embryo culture. *In:* J Reinert, YPS Bajaj, (eds.). *Plant Cell, Tissue, and Organ Culture*. Springer-Verlag, Berlin, pp. 375–397.

Raghavan V (1980) Embryo culture. Int. Rev. Cytol. Suppl. 11B: 209–240.

Raghavan V (1985) The applications of embryo rescue in agriculture. *In:* Biotechnology in International Agriculture. (IRRI), Manila, pp. 189–197.

Raghavan V (1986a) Variability through wide crosses and embryo rescue. *In:* IK Vasil, (ed.). *Cell Culture and Somatic Cell Genetics of Plants*. Academic Press, Orlando, FL, Vol. 3. pp. 613–633. Raghavan V (1986b) *Embryogenesis in Angiosperms. A Developmental and Experimental Study*. Cambridge Univ. Press, New York, NY.

Raghavan V (1993) Embryo culture: methods and applications. *In:* J Prakash, RLM Pierik, (eds.). *Plant Biotechnology. Commercial Prospects and Problems.* Oxford & IBH Publi. Co., New Delhi, pp. 143–168.

Raghavan V (1997) *Molecular Embryology of Flowering Plants*. Cambridge Univ. Press, New York, NY.

Raghavan V (2000a) The coming of age of plant embryology. Curr. Sci. 80:244–251.

Raghavan V (2000b) Pattern formation in angiosperm embryos. Botanica 50:33–47.

Raghavan V (2001) Life and times of the suspensor of angiosperm embryos. Phytomorphology. 51:251–276.

Raghavan V (2002) Induction of vivipary in *Arabidopsis* by silique culture: implications for seed dormancy and germination. Amer. J. Bot. 89:766–776.

Raghavan V (2004) Somatic embryogenesis. *In:* PK Saxena, SJ Murch (ed.). *The Journey of a Single Cell to Plant.*

Raghavan V, Sharma KK (1995) Zygotic embryogenesis in gymnosperms and angiosperms. *In:* TA Thorpe, (ed.). *In Vitro Embryogenesis in Plants*. Kluwer Acad. Publ., Dordrecht, Neitherlands pp. 73–115.

Raghavan V, Srivastava PS (1982) Embryo culture. *In:* BM Johri., (ed.). *Experimental Embryology of Vascular Plants.* Springer-Verlag, Berlin, pp. 195–230.

Raghavan V, Torrey JG (1963) Growth and morphogenesis of globular and older embryos of *Capsella* in culture. Amer. J. Bot. 50:540–551.

Raghavan V, Torrey JG (1964) Effects of certain growth substances on the growth and morphogenesis of immature embryo of *Capsella* in culture. *Plant Physiol.* 39:691–699.

Randolph LF (1936) Developmental morphology of the caryopsis in maize. J. Agr. Res. 53:881–916.

Raz V, Bergervoet JHW, Koornneef M (2001) Sequential steps for developmental arrest in *Arabidopsis* seeds. Development 128:243–252.

Rietsema J, Satina S, Blakeslee AF (1953) The effect of sucrose on the growth of *Datura stramonium* embryos *in vitro*. Amer. J. Bot. 40:538–545.

Rijven AHGC (1952) *In vitro* studies on the embryo of *Capsella bursa-pastoris*. Acta Bot. Neerl. 1:157–200.

Rijven AHGC (1956) Glutamine and asparagines as nitrogen sources for the growth of plant embryos *in vitro*. A comparative study of 12 species. Aust. J. Biol. Sci. 9:511–527.

Robichaud CS, Wong J, Sussex IM (1980) Control of *in vitro* growth of viviparous embryo mutants of maize by abscisic acid. Dev. Genet. 1:325–330.

Ryczkowski M (1960) Changes of the osmotic value during the development of the ovule. *Planta* 55:343–356.

Sanders ME, Burkholder PR (1948) Influence of amino acids on growth of *Datura* embryos in culture. Proc. Natl. Acad. Sci. USA 34:516–526.

Schel JHN, Kieft H (1986) An ultrastructural study of embryo and endosperm development during *in vitro* culture of maize ovaries (*Zea mays*). Can. J. Bot. 64:2227–2238.

Scheres B, Wolkenfelt V, Willemsen V, Terlouw M, et al. (1994) Embryonic origin of the *Arabidopsis* primary root and root meristem initials. Development 120:2475–2487.

Schulz P, Jensen WA (1977) Cotton embryogenesis: the early development of the free nuclear endosperm. Amer. J. Bot. 64:384–394.

Schulz R, Jensen WA (1968) *Capsella* embryogenesis: the egg, zygote, and young embryo. Amer. J. Bot. 55:807–819.

Schwartz BW, Yeung EC, Meinke DW (1994) Disruption of morphogenesis and transformation of the suspensor in abnormal suspensor mutants of *Arabidopsis*. Development 120:3235–3245.

Smith JG (1973) Embryo development in *Phaseolus vulgaris*. II. Analysis of selected inorganic ions, ammonia, organic acids, amino acids, and sugars in the endosperm liquid. Plant Physiol. 51:454–458.

Souter M, Lindsey, K (2000) Polarity and signaling in plant embryogenesis. J. Exp. Bot. 51:971–983.

Stefaniak B (1987) The *in vitro* development of isolated rye proembryos. Acta Soc. Bot. Polon. 56:37–42.

Stewart JM, Hsu CL (1977). Ovulo embryo culture and seedling development of cotton (*Gossypium hirsutum* L.). Planta 137:113–117.

Sumner MJ (1992) Embryology of *Brassica campestris*: the entrance and discharge of the pollen tube in the synergid and the formation of the zygote. Can. J. Bot. 70:1577–1590.

Sumner MJ, van Caeseele L (1989) The ultrastructure and cytochemistry of the egg apparatus of *Brassica campestris*. Can. J. Bot. 67:177–190.

Tilton VR, Wilcox LW, Palmer RG (1984) Postfertilization wandlabrinthe formation and function in the central cell of soybean, *Glycine max* (L.) Merr. (Leguminosae). Bot. Gaz. 145:334–339.

Töpfer R, Steinbiss HH (1985) Plant regeneration from cultured fertilized barley ovules. Plant Sci. 41:49–54.

Torres-Ruiz R, Jürgens G (1994) Mutations in the *FASS* gene uncouple pattern formation and morphogenesis in *Arabidopsis* development. Development 120:2967–2978.

Traas J, Bellini C, Nacry P, Kronenberger J, Bouchez D, Caboche M (1995) Normal differentiation patterns in plants lacking microtubular preprophase bands. Nature 375:676–677.

van Lammeren AAM (1986) A comparative ultrastructural study of the megagametophytes in two strains of *Zea mays* L. before and after fertilization. Agric. Univ. Wageningen Papers 86-1:1–37.

van Lammeren AAM (1988) Observations on the structural development of immature maize embryos (*Zea mays* L.) during *in vitro* culture in the presence or absence of 2,4-D. Acta Bot. Neerl. 37:49–61.

van Overbeek J, Conklin ME, Blakeslee AF (1942) Cultivation *in vitro* of small *Datura* embryos. Amer. J. Bot. 29:472–477.

van Overbeek J, Siu R, Haagen-Smit AJ. (1944) Factors affecting the growth of *Datura* embryos *in vitro*. Amer. J. Bot. 31:219–224.

Veen H (1963) The effect of various growth-regulators on embryos of *Capsella bursa-pastoris* growing *in vitro*. Acta Bot. Neerl. 12:129–171.

Vroemen C, de Vries S (1999) Flowering plant embryogenesis. *In:* VEA Russo, DJ Cove, LG Edgar, R Jaenisch, F Salamini, (eds.). *Development, Genetics, Epigenetics and Environmental Regulation*. Springer-Verlag, Berlin, pp. 121–132.

Vroemen CW, Langeveld S, Mayer U, Ripper G, et al. (1996) Pattern formation in the *Arabidopsis* embryo revealed by position-specific lipid transfer protein gene expression. Plant Cell 8:783–791.

Webb MC, Gunning BES (1991) The microtubular cytoskeleton during development of the zygote, proembryo and free-nuclear endosperm in *Arabidopsis thaliana* (L.) Heynh. Planta 184:187–195.

Williams EG, Maheswaran G, Hutchinson JF (1987) Embryo and ovule culture in crop improvement. Plant Breed. Rev. 5:181–236.

Yadegari R, de Paiva GR, Laux T, Koltunow AM, et al. (1994) Cell differentiation and morphogenesis are uncoupled in *Arabidopsis raspberry* embryos. Plant Cell 6:1713–1729.

Yan H, Yang HY, Jensen WA (1991) Ultrastructure of the developing embryo sac of sunflower (*Helianthus annuus*) before and after fertilization. Can. J. Bot. 69:191–202.

Yeung EC (1980) Embryogeny of *Phaseolus*. The role of the suspensor. Z. Pflanzenphysiol. 96:17–28.

Yeung EC, Brown DCW (1982) The osmotic environment of developing embryos of *Phaseolus vulgaris*. Z. Pflanzenphysiol. 106:149–156.

Yeung EC, Sussex IM (1979) Embryogeny of *Phaseolus coccineus*: the suspensor and the growth of the embryo-proper *in vitro*. Z. Pflanzenphysiol. 91:423–433.

Zenkteler M, Nitzsche W (1985) *In vitro* culture of ovules of *Triticum aestivum* at early stages of embryogenesis. Plant Cell Repts. 4:168–171.

Ziebur NK, Brink RA, Graf LH, Stahmann MA (1950) The effect of casein hydrolyzate on the growth in vitro of immature *Hordeum* embryos. Amer. J. Bot. 37:144–148.

12

Principles of Micropropagation

Alan C. Cassells
Department of Plant Science,
National University of Ireland, Cork, Ireland
e-mail: a.cassells@ucc.ie

INTRODUCTION

Micropropagation is the application of plant cell, tissue, and organ culture research to the clonal multiplication of plants. Micropropagation is a substitution technology in the plant propagation industry where it competes with traditional seed and vegetative propagation methods (Fig. 12.1). Micropropagation involves a number of stages as first defined by Murashige (1974) and subsequently expanded by Debergh and Maene (1981) (see Table 12.1). These steps all involve manual operations and consequently, the production cost of the micropropagule, that is the microplant, microtuber etc., is heavily influenced by local labor costs. It has been estimated that the labor cost is around 60% of the total production cost in traditional micropropagation (George 1996). High labor costs have had two influences on micropropagation; firstly, pressure to compromise quality in favor of quantity (Grunewaldt-Stocker 1997) and secondly, a move of production facilities to low labor cost economies (Prakash 2000).

Micropropagation can only compete with traditional methods of plant propagation in niche market sectors, e.g. with high value hybrid seed and more expensive vegetative propagules. In the developed economies, the cost of producing a bare-rooted microplant is approximately 15 Euro or US cents and of an established microplant approx. 20 Euro/US cents. This means that the global seed market valued 3.9 billion USD (*www.fas.usda.gov*) is unlikely, even with automation, to be eroded significantly by micropropagation. Indeed, it is not competitive with the vegetative propagation

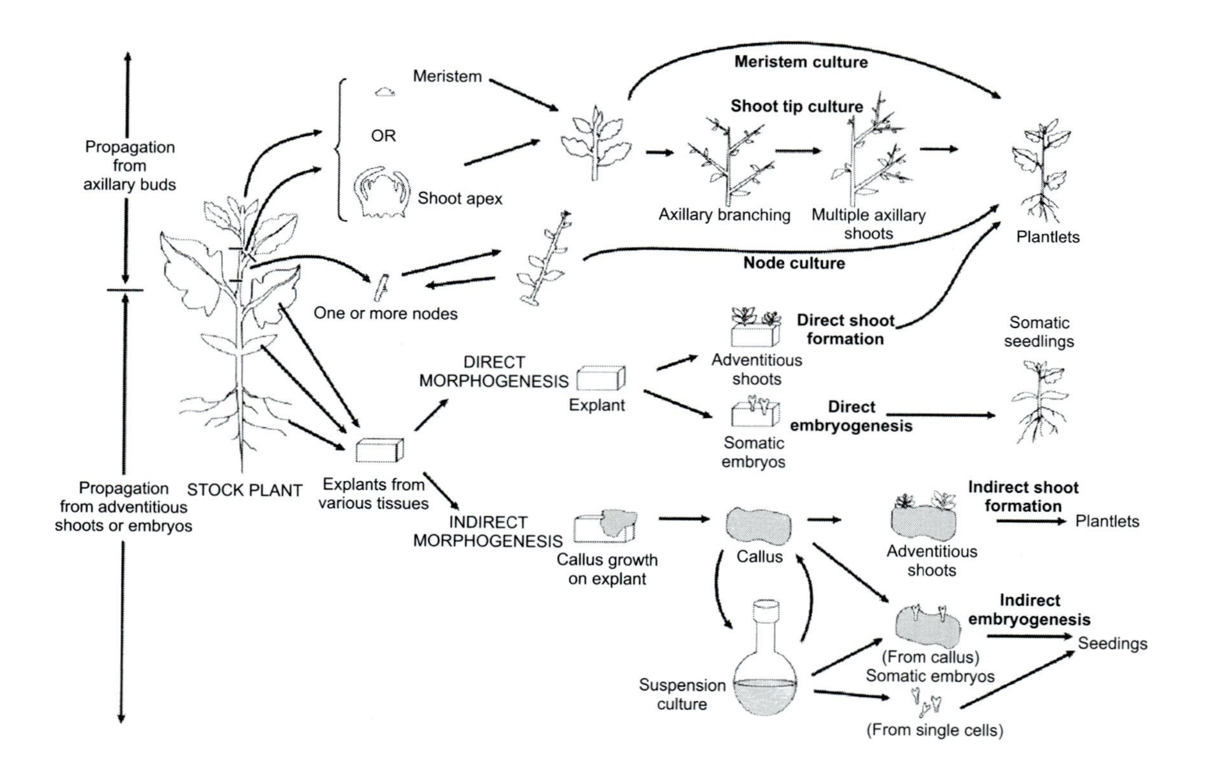

Fig. 12.1 Scheme showing structure of the plant propagation industry. Commer-cial micropropagators compete with seed producers and vegetative propagators. Profit margins can be accumulated at each step in the chain from production to retail distribution.

Table 12.1 Stages in micropropagation. Stages 1-4 after Murashige (1974) and Stages 0 and 3a after Debergh and Maene (1981). Weaning refers to the adaptation of the microplant physiology to ambient environmental conditions. Usually Stages 1-4 are followed; sometimes, however, Stage 1 and/or 3 is omitted. See George (1993, 1996) for strategies used for different crop types.

Stage	*Activity*	*Notes*
Stage 0	Selection of stock plants; pathogen indexing, maintenance under quarantine conditions	For crop certification pathogen indexing should be carried out using official guidelines
Stage 1	Establishment of aseptic culture, confirmation by bacterial indexing of freedom from cultivable microorganisms	It is advisable that the culture be established from meristem explants to eliminate pathogens and bacterial contaminants. It is advised that after explant establishment, basal explants be cultured intact on a range of bacteriological media
Stage 1a	Use of meristem culture and/or chemo- or thermotherapy *in vitro* to eliminate pathogens	It is important to follow correct procedure in selection of antimicrobial chemicals and to use correctly. For certification purposes the plants must be reestablished and returned to Stage 0 for indexing
Stage 2	Clonal mass propagation	It is essential to use a stable cloning strategy
Stage 3(a)	Preparation for rooting	Some *in vitro* shoots may benefit from gibberellic acid treatment to elongate the stems before Stage 3(b)
Stage 3(b)	Preparation for return to the environment (weaning)	Strategies are variable, some microshoots self-root, others need to be transferred to a rooting medium. This is an opportunity to wean the microplants *in vitro* and to introduce biological inoculants.
Stage 4	Microplant establishment	Plants may require shading and misting and special attention to avoid damping-off diseases.

of chrysanthemum, potato, strawberry, plantation and other crops except in the early stages of production of disease-free nuclear stock for subsequent vegetative multiplication (Cassells 1997). The industry has sought to expand its market firstly, by using its potential for continuous production, independent of ambient climatic changes (wet or dry, cold or hot seasons) to obtain premium prices for the early release of new varieties and disease-free stock; secondly, to use its lesser stock dependence and lower stock storage costs compared with traditional vegetative propagation to recycle old varieties. This reflects product life cycles where the popularity of a new variety increases, peaks and declines. Thirdly, the industry has sought less labor intensive, partially automated processes, albeit these will still be relatively expensive, especially compared with seed-production costs.

While production price is important, micropropagators must produce a product that conforms to trade legislation, namely, "fit for the purpose for which it is intended" and so, microplant quality is an important element in competition with the traditional industry product. The aspects of quality important in micropropagation are: genetic stability (clonal or trueness-to-type), epigenetic or developmental status (maturity), physiological quality, and health status. The potential to produce disease-free plants of good genetic, epigenetic, and physiological quality also gives a competitive advantage to the micropropagator although there has been criticism in the past of the industry for failure to supply good quality material (Grunewaldt-Stocker, 1997).

Currently, certification and plant quarantine authorities do not generally accept the results of disease indexing carried out on the tissue *in vitro*; instead they require that the stock plants used for micropropagation, and sometimes the progeny plants, be pathogen indexed (Cassells 1997). This is a major time-and-cost constraint on the industry. In some countries and for some crops, e.g. strawberry in the Netherlands, the phytosanitary authorities will not allow micropropagation in the production of certified stock, primarily on the grounds that genetic stability is not maintained in the multiplication stages. Generally, phytosanitary standards agencies such as the European and Mediterranean Plant Protection Organisation (EPPO: *www.eppo.org*) do not accept the results of indexing carried out on *in-vitro* material. Equally important to the clients of micropropagation companies is the occasional poor physiological quality of micoplants which results in failure or a growth check at establishment and poor epigenetic quality, usually "juvenility", resulting in delayed flowering and poor cropping (George 1993). "Rejuvenation" or "reinvigoration" is common, if not universal, in tissue culture, such that the developmental status of microplants is more mature than seedling material but less mature than vegetative propagules such as cuttings (George 1993). The maxim that "flowers sell plants" is important in the high value ornamental sector where micropropagation is relatively successful, so it is important that microplant flowering not be delayed. In contrast, for the woody plant propagator where developmentally old tissues root poorly if at all, or for those who are using micropropagation to produce healthy stock plants for the production of cutting, rejuvenation *in vitro* is desirable as the rejuvenated plants root readily and can have a multibranched habit, which is desirable in stock plants for further cycles of *in vivo* vegetative propagation. Here, emphasis is given to best production practices for micropropagation.

MICROPROPAGATION PROCESS

Micropropagation should aim not just to produce healthy, true-to-type microplants at an economic price, but should also be concerned about the performance *post vitrum* of the progeny plants, namely, with flower and fruit quality, tuber yield, etc. Here factors contributing to quality are considered. In this regard, while micropropagation is conveniently divided into "Stages", it must be emphasized that it is a process. The implication of this is that care must be taken when optimizing individual stages, critically Stage 2, the multiplication stage, so that product quality is not compromised. The stages in micropropagation are listed in Table 12.1.

Stage 0

The process begins with the choice of stock plants, which should be selected as representative of the genotype (variety) and be of good vigor. These should be placed in an isolation facility, preferably a quarantine facility in which there is no risk of disease or pathogen vector entry. If this is not available, individual plants, depending on size, can be maintained in ventilated plastic bags, e.g. Suncap bags (*www.sigmaaldrich.com*) to facilitate gaseous exchange. Larger plants can be maintained in muslin cages as for virus testing. Care should be taken to avoid spreading infection during maintenance cultivation, e.g. watering should be from the bottom to avoid transmission of a pathogen in water droplets instruments should be sterilized before moving from one plant to the next. Once isolated, individual plants can be indexed for pathogens. Guidelines for pathogen testing are published by international certification authorities and underpin the safe international movement of plants. These have been discussed elsewhere (Cassells 2000a) and further details are provided at *www.fao.org* and *www.eppo.org*. Micropropagators mainly work with high value ornamental crops for which there may be no official crop certification schemes and limited crop intelligence. In these cases it is recommended that micropropagators follow the principles laid down in the above guidelines for the same pathogens in other crops or for the target pathogen in related crops. Generally, the client specifies the diseases of concern and requests that the plants be tested and confirmed free of the target pathogens. It is important to stress that this indexing should be carried out on the stock plant *in vivo* and not on the tissue cultures since little is known about the distribution and amount of pathogens in plant tissues *in vitro* (O'Herlihy and Cassells 2003). Information about plant diseases and their diagnosis can be found on the website of the Americam Phytopathological Society (*www.apsnet.org*) and via its links to other relevant sites.

There are many pitfalls in pathogen indexing including seasonal fluctuations of the pathogen in its host plant, uneven distribution in the plant, interaction between pathogens in plants, and sensitivity and specificity of the diagnostic or diagnostic procedure. Most certification schemes rely on traditional methods to detect pathogen with specific methods for the different pathogen groups (see Table 12.2). Symptoms are very important to the agricultural or horticultural plant pathologist as they narrow down the options for indexing. They then use at least for major pathogens, commercially available diagnostics (see e.g. (*www.agdia.com; www.bioreba.com*) but it is advisable that serological or biochemical diagnostic tests be followed by inoculation of a susceptible host or indicator plants to confirm negative results (Cassells 2000a).

Where the stock plants index positive for viral pathogens, they may be subjected to thermotherapy to eliminate the virus(es), after which they should be rigorously reindexed. It has been widely reported that viruses may reemerge slowly after thermotherapy and so caution should be exercised (O'Herlihy and Cassells 2003)

Stage 1

Plants in nature, in addition to being infected with symptom-producing and latent pathogens, may also be systemically or randomly contaminated intercellularly endophytically or epiphytically/hemiendophytically with cultivable environmental bacteria, and less commonly with fungi, especially yeasts. These are generally not plant pathogenic isolates but may be opportunistic pathogens. They can usually grow on plant tissue culture media and may emerge in Stage 1 at explant establishment or accumulate in the tissues with the risk of emergence late in production when much expenditure has occurred, or even *post vitrum,* to cause economic problems. These may include pathogens of other crops or human pathogens, albeit the risk of either of the latter is probably low (Cassells 1997; Weller 1997).

The common strategy used by micropropagators to eliminate pathogens and endophytes is to introduce their plants into culture by using excised meristem "tips" (meristem culture; George 1993). The basis of this approach is that many pathogenic and nonpathogenic endophytes do not penetrate the tip of the plant beyond the vascular system. The vascular tissues differentiate back from the apical tip. Two factors are important for success, namely that the smallest possible 'tip' is excised

Table 12.2 Application of different indexing methods to various groups of pathogens. Genomic diagnostics are increasingly being used while proteomic methods have been investigated more recently.

				Test Methods				
Pathogen type	*Symptoms*	*Indicator plants*	*Serological diagnostics*	*Microscopy*	*Biochemical test kits*	*Selective culture media*	*Genomic diagnostics*	*Proteomic diagnostics*
Virus	Widely used	Widely used to confirm pathogenicity and identify strains	ELISA is a standard method	Limited use of electron microscopy	Not applicable	Not applicable	Detection ofvariable coat genes/ conserved sequences	Detection of viral coat proteins
Viroid	Widely used	Widely used to confirm pathogenicity and identify strains	Not applicable	Not used	Not applicable	Not applicable	Standard method	Not applicable
Bacterium*	Widely used	Widely used to confirm pathogenicity	Not commonly used	Widely used for Gram and spore staining, etc.	Not commonly used for plant pathogens	Historically used in plant pathology	Detection of rDNA	Detection of bacterial biomarkers
Fungus	Widely used	Widely used to confirm pathogenicity	Not commonly used	Widely for used morphological identification	Not applicable	Widely used	Detection of rDNA	Detection of fungal biomarkers

*Fatty acid profile analysis is used for the detection of bacterial plant pathogens (after Cassells and Doyle 2003).

and that the pathogen or contaminant obeys the hypothesis! In general, it is recommended that the apical dome and first pair of leaf primordial be excised. In difficult cases, these can be grafted to aseptic seedlings *in vitro* to aid in establishment of the small explant (George 1993). There is always the risk in using other explants, lateral buds, internodes, etc. that latent contaminants may be introduced into the culture. It is rare to see pathogen symptom expression in plant tissues *in vitro* and so pathogens may be clonally transmitted in the tissues to result in large-scale release of infected material.

Meristem excision is a stressful process for the plant tissues and often results in the production of phenolics, which may form a barrier between the explant and the culture medium. To avoid this problem the explant can be repositioned on the medium or antioxidants or phenol binding agents are sometimes incorporated into the medium (George 1993). Plant tissue media are discussed in detail in George (1993, 1996). Before proceeding to Stage 2 it is important to index the tissues for cultivable contaminants, which may have been carried over in the explant (Leifert and Cassells 2001). To encourage the expression of bacterial contaminants, components of bacteriological media, e.g. casein hydrolyzate, can be added to the explant medium. Indexing should involve excision of the basal tissues of the microshoot that has developed from the original apical explant. These should be placed intact on a range of bacteriological media (*www.sigmaaldrich.com; www.duchefa.com*). Some bacteria may not grow on plant tissue culture media but can live in cell exudates within the plant tissues. Expressing the sap may result in production of wound phenolics, which inhibit these bacteria. For the latter reason, it is recommended that the intact explant be transferred to bacteriological media for testing. Complicated strategies have been published for sampling the cultures. These tend to assume that bacterial distribution will be uneven in the plants (George 1993). While this may be true of contamination by epiphytic bacteria which may become hemiendophytic in plant tissues *in vivo*, it could be argued that the tendency to hyper-hydricity, that is, cell enlargement and creation of large air spaces in plant tissues *in vitro*, may facilitate systemic endophytic contamination in cultures.

Stage 2

Stage 2 is the multiplication stage in which the objectives are to mass clonally propagate healthy, true-to-type plants. The assumption is that the cultures are free of pathogens (based on indexing of the stock plants and selection of disease-escapes) and that any nonfastidious environmental microorganisms, mainly bacteria but also yeasts, have been detected in Stage 1. Given that the cultures are contaminant-free, the two key objectives of Stage 2 are to choose a stable cloning strategy and to use good working practice to prevent microbial contamination during the mass multiplication (Leifert and Cassells 2001). The choice of cloning strategy depends on the stability of the genotype in adventitious regeneration and on the characteristics of the plant, e.g. African violet and begonia are stably propagated vegetatively in conventional vegetative propagation from leaf cutting and *in vitro* from leaf explants (bypassing Stage 1), while lilies, cyclamen, etc. produce vegetative reproductive structures *in vivo,* which can be induced to form and can be subdivided and multiplied *in vitro*. A comprehensive list of crops and their micropropagation protocols are given in George (1996); for information on vegetative propagation see Hartman et al. (2001). Most plants are propagated *in vitro* either: (i) by induction of adventitious buds (or somatic embryos) on explants; (ii) by stimulation of the apical bud to produce axillary buds (bud clusters) which are repeatedly divided and subcultured; or (iii) by elongation of the apical and nodal buds to form stems (nodal bud culture) which are subdivided into nodes for further cycles of multiplication. These pathways are illustrated in Fig. 12.2. One precautionary example is strawberry, in which initially proliferation of apical bud clusters results in stable progeny but over time adventitious shoots may arise from the callus developing at the base of the proliferating bud cultures, resulting in genetically unstable progeny (Boxus et al. 2000). The preferred cloning option is nodal culture in which the origin of the explant for replication cycles (nodal explants) can be visibly tracked. However, for some species it is difficult to achieve good stem elongation for rapid dissection into nodal explants for subculture. In crops or crop varieties in which the genome is stable in adventitious regeneration, there may be a choice of options, viz. cloning via adventitious shoots or somatic embryos, in addition to nodal bud clusters and nodal cultures.

Regardless of the strategy chosen, it is essential to avoid microbial contamination of the cultures in the laboratory. In some commercial laboratories every stock culture is visually examined before being used as a source of explants for subculture. This however, should be supported by random subculture of basal nodes to confirm the absence of latent contaminants. Management of microbial contamination in plant tissue cultures has been extensively reviewed (see Cassells 1997; Leifert and Cassells 2001; for a list of indicator contaminants see Table 12.3).

Sampling *in vitro* production and carrying out genetic fingerprinting using AFLPs for example, can be used to monitor genetic stability (Karp et al., 1998). But Genomic analysis is expensive and the most practical strategy is to establish a sample production of 50-100 progeny plants and visually examine them for uniformity. Analysis can be made less subjective by using computerized image analysis (Kowalski and Cassells, 1999). Again, if the cloning strategy is based on bud culture, stability is likely to be maintained; but variation may arise since epigenetic adaptation to the

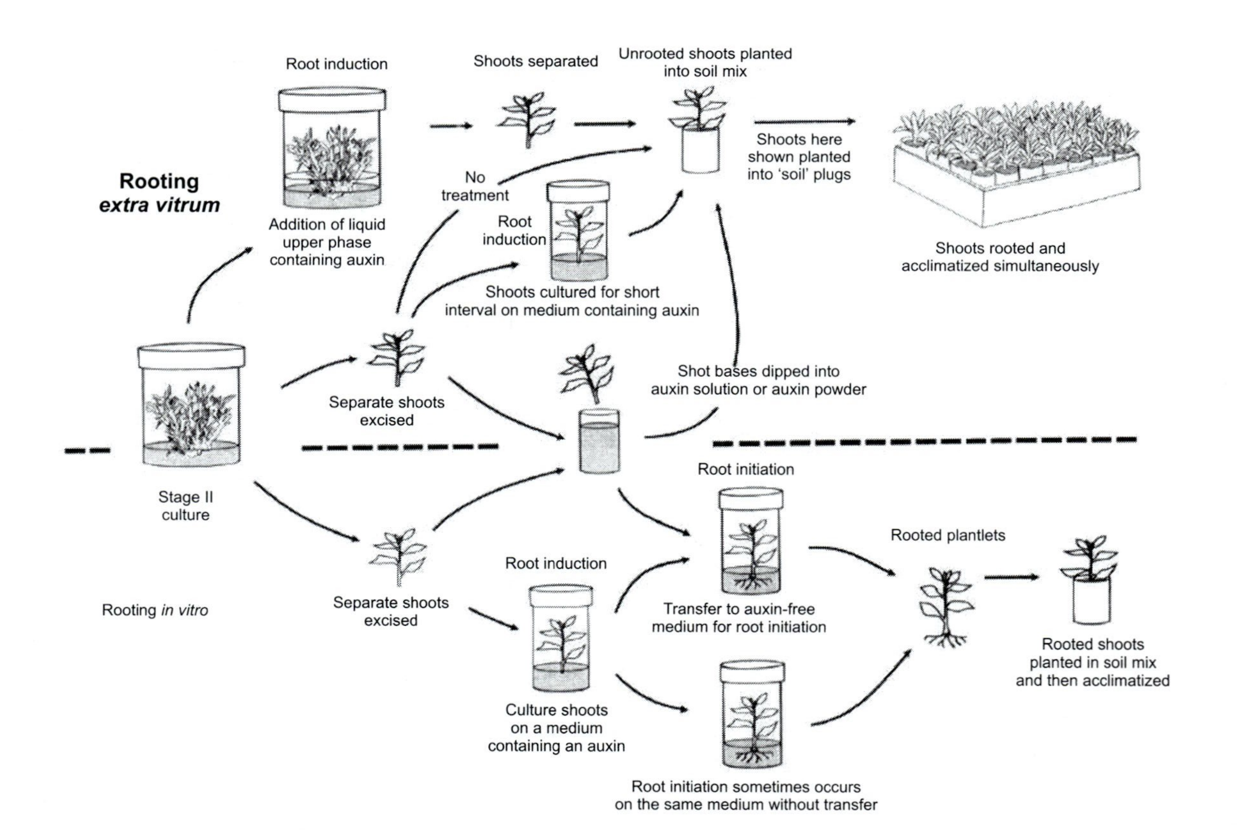

Fig. 12.2 Cloning strategies available to the micropropagator. The cloning strategy of choice is that which is the most profitable, taking into account the genetic stability of the genotype. The choice is also influenced by the amenability of the plant habit *in vitro* to ease of subculturing (from George 1993 with permission).

Table 12.3 Critical control points (CCPs), indicator microorganisms, and equipment and methods used for microbiological production control in plant tissue culture and prevention of contamination (after Leifert and Cassells 2001)

Stages of Plant Tissue Culture	*Parameters (CCPs)to be monitored*	*Indicator microorganisms*	*Equipment needed*
Stage 0 Stock plant treatment	Symptoms of viroid, viral, bacterial, and fungal pathogens. Detection and identification of viroids, viruses, bacteria, and fungi		Hand lens, light microscope, phase contrast microscope; selective bacteriological media; serological tests/DNA probes; indicator plants
Stage 1 Surface sterilization	Active chlorine content in hypochlorite solutions Activity of other biocides	Gram-negative bacteria: *Pseudomonas, Fluorescens, Erwinia* spp., *Klebsiella* spp., *Serratia* spp., *Agrobacterium* spp.	Whatman indicator paper, arsenite solution, titration column, depending on compound
Stage 2 Indexing	Presence of viruses, bacteria, and fungi on explants after surface disinfection		Indexing media Serological tests Nuclei acid hybridization tests

in vitro conditions occurs over subculture cycles (see below). For a review of factors influencing genetic and epigenetic variation in micropropagation see Cassells and Curry (2001) and Joyce et al. (2003).

Stage 3

Stage 3, which may involve transfer to a modified medium, is usually referred to as the 'rooting stage'. This involves preparation of the bud clusters, adventitiously proliferating explants, microshoots, or shoots (from nodal culture) for establishment (Fig. 12.2). While the aim of Stage 2 is to maintain genetic stability, the main aim of Stage 3 is to ensure that the microplants are of good physiological and epigenetic quality, i.e. the plant organs, roots, shoots, and leaves are functional and the microplants developmentally mature, such that on transfer to the growth substrate they will establish and grow without a growth check, are resistant to damping-off pathogens, and grow to flowering, tuberization, etc., giving a satisfactory harvest. The shoots or self-rooted shoots from Stage 2 may be transferred directly to the greenhouse for establishment but sometimes do require an elongation treatment (Stage 3 a) or/and transfer to a rooting medium (see Fig. 12.2). For details of rooting media see George (1993, 1996). Poor physiological quality is associated with the hyperhydricity (syn. vitrification) syndrome *in vitro*. In the extreme, this can be expressed as shoot- and leaf- tip necrosis associated with basal callousing of the explants, death of the shoots, and loss of competence. This is extreme and reflects avoidable stress levels. More commonly, hyperhydricity is subliminal in the *in-vitro* cultures, resulting in malfunctional stomata and underdevelopment of the photosynthetic apparatus (Ziv and Ariel 1994).

While a number of media factors have been implicated in hyperhydricity including hormonal imbalance, excessive nitrogen, ethylene accumulation in the culture vessel, one of the key factors may be high humidity. Tightly sealed culture vessels, while preventing bacterial and microarthropod contamination do not allow a transpiration stream to develop in the microshoots. This is essential for the calcium uptake required for stomatal function and as a counter ion for auxin basipal movement (Taiz and Zeigler 2002). Controlling the gaseous atmosphere in the culture vessel can eliminate signs of hyperhydricity and restore stomatal functionality (Cassells and Walsh 1994).

While the tissues may be green *in vitro*, their photosynthetic activity can be low due to the inhibition of RUBISCO by sucrose in the medium (De Riek et al. 1991). Indeed, while plants *in vitro* are described as heterotrophic, it may be more accurate to describe them as mixotrophic whereby some of the energy for growth is provided by photosynthesis but most by sucrose. Photoautotrophic culture has been shown to improve the growth of plants *in vitro*, advance their development, and improve their performance *post vitrum* (Zobayed et al. 2000). Photoautotrophic culture (syn. aseptic microhydroponics) uses a low phosphate medium free of organics and is compatible with inoculation with biocontrol bacteria and mycorrhizal fungi (Cassells 2000b). Inoculation of microplants *in vitro* and *post vitrum* confers protection against weaning stresses, including damping-off pathogens, and against biotic and abiotic stresses in the nursery and field (Cordier et al. 2000; Vestberg et al. 2002). Inoculation with arbuscular mycorrhizal fungi (AMF) and plant growth-promoting rhizobacteria (PGPR) *in vitro* has been advocated as more cost effective than *post vitrum* inoculation.

Little is known about the functionality of *in-vitro* produced roots (Davies and Santamaria 2000). It is possible that roots developed in aerated substrates may have greater functionality than those produced in agar or liquid culture.

Stage 4

If the above guidelines have been followed, microplants from hetero- and mixotrophic culture should establish at high frequency but, similar to seedlings, they will require protection against desiccation, damping-off diseases, and excessive light. This is provided by fogging or misting and the provision of shading in the first stages of weaning and establishment. Debergh et al. (2000) have discussed the loading of microplants with sucrose in Stage 3 to build up tissue energy reserves in a trade-off with the concomitant decrease in net photosynthetic capacity. They also argue for the importance of a functional scavenging system for reactive oxygen species (see Cassells and Curry 2001). Microplants may react adversely at the weaning stage to commonly used pesticides (Werbrouck et al. 1999). Sensitivity to the latter should be tested with a sample batch of microplants before general application.

MICROPLANT QUALITY

As discussed in the introduction, microplants should be fit for the purpose for which they are intended, that is, they should be pathogen and contaminant-free, genetically true-to-type, and of good developmental status and physiological quality. The health status of microplants has been reviewed extensively elsewhere (see Cassells 1997; Cassells et al. 2000) and is not discussed here. Historically, micropropagators and their clients have mainly concerned themselves with physiological quality, as this is the interface at which the micropropagator and the nursery person meet and not the development status. In the 1960s, some micropropagators believed they could market their plants "without frontiers," i.e., have an unregulated global market. But the concept of allowing tissue cultures unrestricted access across quarantine boundaries was unacceptable to the phytosanitary authorities and the clients found it difficult to establish plants from tissue culture vessels. The former issue remains dependent on health certification of the stock plants from which the cultures were derived. Microplants are now usually shipped bare-rooted, that is, washed free of medium and agar or other substrate.

The physiological quality of microplants is practically difficult to define other than by testing for percentage establishment and time taken in weaning, albeit there have been attempts to develop physiological markers (Cassells et al. 2000). Equally difficult is determination of the developmental (epigenetic) status of the microplants, that is, their development maturity. This can be simply defined as their time to flowering. It has long been recognized that explants taken from the mature phase of woody plants can be "rejuvenated" or "reinvigorated" (the term "partial rejuvenation" may be more accurate (Joyce et al. 2003)) by passage through tissue culture (George 1993). Thus, for example, while stem cutting from hydrangea may be difficult to root and show strong apical dominance, the progeny from micropropagation show prolific rooting and a well-branched habit. That nonwoody plants might also show difference in developmental maturity from *in vitro* compared with conventional propagules such as cutting, depending on protocol variables, has not generally been considered by micropropagators and yet it may have significant implications for both productivity *in vitro* and plant performance *post vitrum*. Here the hypothesis that stress *in vitro* influences genetic, epigenetic, and physiological quality of microplants *in vitro* and *post vitrum* is discussed with emphasis on its economic implications. For background see Cassells and Curry (2001) and Joyce et al. (2003)

Stress in *in vitro* Culture

Plants *in vivo* are exposed to environmental stresses and have well characterized biotic and abiotic stress responses (Lerner, 1999). These stress responses include defenses against excess light, heat, cold, drought, waterlogging, acidity, salinity, etc. Defenses are activated when homeostatic thresholds are exceeded and involve stress perception, signal transduction, activation of transcription, and genomic, proteomic, and metabolic changes resulting in an adaptive response. If the stress has not been too great, resulting in irreversible changes, on its removal the initial ground state is reestablished or a new adapted stage results. There is overlap between some stress responses and an oxidative stress response is a general component of all stress responses. Mammalian response to stress above the tolerance level is to migrate to a less stressful environment. Plants cannot do this; instead they retain greater genomic plasticity (Joyce et al. 2003). This is reflected in the retention of cell totipotency and deferral of germ line differentiation. During ontogeny plants undergo a phase change from the vegetative to reproductive phase and this may be associated with a change in leaf shape and plant habit and with the development of reproductive structures and loss of rooting potential (George 1993). *In vivo,* stress tends to accelerate developmental maturation and reproduction.

The effects of oxidative stress on the eukaryotic genome are an upregulation of the oxidative stress responses and, when homeostasis cannot be maintained, by the synthesis of stress proteins and stress metabolites. These are respectively enzymes associated with the synthesis of antioxidants and involved in the breakdown of reactive oxygen species, and antioxidants. There is also a shutdown of the cell cycle and activation of DNA repair mechanisms (Inze and Van Montagu 2002). Failure to neutralize reactive oxygen species results in base and chromosome damage including DNA demethylation which can result in mutation and altered gene expressions, resulting in epigenetic changes (Cassells and Curry 2001). It is not perhaps unexpected that cells in the vicinity of cut surfaces formed at explant excision mutate or die, as they are exposed to the oxidative cell burst following cell wounding. Mutation in adventitious regeneration ("somaclonal variation"; Larkin and Scowcroft 1981) can be seen as a consequence of the wound damage (generation of reactive oxygen species) mediated by the stress defenses of the host plant and the buffering potential of the genome (Cassells and Curry 2001). There may also be an influence of the culture environment in selecting for adapted mutated cell lineages. It is interesting that apical meristems, also exposed to oxidative damage at excision, do not express high frequency mutation in the buds formed from them. This may be due to a combination of stronger defenses against oxidative damage in apical cells and/or the influence of diplontic selection in displacing mutant cells from the apical initials (Cassells et al. 1993). The cell line instability in callus or in suspension culture is known both from loss of competence or loss of the ability to produce secondary metabolites and again suggests a mutation putative oxidative stress, with selection of adapted cell lines (George 1993).

There arise the questions as to which factors are affecting the physiological and developmental status of the tissues *in vitro* which can vary from hyperhydrated, to juvenile, to relatively mature morphologies. On occasion even flowering may be induced *in vitro* (George 1993). There is some evidence that in oxidative stress, auxin and ethylene may be involved. The evidence for auxin involvement is suggested by the alleviation of tip necrosis by induction of a transpiration stream in the microshoots or by increasing the calcium content of the medium (Cassells and Walsh 1994). The involvement of ethylene is seen by the use of membranes as vessel lids that have differential permeability to ethylene: where ethylene is allowed to accumulate, potato for example may form tubers, a mature character; where ethylene is partially retained, an "etiolated" morphology arises; and where ethylene is vented, a relatively more mature morphology is seen (Cassells et al. 2003). *In vitro* morphology, especially that of sensitive genotypes would seem to be affected by the relative humidity, accumulation of ethylene in the tissues and vessel (Cassells et al. 2003), and disturbed auxin metabolism and/or movement (Cassells and Walsh, 1994). Whether expressed *in vitro* as in potato in aberrant morphology, or not, as in *Dianthus*, the *in vitro* environment can affect the relative developmental status explants with economic consequences for *post vitrum* performance (Cassells et al. 2003; Cassells and Walsh 1998), for potato and *Dianthus* respectively, albeit, once in the natural environment, development maturity is progressively regained (Cassells et al., 1999).

Caution is urged when extrapolating from *in vivo* stress responses to *in vitro* stress responses. First, the physiology of plants *in vitro* differs from that *in vivo* and they may not be receptive to some environmental cues. Second, plants *in vitro* have relatively larger energy resources available from the sucrose in the medium. Third, unusual stress combinations may coincide *in vitro,* e.g. light stress, high humidity with possible waterlogging and entrapment of ethylene in the tissues and ethylene accumulation in the culture vessel, osmotic stress, medium hormone stress, etc. The responses may be adapted with cross interaction between the stress pathways to achieve novel outcomes.

TRENDS IN COMMERCIAL MICROPROPAGATION

Would pioneers of commercial micropropagation from the 1970s see any differences in business plans and practices in laboratories of the 2000s? Certainly. In the high labor-cost countries there has been a move away from contract micropropagation which may involve costly learning curves for new varieties, to in-house applications of the technology with linkage to early release of new, royalty-bearing or protected in-house varieties and/or to increasing profitability by exploiting the added-value chain by growing on the microplants, selling finished plants thereby accumulating profit margins along the chain. The structure of the plant production industry has been shown in Fig. 12.3. Had the early pioneers kept up with the literature, they would have read of artificial seeds based on encapsulation of somatic embryos, of mechanization (Cassells et al. 1993), of plant production in bioreactors (Hoslef-Eide and Preil 2004), and of photoautotrophic plant culture (Zobayed et al. 2000). However, arguably, relatively little has changed in practice in cash-flow dependent microproduction companies, that is, those not subsidized by large parent

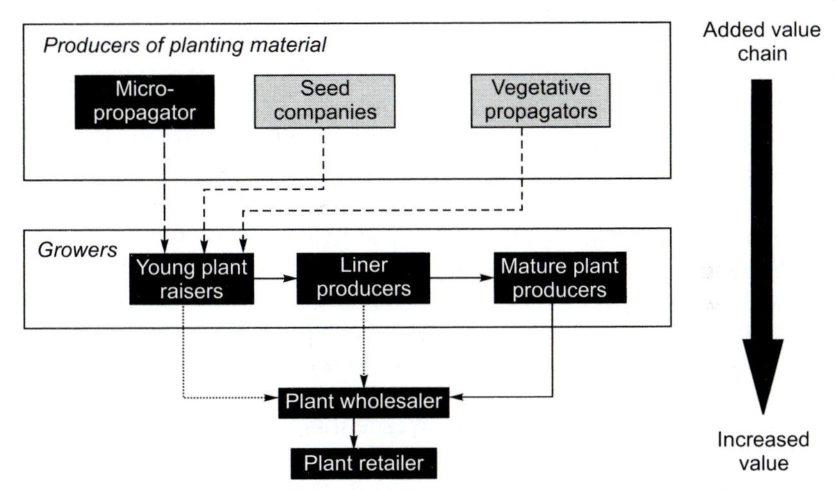

Fig. 12.3 Structure of the plant production industry

companies. Micropropagators generally still use small inexpensive glass food jars (reusable) or cheap plastic food containers (not reusable) as culture vessels. The main reason for this is to contain contamination by environmental microorganisms and microarthropods (see autotrophic culture). Usually the vessel lid is open, normally with a cellulose washer to reduce the risk of contamination, while allowing gaseous exchange. The container may also be wrapped in "cling-film", which allows gaseous exchange while reducing the risk of contamination. One procedure adopted by some micropropagators is the use of bottom cooling (Maene and Debergh 1987). Bottom cooling of the culture vessels is achieved by circulating ground water through the base of the shelves. This promotes development of a transpiration stream in the microshoots as the water vapor condenses on the cooled surface of the medium. Bottom cooling also promotes rooting in many species. Murashige and Skoog (1962) basal medium is still used very extensively, in approximately 95 % of published papers, but specialized media are used for woody plants, rhododendron, etc. (for extensive lists of media and protocols, see George 1996). Agar, in spite of its known variability between batches, is also still widely (Cassells and Collins 2000). While the composition of the minerals, organics, and growth substrate, again almost universally sucrose, remains unchanged (see above), growth regulators have changed. In addition to the traditional auxins, cytokinins, and gibberellins, there is now widespread use of thidiazuron (TDZ; 1-phenyl-3-(1,2,3-thiadiazol-5-yl)urea), particularly in somatic embryogenesis (Murthy et al. 1998).

As mentioned above, labor costs in conventional micropropagation are high and so there is a potential market for innovative technology. Aside from artificial seed and mechanization, two strategies are advocated to improve productivity, namely liquid culture in bioreactors (Hvoslef-Eide and Preil 2004) and photoautotrophic culture (Zobayed et al. 2000). A form of liquid culture, temporary immersion culture (Teisson and Alvard 1995), has attracted much attention in recent years, mainly from academic and "not for profit" production. This is based on culture in conventional sized vessels in which the medium is pumped in/out at experimentally predetermined intervals; it has similarities to misting of plants in that feeding is via all plant surfaces. That the plant tissues are not permanently submerged appears to overcome the problem of hyperhydricity associated with tissue submersion. The temporary immersion system is both more capital intensive and arguably more labor intensive than use of conventional culture vessels. To overcome this there has been some commercial interest and take-up of simple 2-5 L bioreactors for production (Ziv and Shemesh 1996; *see www.osmotech.com*). These are aerated and have been successfully used for the commercial production of some species (Ahroni 2002). As with scale-up in any heterotrophic culture system the problem of excluding microbial contaminants remains (Levin et al. 1996). For a review of the use of bioreactors in micropropagation see Hvoslef-Eide and Preil (2004)

The move to photoautotrophic culture (syn. aseptic microhydroponics) allows scalability from conventional containers to large aseptic growth-rooms (Zobayed et al. 2000). However, the capital costs can be high for optimization as elevated light is required with the concomitant costs of removing the heat generated and supplying CO_2 . The optimal light and CO_2 supplementation are in the range PFF 100-200 µmol m^{-2} s^{-1} and 1,000 ppm respectively (Kozai et al. 1997). If contaminants get into the system they may become systemic but are limited by the nutrient availability from root leakage and so may not cause damping-off. The main problem is that the explants may need to be primed before becoming autotrophic. This can involve a hormonal treatment to achieve bud break and initial exposure to sugar. In spite of the interface between photoautotrophic culture (aseptic microhydroponics) and soilless culture which allows the transfer of extensive technology and expertise to the former, the capital costs, and perhaps the restriction to a limited range of crop species, have limited the uptake of this technology by commercial micropropagators.

Notwithstanding the lack of progress in reducing the labor component of micropropagation, there has been some progress in improving quality of microplants *post vitrum* by inoculating them *in vitro* but mainly *in vivo*, with biological inoculants. Aseptic microplants represent a "biological vacuum" analogous to aseptically sprouted seeds. The root exudates are nutrient rich and will support a vigorous microbial population. The beneficial effects of mycorrhizal fungi and PGPR in promoting plant growth and sensitizing the plants to microbial attack have been demonstrated (Cordier et al. 2000; Vestberg et al. 2002). In an integrated strategy to optimize *in vitro* productivity and microplant quality, it has been demonstrated, using potato as a model, that it is possible to manipulate the plant habit in Stage 2 to facilitate automated cutting, followed by transfer to autotrophic culture in Stage 3 for improvement of physiological quality and inoculation with AMF and PGPR in an *in vitro* weaning strategy (Cassells et al. 2003).

CONCLUSIONS

While it is frustrating for academic and applied tissue culture researchers that sometimes their research is not exploited by commercial micropropagators, it is not surprising when the structure of the industry is considered. Aside from the relatively low profitability of the industry and its rising costs due to wage inflation, the industry is, in its sector, relatively capital intensive and contract micropropagators especially require flexible technology. Advanced processes, e.g. based on the use of bioreactors, may only be applicable to a narrow range of crop species (Ahroni 2002). The greater potential for technological progress rests with the large crop specialist companies, such as forestry companies, or breeders of large volume crops with good profit margins, e.g. some of the ornamental or plantation crops albeit they have to compete with production in low labor cost economies. But in this sector, for many crops that are profitable to micropropagate the market size may be small and the market easily over supplied, resulting in a price collapse.

Micropropagation has established some viable niche markets and in Europe has survived the challenge from competition from lower labor-cost regions (O'Riordain 2002). A better understanding of the physiology and epigenetics of plants *in vitro* will continue to lead to quality improvement. The ability to stabilize plants in adventitious regeneration may lead to new opportunities to automate production but with the caveat that even with significant reduction in labor costs, interest on loans, consumables costs, energy and insurance and other business costs will still restrict the competitiveness of micropropagation in the plant propagation industry. Meanwhile exploitation of biological inocul-ants may contribute to overall *post vitrum* improvement in performance that achieves an value added propagule for the micropropagator.

REFERENCES

Ahroni M (2002) Aspects of commercial plant tissue culture propagation in liquid media. *In*: T Hvoslef-Eide and W Preil (eds). 1st Int. Sym. Liquid Systems for *in Vitro* Mass propagation of Plants. Dept. Horticulture and Plant Science Aas. pp. 76-77.

Boxus Ph, Jemmali A, Terzi JM, Arezki O (2000) Drift in genetic stability in micropropagation: the case of strawberry. Acta Hort. 530: 155-162.

Cassells AC (1997) Pathogen and Microbial Contamination Management in micropropagation. Kluwer, Publ., Dordrecht, Netherlands 370 pp.

Cassells AC (2000a). Contamination detection and elimination. *In*: RE Spier (ed.). Encyclopedia of Plant Cell Biology. John Wiley and sons inc.,Chichester, UK, pp. 577-586.

Cassells AC (2000b) Aseptic microhydroponics: a strategy to advance microplant development and improve microplant physiology. Acta Hort 530: 187-194.

Cassells AC, Collins IM (2000). Characterization and comparison of agars and other gelling agents for plant tissue culture use. Acta Hort. 530: 203-212.

Cassells AC, Curry RF (2001). Oxidative stress and physiological, epigenetic and genetic variability in plant tissue culture: implications for micropropagators and genetic engineers. Plant Cell Tiss. Org. Cult. 64: 145-157.

Cassells AC, Doyle BM (2003) Pathogen indexing. *In*: RN Trigiano, DJ Gray (eds.). Plant Tissue Culture and Development CRC Press, Boca Raton, FL (USA). (in press)

Cassells AC, Doyle BM, Curry RF (2000) Proc. Int. Sym. Methods and Markers for Quality Assurance in Micropropagation. Acta Hort pp. 530.

Cassells AC, Hurley D, Long R, O'Shea A, Perrin AP (1993). Improved equipment used to cut plants being used in micropropagation systems has rotating holder carrying inverted tray containing microplants, across which cutters are passed to cut plant nodes which are collected and sealed in trays wrapped in thin film material. World Patent Applic. No. 9319586

Cassells AC, Joyce SM, O'Herlihy EA, Perez-Sanz MJ, Walsh C (2003) Stress and quality in *in vitro* culture Acta Hort. 625.

Cassells AC, Kowalski B, Fitzgerald DM, Murphy GA (1999). The use of image analysis to study developmental variation in micropropagated potato (*Solanum tuberosum* L.). Potato Res. 42: 541-548.

Cassells AC, Walsh C (1998) Characteristics of *Dianthus* microplants grown in agar and polyurethane foam using air-tight and water-permeable vessels lids. *In*: G Reuther (ed.). Physiology and Control of Plant Propagation *in vitro*. European Communities, Luxembourg, pp. 122-126.

Cassells AC, Walsh C, Periappuram C. (1993). Diplontic selection as a positive factor in determining the fitness of mutants of *Dianthus* 'Mystere' derived from X-irradiation of nodes in *in vitro* culture. Euphytica 70: 167-174.

Cassells, A.C. and Walsh, C. (1994). The influence of the gas permeability of the culture lid on calcium uptake and stomatal function in *Dianthus* microplants. Plant Cell Tiss. Org. Cult. 37: 171-178.

Cordier C, Lemoine MC, Lemanceau P, Gianinazzi-Pearson V, Gianinazzi S (2000) The beneficial rhizosphere; a necessary strategy for microplant production. Acta Hort 530: 259-268.

Davies WJ, Santamaria JM (2000) Physiological markers for microplant shoot and root quality. Acta Hort 530: 363-376.

De Riek J, Van Cleemput O, Debergh PC (1991) Carbon metabolism of micropropagated *Rosa multiflora* L. *in vitro.* Plant 91: 57-63.

Debergh PC, Maene LJ (1981) A scheme for the commercial propagation of ornamental plants by tissue culture. Sci. Hort. 14: 335-345.

Debergh PC, Topoonyanout N, Van Huylenbroerck J, Moreira da Silva H, Oyaert E (2000) Preparation of microplants for *ex vitro* establishment. Acta Hort 530: 269-276.

George EF (1993) Plant Propagation By Tissue Culture 1. The Technology. Exegetics, Basingstoke, UK.

George EF (1996) Plant Propagation by Tissue Culture 2. In Practice. Exegetics, Basingstoke, UK.

Grunewaldt-Stocker G (1997) Problems with the health of *in vitro* propagated *Anthurium* spp. and *Phalaenopsis* hybrids. *In*: AC Cassells (ed.) Pathogen and Microbial Contamination Management in Micropropagation. Kluwer Publ Netherlands, Dordrecht, pp 363-370.

Hartman HT, Kester D, Davies F, Geneve R (2001) Hartman and Kester's Plant propagation: Principles and Practice. Pearson, NJ (USA).

Hvoslef-Eide T, Preil W (2003) Plant Propagation in Liquid Culture. Kluwer Publ., Dordrecht, Netherlands.

Inze D, Van Montagu M (2002) Oxidative Stress in Plants. Taylor and Francis, London, UK.

Joyce SM, Cassells AC, Mohan Jain S (2003). Stress and aberrant phenotypes *in vitro* culture. Plant Cell Tiss. Org. Cult. (in press)

Karp A, Isaac PG, Ingram DS (1998) Molecular Tools for Screening Biodiversity. Chapman and Hall, London, UK.

Kowalski B, Cassells AC (1999). Mutation breeding for yield *and Phytophthora infestans* (Mont.) de Bary foliar resistance in potato (*Solanum tuberosum* L. cv. Golden Wonder) using computerised image analysis in selection. Potato Res. 42: 121-130.

Kozai T, Kubota C, Jeong BR (1997) Environmental control for the large-scale production of plants through *in vitro* techniques. Plant Cell Tiss. Org. Cult. 51: 49-56.

Larkin PJ, Scowcroft WR (1981) Somaclonal variation—a novel source of variability from plant cultures for plant improvement. Theor. Appl. Genet. 60: 197-214.

Leifert C, Cassells A. C. (2001). Microbial hazards in plant tissue and cell cultures. *In vitro* - Plant 37: 133-138.

Lerner HR (Ed) (1999) Plant Responses to Environmental Stresses. Marcel Dekker Inc., New York, NY.

Levin R, Stav R, Alper Y, Watad AA (1996) *In vitro* multiplication in liquid culture of *Syngonium* contaminated with *Bacillus* spp. and *Rathayibacter triciti*. Plant Cell Tiss. Org. Cult. 45: 277-280.

Maene LJ, Debergh PC (1987) Optimalization of the transfer of tissue cultured shoots to *in vivo* conditions. Acta Hort. 212: 335-348.

Murashige T (1974) Plant propagation through tissue cultures Annu. Rev. Plant Physiol. 25: 135-166.

Murashige T, Skoog F (1962) A revised medium for rapid growth and bio-assays with tobacco tissue cultures. Physiol. Plant 15: 473-497.

Murthy BNS, Murch S, Saxena PK (1998) Thidiazuron: A potent regulator of plant morphogenesis. *In vitro* Cell Dev. Biol. 34:267-275.

O'Herlihy EA, Cassells AC (2003). Influence of *in vitro* factors on titre and elimination of model fruit tree viruses. Plant Cell Tiss. Org. Cult. 72: 33-42.

O'Riordain F (2002) COST 843 – the directory of European plant tissue culture. European Commission, Luxemboury.

Prakash J (2000) Factors influencing the development of the micropropagation industry: the experience in India. Acta Hort. 530: 165-172.

Taiz L, Zeigler L (2002) Plant Physiology. Sinauer, Sunderland (3rd ed.).

Teisson C, Alvard D (1995) A new concept of plant *in vitro* cultivation liquid medium: temporary immersion. *In*: M Terzi, R Cella, A Falavigna (eds.). Current Issues in Plant Molecular and Cellular Biology. Kluwer publ., Dordrecht, the Netherlands, pp 105-109.

Vestberg M, Cassells AC, Schubert A, Cordier C, Gianinazzi S (2002). AMF and micropropagation of high value crops. *In:* Mycorrhizal Technology in Agriculture: From Genes to Bioproducts Birkauser, Zurich, pp. 223-233.

Weller R (1997) Microbial communities on human tissue: an important source of contaminants in plant tissue cultures. *In*: AC Cassells (ed.). Pathogen and Microbial Contamination Management in Micropropagation. Kluwer Publ., Dordrecht, the Netherlands, pp. 245-258.

Werbrouck S, Goethals K, Van Montagu M, Debergh P (1999) Surprising micropropagation tools. *In*: A Altman, M Ziv, S Izhar (eds.). Plant Biotechnology and *in vitro* Biology in the 21st Century, Kluwer Publ., Dordrecht, Netherlands, pp. 667-672.

Ziv M, Ariel T (1994) Vitrification in relation to stomatal deformation and malfunction in carnation leaves *in vitro*. *In*: PJ Lumsden, JR Nicholas, WJ Davies (eds.). Physiology, Growth and Development of Plants in Culture. Kluwer Publ., Dordrecht, the Netherlands, pp. 143-154

Ziv M, Shemesh D (1996) Propagation and tuberization of potato bud cluster from bioreactor culture. *In vitro* Cell Dev. Biol. – Plant 32: 31-36.

Zobayed SMA, Afreen F, Kozai T (2000) Quality biomass production via photoautotrophic micropropagation. Acta Hort. 530: 377-386.

13

Micropropagation of Floricultural Crops

G.R. Rout[1] and S. Mohan Jain[2]
[1]*Plant Biotechnology Division, Regional Plant Resource Centre*
Bhubaneswar-751 015, India
[2]*International Atomic Energy Agency, FAO/IAEA Joint Division*
Box-100, A-1400, Vienna, Austria

INTRODUCTION

Flowers are used for all occasions. The worldwide floricultural industry is valued at over 50 billion Euros and is increasing annually by 8-10 percent, especially in the developing countries such as India, Kenya, Colombia, Ecuador and Thailand. However, a few other countries, e.g. Vietnam, Malaysia, the Philippines are trying to enter this lucrative market. Concomitantly a worldwide slow down in floriculture production has been noted. This is mainly due to high labor and production costs; as a result more and more investment is flowing to the developing countries to boost the floriculture industry. Major flower crops such as roses, orchids, carnation, and chrysanthemum are being produced in the developing countries. The cut flower market is valued at 20 billion Euros; half are produced in the Netherlands, the USA, and Japan. Among cut flowers, roses, tulip, mums, gerbera, carnation are major money-makers. In floriculture, introduction of new cultivars on the market is essential (e.g. in the fashion industry, new or unusual traits increase the demand). Furthermore, production of new cultivars is required to meet increasing environmental demands, in particular the reduction of chemicals used for disease and pest control.

Presently, major producers of flowers are a few developing countries, the Netherlands and Japan. The shares of major producers are the Netherlands (33%), Japan (24%), Italy (11%), USA (12%), Thailand (10%) and others (14%). The major exporting countries are the Netherlands (59%),

Colombia (10%), Italy (6%), Israel (4%), Spain (2%), Kenya (1%), and others (18%). The four leading exporters (Holland, Colombia, Italy and Israel) constitute about 80% of the world market. The share of the developing countries of Africa, Asia, and Latin America as well as Thailand, Eucador and India is less than 20% (Janakiram 2000; Rajagopalan 2000). Considering the Indian scenario, in a short span of floriculture trade, roughly seven years or so, floriculture farming has shown a positive trend. India has achieved the distinction of becoming the biggest supplier of flowers to Japan. The per-centage growth of Indian Floriculture to Japan is 353.79% and Spain 587.9%. Indian flower exports are 45% Europe, 35% Japan, 5% USA, 5% Singapore, and 10% other countries (Singh 2001).

The international trade in cut flowers is growing by about 11% each year. The major importing countries for cut flowers are Germany, the USA, France, the UK, the Netherlands, Japan and Switzerland. The world market of cut flower products is worth about US $8 billion. However, growth is mostly concentrated in western Europe. North America and Japan sales through the Aalsmer flower auction in 1998 approached Dfl. 3000 million (US $1657 Million), a 5.1% increase over sales in 1977 which yearly shows that the trend is taking a swing toward marketing, re-engineering business processes, and internationalization (Negi 2000; Singh 2001). Flower cultivation is a good source of foreign revenue. India is in a fortunate position of having a diversity of climate and soil for growing a variety of ornamental crops year round, unlike the severe winters prevailing in the West. However, the trade in cut flowers is negligible compared to a small country like the Netherlands. The main lacunae lie in its postharvest technology of cut flowers, which needs to be strengthened to increase flowers quality.

IN VITRO PROPAGATION

In vitro culture is one of the key components of plant biotechnology that makes use of the totipotent nature of plant cells, a concept proposed by Haberlandt (1902) and unequivocally demonstrated for the first time by Steward et al. (1958). Tissue culture is a blanket term for protoplast, cell, tissue, and organ culture (Bhojwani and Razdan 1996). Plants have been regenerated from different explants, excised from *in vitro*-grown and field-grown plants. *In vitro* propagation can be employed for large-scale production of disease-free clones, mass cloning of selected genotypes, and gene pool conservation. The ornamental industry has intensively applied the *in vitro* propagation approach for large-scale plant multiplication of elite superior varieties. As a result, hundreds of plant tissue culture laboratories have come up worldwide, especially in the developing countries, mainly due to labor intensive and cost-effective technology.

Micropropagation

Micropropagation is an alternative method of vegetative propagation. It involves four distinct stages: shoot initiation, shoot multiplication, rooting of the micropropagated shoots and acclimatization. The first stage of initiation of culture depends on the type and the physiological state of the donor plant at the time of explant excision. Explants from actively growing shoots are generally used for large-scale plant multiplication. Shoot multiplication is the crucial stage and is achieved by adding plant growth regulators (auxins and cytokinins) to the culture medium. Elongated shoots derived from the multiplication stage are subsequently rooted either *ex-vitro* or *in vitro*. In some cases, liquid medium favors *in vitro* root induction from excised shoots. This is usually done either on a filter paper bridge soaked in the liquid medium or by constant shaking of shoot cultures at low speed on a horizontal shaker. Continuous shaking of cultures seems to be more appropriate for shoot multiplication due to better aeration of cultures. Transplantation or hardening of *in vitro* plants is also an important stage in micropropagation. Plants multiplied *in vitro* are exposed to a unique set of growth conditions, which are invariably controlled (high levels of organic and inorganic nutrient ions, growth regulators, a carbon source, high humidity, low light, and poor gaseous exchange). These factors support rapid growth and multiplication, but the conditions induce structural and physiological changes in the plant that may render it unfit for survival under field conditions. Thus a gradual acclimatization from culture to field conditions is necessary to adapt to new different growing conditions, i.e. low humidity/high irradiance.

Photoautotropic micropropagation has recently been proposed as a means of reducing production costs and automation-robotization of the micropropagation process (Kozai 1988). Carbon dioxide is used as a carbon source instead of sucrose. *In vitro* plants are grown in a carbon dioxide enriched environment under high photon flux. Photoautotropic micropropagation of various crops including ornamental plants has been reviewed (Kozai et al. 1988; Kozai 1990a, b; Kozai et al. 1990a,b). The concept of photoautotropic micropropagation is an interesting idea for developing an automated micropropagation system to achieve a drastic reduction in production costs (Kozai 1991a, b). So far, the method has yet to be adopted by commercial micropropagators, even though it does cut down contamination rate as well as some laboratory operational costs.

Liquid medium seems to be more effective for shoot regeneration and root induction, which is due to better aeration. Simmonds and Werry (1987) used liquid medium for enhancing the micropropagation profile of *Begonia* x *hiemalis*. Watad et al. (1997) compared performance of agar-solidified medium and interfacial membrane rafts floating on liquid medium for shoot multiplication and root induction. The results showed shoot multiplication was highest on membrane rafts floating on the liquid medium, and also plants rooted much better. Similarly, Osternack et al. (1999) succeeded in inducing somatic embryogenesis and adventitious shoots and roots from hypocotyl tissues of *Euphorbia pulcherrima* on cytokinin containing medium. Subsequently, Preil (2003) noted that the regeneration potential of isolated cells, tissue or organs and the callus cultures is highly variable. Furthermore, petiole cross sections cultivated on auxin and cytokinin containing medium give rise to adventitious shoots from epidermal cells and subepidermal cortex cells, never from pith cells of the central regions of the petiole.

Somatic Embryogenesis

Somatic embryogenesis is another method of micropropagation. Under suitable *in vitro* conditions, somatic cells undergo embryogenic development, almost comparable to that exhibited by the zygote, which is a result of gametic fusion. Somatic embryos are bipolar structures that arise from individual cells and have no vascular connection with the maternal tissue of the explant (Haccius 1978; vonArnold et al. 2002). Embryos may develop directly from somatic cells (direct embryogenesis) or development of recognizable embryogenic structures is preceded by numerous organized, nonembryonic mitotic cycles (indirect embryogenesis). Considerable success has been achieved using this technology in floricultural crops, e.g., chrysanthemum (*Dendranthema grandiflorum*) (May and Trigiano 1991; Tanaka et al. 2000), carnation (*Dianthus*) (Choudhary and Kokchin 1995), rose (*Rosa hybrida*) (Rout et al. 1991; Rout et al. 1999), anthurium (*Anthurium andreanum*) (Kuehnle and Sugii 1992), petunia (Handro et al. 1973; Rao et al. 1973a; Sangwan and Harada 1976; Gupta 1982), orchids (Chen et al. 1999; Chen and Chang 2000a), and gladiolus (Remotti 1995). Kunitake et al.(1993) noted the ability of the zygotic embryo-derived calli of *R. rugosa* to develop somatic embryos on media containing no exogenous growth regulators; however, their embryogenic potential lasted no longer than 6 months. Induction of somatic embryogenesis is genotypic dependent, e.g. roses (Hsia and Korban 1996), type of growth regulators (Roberts et al. 1990), explant type in chrysanthemum (*Dendranthema grandiflora)* (May and Trigiano 1991), and sucrose concentration of the culture medium, e.g. 9 –18% sucrose. Twelve of the 23 cultivars evaluated produced somatic embryos, but complete plants were recovered only from five cultivars (May and Trigiano 1991). Remotti (1995) was successful in obtaining primary and secondary somatic embryogenesis from cell suspension cultures of *Gladiolus* x *grandiflorus* and 60% germination rate of somatic embryos in liquid medium with a filter paper bridge without growth regulators. Plants transferred to soil showed a 100% recovery and developed normally. Similarly, Choudhary and Kokchin (1995) were successful in inducing somatic embryogenesis and plantlet regeneration in *Dianthus caryophyllus*, which grew well in the field. There are advantages and disadvantages in using somatic embryogenesis in large-scale plant multiplication (Jain 2002). The major advantages are large-scale somatic embryo production on bioreactors, and clonal propagation. The major limitation is genotypic dependence of somatic embryo production and germination.

Bioreactor Production

Plant propagation by tissue culture techniques can be encumbered by the intensive labour requirement for the multiplication process. Thus, scaling-up systems and automation of unit operations are necessary to reduce production costs (Aitken-Christie 1991; Ziv 1991, 1992; 1995; Preil 1991). The various propagation aspects of several plant species in bioreactors, application and some of the problems associated with the operation of bioreactors have already been reviewed (see Takayama and Akita 1998; Ziv 2000; Paek et al. 2001). The major problems facing the bioreactor system are sensitivity to shear force, mechanical damage, and foam formation in bubble aerated bioraectors. Mechanization and automation of the micropropagation process can greatly overcome the limitation of labor costs (Paek et al. 2001). Several researchers have worked on automation of plant multiplication by dealing with repeated cutting of shoots, separation, subculture, and transfer of buds, shoots, or plantlets during the multiplication and transplanting phases (Levin et al. 1988; Aitken-Christie 1991; Aitken-Christie et al. 1995; Vasil 1994). Progress in tissue culture automation will depend on the use of liquid cultures in bioreactors, which will allow fast proliferation, mechanized cutting, separation, and automated dispensing (Sakamoto et al. 1995). Adapting bioreactors with liquid media for micropropagation is favorable given the ease of scaling-up (Preil 1991) and precluding physiological disorders of shoot and leaf hyperhydricity (Ziv 1999). Liquid media have been used for somatic embryos and cell suspension cultures in both agitated flasks and various types of bioreactors (Smart and Fowler 1984; Tautorus and Dunstan 1995; Ziv 2000; Paek et al. 2001). Although bioreactors have been used for cell suspension cultures and somatic embryogenesis (Takayama and Akita 1998; Ziv 2000), shoot bud proliferation of *Ornithogalum dubium* was done by using column bubble bioreactors (Ziv et al. 1997). The proliferation rate was enhanced about 8-fold during the second 50-day-long subculture. Further, Lim et al. (1998) developed a 2-stage culture process on a pilot scale mass production of *Lilium* bulblets *in vitro*. Several thousand bulblets were produced in 5-20 litre batches in nonstirred bioreactors within 60 days. Young et al.(2000) reported mass multiplication of *Phalaenopsis* through bioreactor systems. About 18,000 protocorm-like bodies were harvested from 20 g inoculum an after 8 week incubation. A partial oxygen pressure in bioreactors assisted cell proliferation and subsequent differentiation of somatic embryos from suspension cultures of *Cyclamen persicum*. A significantly higher number of germinating embryos was obtained from cultures grown at 40% pO_2 than from those grown in flasks or in bioreactors at 5, 10, and 20% pO_2 (Hohe et al. 1999). Seon et al.(2000) reported that mass production of *Lilium* bulblets was achieved using bioreactor culture with periodic medium supplementation in the cultures. Paek et al. (2001) compared the application of different bioreactor culture techniques for micro-propagation of *Lilium*. Periodic immersion liquid culture using the ebb-and-flood system and column–type bubble bioreactors equipped with a raft support system to maintain plant tissues at the air and liquid inter-face, proved suitable for micro-propagation of plants via organogenesis. Balloon-type bubble bioreactors were suitable for somatic embryo production due to less shear on cultured cells. These experiments established the two-stage culture method of bioreactor production for several cultivars of *Lilium,* resulting in a nearly 10-fold increase in bulblet growth.

Recently, Etienne and Berthouly (2002) reviewed work on the temporary immersion system for plant propagation. Among ornamental plants, this system has to date been used with orchids. It has the advantage of combining aeration and the positive effect of liquid medium. It avoids continuous immersion of cultures, provides adequate oxygen supply, limits shear levels, enables sequential medium changes and automation, reduces contamination, and offers a low production cost. The use of bioreactors for plant propagation of floricultural crops is summarized in Table 13.1.

Table 13.1 Micropropagation of floricultural crops through bioreactor.

Species	*Response*	*References*
Amaryllis hippeastrum	buds, plants, bulblets	Takayama and Akita 1998
Cyclamen persicum	callus, somatic embryos	Hvoslef-Eide and Munster 1998 Hohe et al. 1999
Dianthus caryophyllus	shoots, plant	Chatterjee et al. 1997
Gladiolus grandiflorum	bud clusters, plants, corms	Ziv 1990a,b; 1991, Ziv et al. 1998
Lilium spp.	Plants, bulblets, plants	Takayama 1991, Takayama and Akita 1998, Lim et al. 998; Seon et al. 2000; Paek et al. 2001
Phalaenopsis	protocorm-like bodies	Young et al. 2000

MICROPROPAGATION VIA THIN CELL LAYER

Thin cell layer is the model systems and applications in higher plant tissue and organ culture and genetic transformation (Teixeira da Silva, 2003). Moreover thin cell layer technology is a solution to many of the issues currently hindering the efficient progress of ornamental and floricultural crop improvement, since it solve the initial step i.e. plant regeneration problem. This technology has also been effectively used in the micropropagation of various crops including floricultural crops (Tran Thanh Van & Bui, 2000; Fiore *et al.,* 2002; Nhut *et al.*, 2003a,b; Teixeira da Silva and Nhut, 2003). Recently, Teixeira da Silva (2003) published a detailed review on the use of thin cell layer technology in ornamental plant micropropagation and biotechnology, which highlights organogenesis and somatic embryogenesis for plant regeneration and genetic improvement via transformation. Mulin and Tran Thanh Van (1989) indicated that *in vitro* shoots and flowers were formed from thin epidermal cells excised from the first five internodes of basal flowering branches in *Petunia hybrida.* Explants (1x10 mm^2) consisted of 3-6 layers of subepidermal and epidermal cells produced vegetative buds within 2 weeks of culture. Ohki (1994) reported that 100-200 shoots per tTCL (transverse thin cell layer) explants were obtained from 03-0.5 mm petiole or 3 x 3 mm lamina sections, respectively of *Saintpaulia ionantha* within 4 weeks of culture. Over 70,000 plants were produced from a single leaf within 3-4 months. Gill *et al* (1992) used tTCL hypocotyl explants (1x10mm) of 1-week-old geranium *(Pelargonium* x *hortorum)* hybrid seedlings for induction of somatic embryogenesis. They observed that the development of somatic embryos was rapid and the number of embryos was about 8-fold higher than in the culture of whole hypocotyl explants. Hsia and Korban(1996) achieved organogenic and embryogenic callus and subsequently regeneration from ITCL (longitudinally thin cell layer) explants derived from dormant bud floral stalks *of Rosa hybrida* cv. Baccara.

Thin cell layer systems could be used as a tool for *in vitro* regeneration and micropropagation. Recent progress in thin cell layer technology has opened new possibilities for improvement of ornamental and floricultural crops.

IN VITRO PROPAGATION FOR NEW VARIABILITY

Protoplast culture

Protoplasts are important for creating genetic variability, somatic cell fusion and genetic transformation for genetic improvement of plants (Binsfeld et al. 1999). For plant regeneration from protoplasts the choice of donor tissue plays an important role in combination with the culture technique. Progress and understanding of protoplast regeneration requires thorough investigation of the cellular processes after transfer of isolated protoplasts to the cultural medium, the process of cell dedifferentiation, cell cycle induction, and programming cells to meristematic competence. Considerable progress has been made with regard to plant regeneration from protoplast of a number of important crops including horticultural plants (Kirby 1988; Puite 1992). Protoplast regeneration was reported for commercial clones of chrysanthemum (Sauvadet et al. 1990), *Saintpaulia* (Winkelmann and Grunewaldt 1995a,b; Hoshino et al. 1995), *Rosa persica* x *xanthina* (Matthews et al. 1991, 1994), and several rose species (Schum et al. 2002). Winkelmann and Grunewaldt (1995a) obtained completely altered flower colurs in plant saintpaulia regenerated from protoplasts.

Protoplast fusion is useful for developing somatic hybrids or cybrids for transfer of traits from one plant to another, especially in sexually incompatible plants and intergeneric plant species (Jain and Maluszynski 2002). The first petunia somatic hybrid was produced in a combination between *Petunia parodii* and *P. hybrida* (Power et al. 1976). Subsequently, somatic hybrids were produced between *P. parodii* and *P. inflata*, two species which exhibit unilateral incompatibility, with pollination of the former species mandatory at the bud stage to obtain the hybrid (Power et al. 1979; Sink et al. 1978). The somatic hybrid plants were confirmed by their intermediate flower morphology, presence of anthocyanin pigmentation, and segregation for parental traits in the offspring. Oh and Kim (1994) achieved regeneration of plants from petal protoplasts of *Petunia hybrida*. More than hundred plants with morphologically normal shoots and roots have been established. Schum et al. (2002) exploited protoplast fusion for introgression of disease resistant genes in rose breeding. Putative somatic hybrid calli were obtained by fusing protoplasts of *Rosa* × *hybrida* and *R. wichuraiana* or *R. multiflora* or *R. roxburghii*. Shoots were regenerated from putative somatic hybrid calli of *R.* × *hybrida* and *R. wichuraiana*. The hybrid nature of some selected regenerants was confirmed by flow cytometry and AFLP analysis. Therefore, somatic cell fusion technology study is certainly helpful in production of hybrids of commercially important floricultural crops, especially for flower color modification.

In vitro Mutagenesis

Most of the available genetic variation used in breeding program has occurred naturally and exists in germplasm collections of new and old cultivars, land races and genotypes. This variation through crosses is recombined to produce new and desired gene combinations (Maluszynski et al. 1995). When existing germplasm fails to provide the desired recombinant, it is necessary to resort to other sources of variation. Since spontaneous mutations occur with extremely low frequency, mutation induction techniques provide tools for the rapid creation and increase in variability in crop species (Jain 1997b, 2000). Mutagens cause random changes in the nuclear DNA or cytoplasmic organelles resulting in gene, chromosomal or genomic mutations (Jain, 2002). The International Atomic Energy Agency (IAEA), Vienna maintains a mutant database of officially released mutant varieties in different parts of the world (*www.iaea.org*). Over 552 mutant varieties of ornamental and decorative plants have been officially released wordwide (*www.iaea.org*). In many vegetatively propagated crops mutation induction in combination with *in vitro* culture techniques may be the only effective method for plant improvement (Novak 1991). Mutations have been induced by chemical (e.g. ethylmethanesulfonate) and physical (e.g. gamma radiation) mutagen treatment for breeding new ornamental plants because ornamental traits are readily monitored. Moreover, ornamental characteristics are highly suitable for improvement by mutation-assisted breeding. Plants growing in a natural environment have been under selection pressure for performance in their environment and not for ornamental value. Some traits of ornamental plants selected from induced mutations are flower characters (color, size, morphology, fragrance), leaf characters (form, size, pigmentation), growth habit (compact, climbing, branching, dwarf), and physiological traits such as photoperiodic response, early flowering, flower shelf life and tolerance to biotic and abiotic stresses.

Considerable work has been done on induced mutations in roses and chrysanthemum using ethylmethanesulfonate (Kaicker 1982), ionizing irradiations (Chan 1966; Broertjes and Van Harten 1978; Smilanksy et al. 1986; Benetka 1985; Jang et al. 1987; Walther and Sauer 1986; Matsumoto and Onozawa 1989). James (1983) reported two induced mutants from an unnamed *Rosa floribunda* by λ-irradiation of actively growing terminal buds. In chrysanthemum, environmental stress tolerant mutants were isolated, e.g. low temperature tolerant (Huitema et al. 1986) salt resistant types (Dalsou and Short 1987). Nikaido and Onogawa (1989) were also able to isolate mutants that contained higher levels of flavonoids and carotenoids. Mandal et al. (2000) induced sectorial somatic mutations in flower color of chrysanthemum by λ-irradiation of root cuttings. Under the FAO/IAEA funded project, MAL/5/024, a new radiation induced mutant variety of orchid, Dendrobium "Sonia Keena Ahmad Sobri" was released in Malaysia. This mutant has diamond-shaped petals, flowers of the parental types and shelf life 15 days (Jain and Maluszynski 2004). In Thailand, under the THA/5/045 project, floriculture improvement using radiation has been successful, e.g. Portulaca, Canna, and chrysanthemum. Mutant varieties of these flowers have been released to the growers. The regenerated plants flowered and exhibited true to type in two successive generations in the field. The combination of micropropagation and induced mutations can develop and multiply elite mutants in a short period of time in most of the ornamental plants.

Somaclonal Variation

Tissue culture-derived variation or somaclonal variation is very common, especially in vegetatively propagated plants, e.g. saintpaulia, begonia, chrysanthemum, gerbera, and others. Somaclonal variation in breeding involves: a) induction and growth of callus or cell suspension cultures for several cycles; b) regeneration of large numbers of plants; c) screening of desirable traits; d) stability testing of selected variants in subsequent generations; and e) multiplication of stable variants and their use in breeding programs. Somaclones showing differences in quantitative traits have been observed in ornamental plants, e.g. shelf-life duration, number of flowers per plant, leaf shape. However, breeders have been reluctant to use somaclonal variation given its unpredictable nature and genetic instability, Despite the fact that it has proven a useful additional tool for crops with a narrow genetic base, providing a rapid source of variability for crop improvement. The combination of induced mutations and tissue culture would greatly benefit the floricultural industry and as a result export earnings and employment could go up. Somaclonal variation in plants was recently reviewed (Jain 2001) and manifested as cytological abnormalities, frequent qualitative and quantitative phenotypic mutation, sequence change, and gene activation and silencing (Kaeppler et al. 2000). Somaclonal variation has proved useful for plant improvement (Skirvin et al. 1993; Jain 1997b, Jain et al. 1997) but strongly depends on the route of regeneration. Most of the variation is observed in plants originating from protoplasts, also termed protoclonal variation (Kawata and Oono 1997; Jain 1997b; Jain and de Klerk 1998). Plants regenerated from unorganized callus varied more than from organized callus, whereas hardly any variation was seen when plants were regenerated directly without an intermediate callus phase (Bouman and deKlerk 1996). Somaclonal variation occurred in plants regenerated from ray-florets of *Chrysanthemum morifolium*. Morphological changes were also observed in plants derived from petal segments (Malaure et al. 1991; Ohishi and Sakurai 1988). Subsequently, Ahloowalia (1992) developed 20 new variants, which differed in height, leaf, flower shape, petal size, and curvature. Somaclonal variants differed from the original cultivar in number, shape and color of petals or growth habit. Increased variability in flowering date, plant height, plant width, number of flowers, and flower morphology was also reported for *Rosa hybrida* (Arene et al. 1993), *Chrysanthemum* (Votruba and Kodyteck 1988), *Begonia* and *Saintpaulia*

(Jain 1993a, b, c) at different frequencies, viz. *Saintpaulia* (2–10%), *Dianthus* (0.8%), *Dracaena* (10%) and *Chrysanthemum* (60%) (Jain et al. 1997). Exploitation of somaclonal variation through callus culture might become a source for new cultivars were this method combined with strategic and efficient *in vitro* selection pressures (Gudin and Mouchotte 1996). The selected somaclones should be genetically stable in seed and vegetatively propagated crops for routine induction of genetic variability through tissue culture, and this aspect should be thoroughly checked before using them in regular crop improvement programs.

Genetic Transformation

Plant genetic transformation has contributed further development of plant breeding techniques and a better understanding of the basic mechanism involved in plant gene regulation (Wising et al. 1988). Gene transfer enables the introduction of foreign genes, or specifically designed hybrid genes, into host plant genomes, thus creating novel varieties with specifically designed characters including resistance to environmental stress, pests and disease (Ahmed and Sagi 1993). While abundant literature is available on transformation, information on progress of transformation studies in floricultural crops is scant (Hutchinson et al. 1989; Hutchinson et al. 1992; Robinson and Firoozabady 1993). Success of *Agrobacterium*-mediated transformation depends on the cultivars (Robinson and Firoozabady 1993), choice of explants (Robinson and Firoozabady 1993; Jenes et al. 1993), delivery system, *Agrobacterium* strain, cocultivation condition, selection method, and mode of regeneration (Mathis and Hinchee 1994). Firoozabady et al. (1991a) succeeded in *Agrobacterium*-mediated transformation of *Rosa hybrida* cv. Royalty. Transgenic rose plants were developed using embryogenic callus line of hybrid tea rose (Noriega and Sondahl 1991). Friable embryogenic callus was cocultivated with *A. tumefaciens,* strain LBA 4404, carrying binary vector pJJ3931 (pnos NPTII/p35SLUC) or *A. rhizogenesis* strain 15834 carrying vector pJJ 3499 (pnos NPTII/p35SGUS), both of which contained an npt II and a GUS gene followed by kanamycin selection. The transformation efficiency was very high; between 40-60 independent kanamycin-resistant calli were produced for each gram of rose calli inoculated with *Agrobacterium* (Robinson and Firoozabady 1993). Almost 100% calli were transformed based on GUS assay and Polymerase Chain Reaction (PCR). From the transformed calli, transgenic somatic embryos were continuously produced. Matthews et al. (1991) also reported *Agrobacterium*-mediated transformation of *Rosa persica* × *xanthina* using protroplasts of embryogenic cell lines. GUS expression of a transformed callus and subsequent regeneration of shoots was reported. Firoozabady et al. (1994) regenerated transgenic rose (*Rosa hybrida* cv. Royalty) plants from friable embryogenic calli inoculated with

Agrobacterium. More than 100 transgenic plants were established in soil and flowered in the greenhouse. Further successful reports appeared on transgenic plants of *Rosa hybrida* (van der Salm et al. 1996; Marchant et al. 1998a; Marchant et al. 1998b). In addition, introducing chitinase transgene in rose (*R. hybrida* cv. Glad Tidings) reduced the blackspot disease by 13 – 43%. The gene expression in the transformed plants was confirmed by enzyme assay (Marchant et al. 1998a; Marchant et al. 1998b).

All transgenic plants were grown in the greenhouse and morphologically true-to-type, with reduced the blackspot disease by 13 to 43%. Subsequently, floribunda roses were transformed with genes for antifungal proteins to reduce their susceptibility to fungal diseases (Dohm *et al.*, 2002). Li *et al.* (2003) developed transgenic rose lines harbouring an antimicrobial protein gene (*Ace-AMP 1*) to enhance resistance to powdery mildew. They have confirmed the stable integration of *Ace-AMP 1* and NPT II genes by Southern blotting. Condliffe *et al* (2003) reported the optimised protocol for transformation (*Agrobacterium-mediated*) in different rose cultivars by using GUS (uidA) gene. They found stable integration of the transgene was confirmed at each stage of somatic embryogenesis and in regenerated plants.

In chrysanthemum, genetic transformation using *Agrobacterium*-based gene vectors has been reported (Ledger et al. 1991; Lemieux et al. 1990; Firoozabady et al. 1991b), which is cultivar dependent and susceptible to wild type *Agrobacterium* strains (Van Wordragen et al. 1992). Use of GUS intron reporter gene revealed that low efficiency gene transfer and transient gene expression took place. Lowe et al. (1993) developed a system for producing transformed plants from explants of *C. morifolium*. In *D. grandiflora*, leaf explants were used for *Agrobacterium*-mediated transformation (Urban et al. 1992 ,1994; de Jong et al. 1993). Kudo et al. (2002) studied various parameters on transformation of *D. grandiflora* with *Agrobacterium*-mediated transformation system, and suggested that differences in promoters or introduced genes might affect gene expression. Vegetatively propagated progeny of transformed plants were identified which expressed GUS activity and contained multiple copies of the TSWV (*Tomato Spotted Wilt Virus*) nucleocapsid gene (Urban et al. 1994).

Literature is plentiful on carnation genetic transformation with different explants such as petal base (Woodson and Goldsbrough 1989), stem sections (Lu et al. 1991), and whole leaves (Firoozabady et al. 1995; Lu et al. 1991) by using *Agrobacterium*-mediated transformation system (Woodson and Goldsbrough 1989; Cornish et al. 1991; Firoozabady et al. 1995). Reported a single transformed plant from petal bases cocultivated with *Agrobacterium* followed by kanamycin selection. A repeateable transformation and regeneration of "Improved White Sim" has been developed (Firoozabady et al. 1991b). Lu et al. (1991) reported recovery of transformed carnation plants (cvs. White Sim, Crowley and Red Sim = syn. Scania) from stem sections with *A. tumefaciens* wild type ICMP802. Hundreds of transgenic plants have been generated. Transgenic carnation plants with some reduced senescence response have been obtained by the introduction of carn363, a carnation gene homologous to the tomato ethylene-forming enzyme (pTOM13) (Cornish et al. 1991).

Choudhary et al. (1994) studied the *Agrobacterium*-mediated transformation of *Petunia hybrida* with a delta-9 fatty acid desaturase gene from *Saccharomyces cerevisiae*. They found that the level of monounsaturated 16:1 and 18:1 fatty acids increased in transformed leaves compared to control plants.

Graves and Goldman (1987) established the *Agrobacterium*-mediated transformation system in *Gladiolus* by using corm slices and found evidence of transformation based on the presence of opines in the resultant tissues. Kamo (1997) reported opine synthesis in different explants of corms, cormels, and seedlings when inoculated with tumorigenic strains of *A. tumefaciens*. Babu and Chawla (2000) compared influence of wounding caused by scalpel and particle bombardment on *Agrobacterium*-mediated transformation in gladiolus. GUS expression frequencies of 5.3% and 23% respectively were obtained from scalpel and particle bombardment wounded explants. Probably particle bombardment generates microwounds in the tissue, resulting in enhanced transformation rate (Jain et al. 1992).

Transformed gerbera plants have been generated by *Agrobacterium* cocultivation of fragmented shoot apices. The *Agrobacterium* LBA 4404 contained a plasmid with kanamycin-resistance and GUS genes. Subsequently, Elomaa et al. (1993) established the *Agrobacterium*-mediated transformation system in *Gerbera hybrida*. The pieces of petioles of the red variety Terra Regina were cocultivated with a disarmed *Agrobacterium tumefaciens* vector containing a nearly full-length antisense cDNA encoding *Gerbera* chalcone synthase under the control of CaMV 35S promoter and a nos-nptII marker gene. The transformed cells were kanamycin resistant and regenerated into flowering plants and antisense cDNA blocked anthocyanin synthesis, resulting in dramatically altered flower pigmentation in some transformants. Recently, Korbin et al. (2002) produced gerbera transgenic plants resistant to tomato spotted wilt virus (TSWV) by inserting TSWV nucleoprotein gene in gerbera explants (shoots, bases of shoot clumps, and leaves with 2-3 mm lamina length) using *Agrobacterium*-mediated transformation system.

Griesbach (1992) transformed Calanthe orchid by exposure of the apical meristem to DNA in the presence of an electric field. Transformation was confirmed by the observation of GUS staining patterns in leaves of 6-month-old plants. Using the particle bombardment method, many ornamental plants were transformed such as *Dendrobium* (Kuehnle and Sugii 1992) and the Cymbidium orchid (Yang et al. 1999). These authors recovered transgenic plants, which were established in soil and acclimatized in the greenhouse. Recently, Belarmino and Mii (2000) transformed *Phalaenopsis* orchid by cocultivation of cell clumps with *A. tumefaciens* carrying hygromycin selective marker gene. Hygromycin-resistant cell cultures proliferated into green callus, which produced protocorm-like bodies on phytohormone-free medium. Genetic transformation studies in some of the major floricultural crops are summarized in Table 13.8.

GERMPLASM CONSERVATION

Clonal Stability Through Tissue Culture

Genetic stability of the micropropagated plants is essential for *in vitro* germplasm conservation (Jain 1997a). Many researchers reported that the meristem-derived plants were more stable than the adventitious shoots derived from callus. Somaclonal variation was more common among adventitious-shoot-derived plants in many species including chrysanthemum (Skirvin 1978). Skirvin and Janick (1976) were among the first to emphasize the importance of clonal variation in genotype improvement of horticulture species. Various types of changes were reported in cell cultures at phenotypic, karyotypic, physiological, biochemical or molecular level. Further, Larkin and Scowcroft (1981) reviewed extensively and reported the phenotypic variations among plants regenerated after a passage through tissue and cell culture. Hasegawa (1980) reported one "abnormal looking" plant among 600 tissue-culture-propagated plants of *Rosa hybrida*. Martin et al. (1981) noted no variation among 2,125 rose plants raised in the field for three years. The lack of somaclonal variability suggests that rose is relatively stable when propagated via axillary buds. However, Lloyds et al. (1988) reported that callus-derived shoots of *R. persica* x *xanthina* exhibited considerable variations in leaf shape. Similarly, callus-regenerated plants of *R. hybrida* showed considerable somaclonal variation in number of petals, dwarf growth habit, round-shaped petals, and color (Arene et al. 1993). Interestingly, Malaure et al. (1991) reported that shoots derived from ray florets of 16 cultivars of chrysanthemum showed more variation than plants regenerated from vegetative parts. However, *in vitro* selection and somaclonal variation are random processes and have yet to be used to achieve specific goals in chrysanthemum improvement. Somatic hybrids resulting from protoplast fusion also show variation in morphology, cytology, fertility and other aspects. Ploidy level of parent is very important in somatic cell fusion work. Izhar and Tabib (1980) showed that the regenerated plants derived from leaf mesophyll protoplasts of petunia were diploid ($2n = 2x = 14$). Further, Izhar et al. (1983) observed that over 1,000 fertile somatic hybrids derived from a fusion product of

P. parodii and *P. hybrida,* were tetraploids. However, when protoplasts, isolated from cell suspension cultures were used as one of the fusion parents, the somatic hybrid plants were of a higher ploidy level (2n = 28) (Clark et al. 1986). Selection of explant, age of culture, genotype, culture conditions and method of plant regeneration are very important features for genetic stability of regenerated plants. Since most of somatic embryos originate from single cells, somaclonal variation among regenerated plants can be minimized. Of course, there is always a limit to number of subcultures before plants show variation.

Determination of Genetic Fidelity with Molecular Markers

Molecular markers have facilitated research on genetic variation at the DNA level. The numerous potential applications of DNA fingerprinting have been tested in population genetics, parentage testing, individual identification, and for shortening breeding programs (Ben-Meir et al. 1997). Markers such as restriction fragment length polymorphism (RFLPs) have recently been used for molecular characterization of tissue culture-derived plants. Since its development, polymerase chain reaction (PCR) has revolutionized many standard molecular biological techniques, with modifications of the original procedure designed to suit a number of needs. Random Amplified Polymorphic DNA (RAPD), arbitrarily primed PCR (AP-PCR), DNA amplification fingerprinting (DAF), intersimple sequence repeat (ISSR), sequence-tagged sites (STSs), amplified fragment length polymorphism (AFLP), and many others generate special classes of markers highly sensitive for genetic analysis of tissue culture-raised plants (Rani and Raina 1996, 2000). RAPD markers have also been used to identify cultivars, to map important agricultural traits, and to construct genetic maps (Williams et al. 1990). Varietal identification is now benefiting from the use of molecular markers, including RFLP and AFLP (Rajapakse et al. 1992). Bouman et al. (1992) also found RAPD polymorphism among micropropagated plants of *Begonia* species. RAPD markers were used for the construction of a chromosome linkage map, using crosses between *Rosa multiflora* derived genotypes that differed in a range of floral and vegetative characters (Debener and Mattiesch 1996), and for parentage analysis in interspecific crosses between various wild rose species (Debener et al. 1997). Ben-Meir et al. (1997) screened carnation (*Dianthus caryophyllus*) and rose (*Rosa hybrida*) with seven different horticultural traits with RAPDs. In chrysanthemum, Huang et al. (2000) did genetic analysis of *Chrysanthemum* hybrids with RAPD markers and classified them into seven types. These were: a) markers shared bands in both parents and offspring, b) markers shared bands in male and female parents, c) markers shared bands in male parent and offspring, d) markers shared bands in female parent and offspring, e) markers indicated in the male parent only, f) markers present in the female parent only, and g) markers present only in offspring. Only type-c markers were suitable for identifying the true male parent. It was concluded that there were no definite rules as to whether markers in offspring were more similar to female or male parents according to similarity analysis.

Cold or Low-Temperature Storage

An effective method for *in vitro* storage for commercial flowers would allow breeders and nurseries to maintain a greater number of varieties for a much longer time. At low temperature, shoot or somatic embryo culture growth slows down and can be maintained for several months with no subculture on the fresh medium. This way, *in vitro* cultures of elite genotypes and germplasm can be maintained. Techniques for the induction of slow growth of embryos in chrysanthemum are known (Preil and Hoffmann 1985; Bajaj 1986; Fukai et al. 1988). Corbeneau et al. (1990) reported that the somatic embryos of oil palm (*Elaeis guineensis*) cannot tolerate even a relatively short exposure to temperature lower than 18 °C but Scorza et al. (1990) reported that the embryogenic cultures of *Prunus persica* stored at 4 °C for 55 days retained their potential for embryogenesis. Later, Janeiro et al. (1995) indicated that the embryogenic callus of *Camellia japonica* stored at 3 – 4°C for 8 weeks significantly improved the plantlet conversion ability of the embryos from 76-100%. These approaches can be used for storage and conservation of elite germplasm, which is beneficial for industry and germplasm banks. More work is needed in ornamental plants.

Cryopreservation

The first report on survival of plant tissues after exposure to ultra low temperatures of –196 °C by Sakai (1960) and the significance of using cryoprotectants such as dimethyl sulfoxide (DMSO) by Quatrano (1968) have revolutionized *in vitro* conservation of germplasm. Preservation of cultured plant tissues in liquid nitrogen is the most efficient method of germplasm conservation. Fukai and Oe (1990) reported cryopreservation of chrysanthemum (*C. morifolium*) shoot tips. Shoot tips were placed into 0.1 mg L^{-1} venzylaminopurine (BA) 1.0 mg L^{-1} NAA 2% sucrose and 5% dimethyl sulfoxide (DMSO) for 2 days, slowly cooled with a cryoprotectant solution (10% DMSO and 3% sucrose) at a rate of 0.2 °C per minute from 0 to 40 °C and then immersed and stored in liquid nitrogen. After thawing in warm water, over 87% shoot tips regenerated shoots. Fukai et al. (1988) noted the application of cryopreservation techniques and the survival rate of 12 species and two interspecific hybrids of chrysanthemum. Shoot regenera-tion rates of the frozen shoot tips varied from 94 to 100%

depending on the species. They also reported that the shoot tips of chrysanthemum showed high viability even after 8 months of storage in liquid nitrogen. Recently, Hitmi et al. (2000) described a simple method for efficient cryopreservation of *Chrysanthemum cinerariaefolium*. The shoot tip explants were treated with 0.55 mmol L^{-1} sucrose plus 4mmol L^{-1} abscisic acid for 3 days and immersed in liquid nitrogen; rapid warming was subsequently done at 40 °C and 75% shoot tips proliferated into plantlets. Cryopreservation is effective by avoiding formation of intracellular ice formation during the stage of freezing. Desiccation by air-drying treatments both before and after immersion in liquid nitrogen is effective to prevent intra-cellular freezing. Bian et al. (2002) successfully cryopreserved protocorm-like bodies of *Dendrobium candidum* by the air-drying method. This and other techniques greatly help long-term *in vitro* conservation of germplasm and could could lead to establishment of a cryo-storage bank.

Somatic Seed Production

Somatic seed production technologies make for easy handling, regular supply of planting material throughout the year, storage and transportation in several crops (Ganapati et al. 1994). Sharma et al. (1992) regenerated plants from encapsulated embryos, stored at 4 °C, up to 180 days, thereby suggesting cold or low temperature storage of plant tissues. Usually, somatic embryos or cells are ideal for cryopreservation. At low temperature explants such as shoot tips and shoot cultures are suitable for storage, e.g. encapsulated shoot tips with calcium alginate. This concept was originally tried in somatic embryos. Corrie and Tandon (1993) studied aspects of *in vitro* and *in vivo* development and reported encapsulated beads converted into plants with a success rate of 64% and 88% in sand and soil mixtures respectively. Similar results were obtained in *Rosa hybrida* (Jayasree and Devi 1997). However, plant regeneration rate was poor. The shoot tips of *chrysanthemum* were encapsulated in different concentrations of sodium alginate to determine the optimum concentration for encapsulation. Among the different gel matrices tested, MS + 0.1 mg L^{-1} IBA showed better increased germination. About 95% of cultures germinated on media containing MS + 0.1 mg L^{-1} IBA within a week. The developing shoots and roots emerged through the alginate matrix (2%) and grew into plantlets in 2 weeks (Rout and Das, unpubl. data). The low permeability of calcium alginate gel to oxygen impairs respiration in the encapsulated propagules and thus affects their viability. Sakamoto et al. (1995) developed novel self-breaking gels to overcome this limitation by using substances that interfere with the hardening reactions; the modification (partial replacement of Ca^{++} with Mg^{++} and K^{++}) also improves the conversion frequency of synthetic seeds. Synthetic seeds are highly susceptible to bacterial, fungal, and other infections (due to their high nutritional status) under glasshouse or field conditions. This problem has been successfully overcome by using antimicrobial agents (bavistin/streptomycin) in the nutrient mix. The problem with this system is that synseeds have to be regenerated in *in vitro* conditions before transferal to the field. This area of research requires more work for overcoming problems facing this technology that could lead to automation of plant production.

PROPAGATION OF ECONOMICALLY IMPORTANT FLORICULTURAL CROPS

Rose

Rose, universally known as the Queen of Flowers, is the most sought after ornamental plant. Apart from its popularity as a garden plant, it is the most important cut flower and is also grown for extraction of its essence and use in herbal medicine. Roses can be grouped into four categories i.e. exotic varieties, indigenously evolved varieties, native species, and exotic species. Rose-breeding programs are normally focused on the enhancement of ornamental value, including flower color, flower size, keeping quality of the bloom, or plant response against environmental stresses. Conventional breeding is faced with limitations arising from the limited gene pool, hybrid cross incompatibility, and polygenic characteristics such as uniform growth and synchronous flowering. An alternative approach is to develop protocols for *in vitro* regeneration and manipulation.

Abbreviations

ab = adventitious shoot bud, adr = adventitious root, c = callus, cm= chimera, ct = cotyledon, e = excised embryos, ims = immature seed, is = internodal segment, l = leaf segment, lb = lateral bud, mb = mid-rib, ms-multiple shoot, ns = nodal segment, p = petiole, ps = pith segment, pt = plantlet formation, r = rooting, sa = shoot apices, sbr = shoot bud regeneration, se = somatic embryogenesis, si = stem internode, spr = shoot primordia, ss= stem segment, st = shoot tip, tb = terminal bud.

There are several reports on *in vitro* mass multiplication of rose (Table 13.2). A single bud produced a single shoot or multiple shoots depending on the species and composition of the medium. Micropropagation of rose through axillary bud culture is shown in Figure 13.1a-d. Shoot proliferation and multiplication *in vitro* was largely based on media formulations containing cytokinins rather than auxins (Rout et al. 1989). Bressan et al. (1982) observed that low concentration of BA stimulated development of

Table 13.2 *In vitro* studies of Rose

Species/ Cultivars	*Explant source*	*Morphogenetic response*	*References*
Miniature cvs. (Sunburst Red, Toy Clown), Ground Cover (cv.Fiona)	st, ns	ms, r, pt	Douglas et al. 1989
Rosa hybrida cv. Landora	ns	ms, r ,pt	Rout et al. 1989
Rosa hybrida cv. Forever Yours	st	ms, r, pt	Skirvin and Chu 1979 Skirvin et al. 1984
Rosa persica × *Rosa laevigata*, **Rosa wichuraiana**	is, st	c, ad	Lloyd et al. 1988
Rosa indica major	ss	c	
Rosa canina inermis cvs. Sonia, Golden Times			Zieslin et al. 1987
Rosa hybrida cv.Bridal Pink	st,	ms, r	Khosh-Khui and Sink 1982a
Rosa manetti, **Rosa hybrida** cv.Tropicana	l, ss	c	Khosh-khui and Sink,1982b
Rosa hybrida cvs.Kings Ransom, Plentiful, Parade, Fragrant Cloud, Lili, Marlene, Garnet Yellow	st, ns	ms, r, pt	Davies 1980
Rosa hybrida cvs. Tropicana, Bridal Pink	st	ms, r	Khosh-Khui and Sink 1982c
Rosa hybrida cv.Super Star	ps	c, ab	Jacobs et al. 1968
Rosa hybrida cv. Improve Blaze	st, ns	ms, r, pt	Bressan et al. 1982 Hasegawa,1980
Hybrid tea rose "The Dioctor"	ss	spr	Hill 1967
Rosa hybrida cv.Bridal Pink	e	ab	Burger et al. 1990
Rosa hybrida cv.Amanda	ns	r	DeVries and Dubois,1988
Rosa hybrida cvs.	ns	ms, r	Hyndmann et al. 1982
Rosa hybrida	st, lb	ms, r	Scotti Campos and

Contd.

Table 13.2 Contd.

Species/ Cultivars	*Explant source*	*Morphogenetic response*	*References*
cv.Rosamini			Pais 1990
Rosa wichuraiana cvs.Dame of Sark, Clarissa	st	ms, r	Tweddle et al. 1984
Rosa hybrida cvs.Crimson Glory, Glenfiditch	lb	ms, r	Barve et al. 1984
Rosa hybrida	st	c, sb	Valles and Boxus 1987
Rosa hybrida cvs.Domingo, Vickey Brown, Tanja, Azteca	l, p	se	deWit et al. 1990
Rosa damascena	is	sbr	Ishiooka and Tanimoto, 1990
Rosa hybrida cvs. Landora, Virgo Happiness, Sea Pearl, Super Star, Queen- Elizabeth	ns	ms, r	Rout et al. 1990
Rosa hybrida cv. Landora	l, ss	sbr	Rout et al. 1992
Rosa chinensis var.minima. (cvs.Debut,Ginny, Red Ace,Tipper)	st	ms, r	Rogers and Smith,1992
Rosa chinensis var. minima (cvs.Baby Katie, Lavender Jewel, Red Sunblaze, Royal Sunblaze)	ns	ms	Chu et al. 1993
Rosa rugosa	ims	se	Kunitake et al.1993
Hybrid tea 'Dr.Verhage'	st	ms, r	Voyiatzi et al. 1995
Rosa multiflora	cm	sbr, r	Rosu et al. 1995
Rosa hybrida	l, p	c, ab, r	Dubois and deVries,1995
Rosa hybrida	adr	c, ab,r	van der Salm et al. 1996
Hybrid tea rose cv. Peace	st, lb	ms, r	Ara et al. 1997
Rosa sp. cvs. Baccara, Mercedes, Ronto, Soraya	l, st	se,pl	Kintzios et al. . 1999

Fig. 13.1 a and b-Shoot proliferation from axial meristems of *Rosa hybrida* cv. Landora on MS medium supplemented with 2.0 mg L^{-1} BA + 0.01 mg L^{-1} IAA + 3% sucrose. c-Initiation of roots from elongated shoot of *Rosa hybrida* cv. Landora on MS medium supplemented with 0.25 mg L^{-1} IAA + 2% sucrose after 14 days of culture. d-Micropropagated plantlets established in the greenhouse.

the axillary buds in roses, which was dependent on the cultivar used. Further, substantial differences were seen in shoot multiplication of rose cultivars taken from different node positions on the stem. Khosh-khui and Sink (1982a, b) observed that the shoot multiplication rate of *Rosa hybrida*, *R. damascena*, and *R. canina* varied significantly during different subculture periods. By decreaseing sucrose concentration in the culture medium, the number of multiple shoots of *R. hybrida* increased (Langford and Wainwright 1987). Similarly, the size of the meristem (both shoot tip and nodal explant) of floribunda, ground cover, and miniature roses had significant effects on shoot multiplication; on average 5.0, 3.1, 1.3, and 2.5 shoots per culture cycle were recorded (Douglas et al. 1989). Several other researchers did studies on increasing shoot multiplication rate, such as ethylene (Kevers et al. 1992), liquid medium (Kevers et al. 1992), and extension of culture period (Chu et al. 1993). The success of rose micropropagation also depended on agar concentration (Ghashghaie et al. 1991), growth room, vessel humidity (Sallanon and Maziere 1992), different types of gelling agents (Podwyszynska and Olszewski 1995; Ara et al. 1997), different types and concentrations of carbohydrates, and explant length and diameter (Salehi and Khosh-khui 1997). Thidiazuron (TDZ) 1.0–2.5 μM was highly effective in *R. damascena* micropropagation (Kumar et al. 2001), whereas microshoots of *R. kordesii* rooted well in the medium with low or no auxin and high salt concentration (Arnold et al. 1995). The quality of light is also crucial in rose micropropagation, especially on rooting component. Skirvin and Chu (1984) highlighted the effect of light quality on root development of miniature roses. Rogers and Smith (1992) reported the consequences of *in vitro* and *ex-vitro* root initiation for miniature rose (*R. chinensis)* and found that after 18 day light treatments either *in vitro* or *ex vitro*, 95-100% rooting was obtained.

Somatic embryogenesis in rose has been reported by several researchers in explants such as leaf, stem segment, internode, stamen filament, root and zygotic embryo (deWit et al. 1990; Rout et al. 1991; Roberts et al. 1995; Kintzios et al. 1999). Kunitake et al. (1993) induced somatic embryogenic calli and somatic embryos in immature seed-derived calli of *R. rugosa* on MS medium without growth regulators. Rout et al. (1991) reported secondary somatic embryos and their maintenance up to 16 months by regular subculture of embryogenic callus mass at 4-week intervals. The growth rate of secondary embryogenic callus grown on ABA increased 36-fold, while the germination rate of somatic embryos was more than five times higher than those derived from embryogenic callus grown on BA and TDZ (Li et al. 2002). Medium supplemented with 2,4-D helped in the long-term maintenance of embryogenic callus potential (Roberts et al. 1990; Mathews et al. 1991; Noriega and Sondahl 1991). However, when somatic embryos of *R. rugosa* were developed on a growth hormone-free medium, the embryogenic potential did not last beyond 6 months of culture (Kunitake et al. 1993). The type and concentration of sugars and hormones are important for somatic embryo quality and development (Hsia and Korban 1996; Noriega and Sondahl 1991;Roberts et al. 1990; Kunitake et al. 1993). Li et al. (2002) found that ABA effectively promoted proliferation and germination of somatic embryos in *Rosa* species. Rout et al. (1991) reported that the inclusion of L-proline in the primary culture, followed by its exclusion in the regeneration medium, stimulated embryo development and reduced the frequency of developmental abnormalities. It seems that exposure of somatic embryos to low temperature at 8 °C for 4 days enhanced germination rates (Rout et al. 1991). However, chilling at 4 °C for 2 weeks was stimulatory and germination rates rose from 12 to 24% (Roberts et al. 1995). An optimized protocol for somatic embryogenesis of *Rosa hybrida* cv. Landora is shown in Figures. 13.2a-e. Somatic embryo-genesis is dependent on genotype, explant type, growth

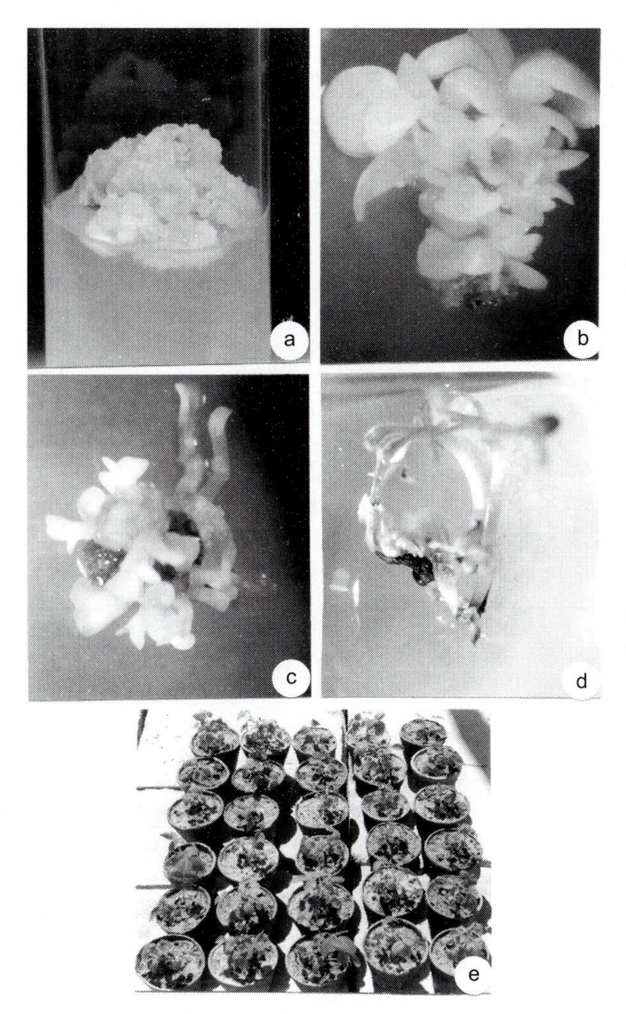

Fig. 13.2 a-Embryogenic callus of *Rosa hybrida* cv. Landora. b-Proliferation of group of somatic embryos with bipolar structures. c-Development of cotyledons from a group of somatic embryos. d-Germination of somatic embryos. e-Somatic embryo-derived plantlets established in soil.

hormones, type of sugar, and light quality. Moreover, somatic embryo germination rate is generally poor, which explains its minimal use in large-scale plant production.

Chrysanthemum

Chrysanthemum (*Dendranthema grandiflora* syn. *Chrysanthemum morifolium*) is extensively grown both as a cut flower and a pot plant worldwide. It is usually propagated vegetatively through cuttings and suckers. Small-flowered types are often raised from seeds producing heterogeneous plants and are used for bedding. Breeding programmes have focused on improving various characteristics to enhance ornamental values, including flower color, flower size and form, and production quality. Although desirable traits have been introduced by classical breeding, there are limitations to this technique. Recent developments in biotechnology of chrysanthemum have been reviewed (Table 13.3; Rout and Das 1997), indicating that shoot multiplication rate is cultivar dependent. Prasad et al. (1983) studied shoot multiplication rates of four cultivars of *C. morifolium*. On average 31 shoots were formed per explant of cv. "Otome Zakura" in 60 days of culture and it was estimated that about 9M plants could be produced from a single explant in one year using this method. Shoot tip growth differed depending on the growth regulator and composition of the medium. Wang and Ma (1978) reported that shoot tips between 0.2–0.5 mm and shoot meristems between 0.1–0.2 mm

Table 13.3 *In vitro* studies of chrysanthemum

Species/cultivars	*Explant source*	*Morphogenic response*	*Reference*
C.morifolium	s	sb	Hill 1968
C.morifolium	st	or,pl	Ben-Jaacov and Langhans 1972
C.cinerariaefolium	ca	sb, r	Roest and Bokelmann1973
C. morifolium	st	pl	Earle and Langhans 1974a,b
C.morifolium	s	or	Roest and Bokelmann 1976
C. morifolium	s	sb, r	Miyazaki et al. 1976
C.morifolium cv. Indianapolis	p, pe, st	sb, pl	Bush et al. 1976
C.morifolium cvs. Blue Bird, Montana, Meladion Delaware	st, inf	sm, r, pl	Wang and Ma 1978
C. morifolium cv. Shin Dong	st	sf, rf	Lee et al. 1979

Contd.

Table 13.3 Contd.

Species/cultivars	*Explant source*	*Morphogenic response*	*Reference*
C.morifolium	pt	c	Schum and Preil 1981
C.morifolium cv. Super Yellow	c	pl, or	Lazar and Cachita 1983
C.morifolium	lm	sml, pl	de Donato and Peruceo 1984
C.morifolium	l	or	Chen et al. 1985
C. hortorum cvs. Pink Camino, Super Yellow, Spider	am	sf, rf	Gertsson and Andersson 1985
C.morifolium cvs. Winter Westland, Yellow Westland, Dark Westland, Snowdon, Yellow Snowdon, Altis and Blanche	st	sml, rf	Ahmed 1986
C.morifolium cv. Birbal sahni	l, s,r	c	Prasad and Chaturvedi 1988
C.morifolium C.coccineum	st, l	sb	Kaul et al. 1990
C.morifolium	l	c, sb	Rademaker et al. 1990
C. coccineum	p	c, sb, pl	Fujii and Shimzu 1990
C.morifolium cv. Royal Purple	s	or, pl	Lu et al. 1990
C. cinerariaefolium	l	c	Zito and Teo 1990
C.morifolium	l, s	or, pl	Bhattacharya et al. 1990
C.morifolium cv. Yellow Spider	l	se	Sauvadet et al. 1990
C.morifolium	mb	se	May and Trigiano 1991
C.hortorum	st	or	Corneanu and Corneanu 1992
C. morifolium cv. Yellow Spider	l	se	Pavingerova et al. 1994
C.maximum	s	sb, r	Kumar and Kumar 1995
C.morifolium cv. Deep Pink	l, s	or, r, pl	Rout et al. 1996
C.morifolium	st,n	sml, pl	Mandal et al. (2000)

Abbreviations: ab-axillary bud; c-callus formation; ca-capitulum; fp-floral peduncle; inf-inflorescences; l-leaf; lm-lateral meristem; md-mid rib; or-organogenesis; p-petal; pd-pedicel; pe-petal epidermis; pl-plantlet formation; r- root; rf-root formation; s- stem; sb-shoot bud regeneration; se-somatic embryogenesis; sf-shoot formation; sm-shoot meriste; sml-shoot multiplication; st-shoot tip.

produced only a single shoot. Larger explants (0.5–1.55 mm) formed multiple shoots.

Low concentrations of auxin are enough for root induction in regenerated shoots. Efficient rooting is dependent on type and concentration of growth hormones, and light intensity as well as strength and type of culture medium (liquid or solid) (Roest and Bokelmann 1975; Rout et al. 1996; Roberts et al. 1992; Roberts and Smith 1990). Rooting was achieved in 90% cultures of "Deep Pink" under light intensity of 2.0 Klux; and rooting rate decreased at higher light intensity [(3.0 Klux) (Rout et al. 1996)].

Somatic embryogenesis in chrysanthemum (*Dendranthema grandiflora)* has been successful by using leaf midrib explants cultured on modified MS basal medium supplemented with 1.0 mg L^{-1} 2,4-D and 0.2 mg L^{-1} BA; and appropriate photoperiod and sucrose concentration (May and Trigiano 1991). The highest production of somatic embryos was achieved on medium containing 9–18% sucrose and the cultures were incubated initially in darkness for 28 days, followed by 16 h photoperiod under the light. Twelve of the 23 cultivars evaluated produced somatic embryos but complete plants were recovered from only five of them. The regenerated plants were phenotypically similar to the parent plants in growth habit, leaf morphology and flower color.

Gerbera

Gerbera jamesonii is an important commercial flower worldwide. It is a perennial herb, native to South Africa and Asia. Rapid clonal propagation of gerberas is done with explants such as shoot tip and capitulum (Table 13.4), and has a great potential for genetic improvement with biotechnology and induced mutations (Jain et al. 1998). *In vitro* , gerbera shoot multiplication was successful by culturing explants including capitulum (Pierik et al. 1973, 1975; Laliberte et al. 1985; Topoonyanont and Dillen 1988) and shoot tips (Murashige et al. 1974; Huang and Chu 1985). By lowering the concentration of cytokinin BA, shoot multiplication was reduced. Hempel et al. (1985) observed that higher cytokinin (10- 20 mg L^{-1}) induced the highest rates of shoot multiplication and rooting of excised shoots of gerbera. Blakesley et al. (1991) demonstrated the metabolism of BA in shoot multiplication of *G. jamesonii;* 2.22 μM BA induced higher shoot multiplication rate.

Besides growth hormones, organic compounds in the culture medium also influence shoot regeneration in gerbera. Soczek and Hempel (1988) reported that the organic compounds in the media increased shoot multiplication rate in gerbera, which was cultivar dependent. Multiplication rate was minimal in cvs. "Clementine" and "Terravisa" on media supplemented with various concentrations of nicotinic acid but was significantly

Table 13.4 *In vitro* studies of Gerbera

Species/ Cultivar	*Explant source*	*Morphogenic response*	*Reference*
Gerbera jamesonii	md	r, c	Pierik and Segers 1973
Gerbera jamesonii	st	ml, r	Murashige et al. 1974
Gerbera jamesonii	inf	ml, r	Pierik et al. 1975
Gerbera jamesonii	cap	ml, r	Pierik et al. 1973
G. jamesonii cv. Fresultane	ov	hap, r	Meynet and Sibi ,1984
Gerbera jamesonii "Pastourelle"	cap	ml, r	Laliberte et al. 1985
G. hybrida "Arendsoog" "Super Giant Yellow"	st	ml, r	Huang and Chu 1985
Gerbera jamesonii hybrids	cap	ml, r	Huang et al. 1987 Harti et al. 1993
Gerbera jamesonii hybrids	cap	sb, r	Arello et al. 1991
Gerbera jamesonii cvs. Clementine, Saskia, Terravisa	st	st, r, pl	Soczek and Hempel 1988
"Super Gerbera"	ov	hap	Cappadocia et al. 1988
Gerbera jamesonii	cap	ml, r	Topoonyanont and Dillen, 1988
G. aurantiaca	st	ml, r	Meyer and Van-Staden 1988
Gerbera jamesonii	mt	ml, r	Blakesley et al. 1991
Gerbera jamesonii G. hybrida	l	sb, r	Jerzy and Lubomski 1991
(G. jamesonii × G. viridifolia) G. viridifolia, *G. piloselloides*	l	sb, r	Reynoird et al. 1993
G. jamesonii cv. Appelbloesem	st	ml, r	Barbosa et al. . 1993
Gerbera jamesonii	pe, l	sb, r, pl	Orlikowska et al. 1999

Abbreviations: c-callus formation; cap-capitulum; hap-haploid plant formation; inf-inflorescences; l-leaf; md-midrib; ml-multiple shoot, ov- ovule; pe-petal; pl- plantlet formation; rf-root formation; sb-shoot bud regeneration; sf-shoot formation; sm-shoot meristem; st-shoot tip.

higher in "Saskia" shoots cultured on medium containing 10 mg L^{-1} each of thiamine, pyridoxine, and nicotinic acid. Woltering (1990) described the beneficial effects of CO_2 on growth and development of gerbera grown *in vitro* . Furthermore, 5% CO_2 supplemented with low light intensity had a positive impact on plant development *in vitro*. Further research is needed to study in detail the impact of CO_2-enriched conditions on growth and development of shoots as well as control of culture contamination by excluding or minimizing sugars in the culture medium.

Gladiolus

Gladiolus is quite popular flower worldwide for its beauty and vaselife. In addition to cut flowers, corm production as planting material is also of commercial importance. *Gladiolus* spp. are native to South Africa and belong to the family Iridaceae. They are propagated by an underground corm that develops at the base of the shoot. To increase multiplication, the corms are cut into small pieces with a bud and a root zone. But clonal multiplication is slow through asexual propagation for meeting the demand of disease–free, quality planting materials. Hence *in vitro* micropropagation of gladiolus has been widely studied as an alternative method for mass-scale production of quality planting material. *In vitro* propagation of *Gladiolus* has been extensively investigated (Table 13.5), and several explants were used for plant regeneration including inflorescence stem (Ziv et al. 1970; Hussey 1975), cormel stem tip (Simonsen and Hildebrandt 1971), and shoot tip (Wilfret 1971; Logan and Zettler 1985). Bajaj et al. (1983) established shoot bud regeneration from callus derived from various explants (inflorescence, stem, flower, cormel, bud and section, leaf) of "Oscar" and "Snow Princess" cultivars. Four gladiolus cultivars were easily regenerable by the two-step bud culture technique for a high rate of shoot regeneration (Sen and Sen 1995) by culturing apical and nodal explants on MS medium supplemented with 1.0 mg L^{-1} BA and 3% sucrose. Activated charcoal and NAA have been widely reported to be useful in rooting. Ziv (1979) reported that addition of 0.5 mg L^{-1} NAA to a pre-transplanting medium increased root growth. For root initiation, plantlets were transferred to ½ MS medium containing 1.0 mg L^{-1} NAA. Also, the impact of temperature is critical on rooting of gladiolus. Ziv and Kipnis (1990) found that high temperature (25 °C) had no affect on root growth. At 17 °C, the roots elongated considerably, branched profusley and contained abundant hairs.

Reports on somatic embryogenesis in gladiolus are scant. Remotti (1995) succeeded in inducing primary and secondary somatic embryogenesis from cell suspension cultures of *Gladiolus* x *grandiflorus*. Plantlets were regenerated on hormone free medium. Germination of somatic embryos was increased to 60% in liquid medium with filter paper bridge without growth regulator. Rooted plants were transferred to fresh medium supplemented with paclobutrazol to reduce stress and premature death after transplanting. Plantlets grown under these conditions developed shorter

Table 13.5 ***In vitro* micropropagation of Gladiolus**

Cultivars	*Explant Source*	*Morphogenic response*	*Reference*
Sans Souci	inf	c, r, sb	Ziv et al. 1970
Jo Wagonar	b	sb	
Firmament	co	c, r, sb,	Simonsen and
Hit Parade	st	c, r, sb	Hildebrandt 1971
Spic , Span	st	c,r,sb	Wilfret 1971
White Friendship	s	c, sb	Ginzburg and Ziv 1973
Blue Conqueror, Fiat Lux, Forest Fire, Oscar, Peter Pears	cs, inf	pl	Hussey 1975
Forest Fire, Elvira, 488/2	cb	axs	Hussey 1977
Eurovision	ab	ms,r,pl	Ziv 1979
Oscar, Snow Princess	Inf, f	c, sb	Bajaj et al. 1983
Manatee-Orange, Red, Pink, White-18	mt, ab	s, r, pl	Logan and Zettler, 1985
Kinneret, Yamit Adi, Nirit	ab	s,r,pl	Lilien-Kipnis and Kochba 1987
Peter Pears, White Prosperity	co	c, sb	Remotti and Loffler,1995
Jenny Lee	cs	sb,pl	Kamo et al. 1990
Jenny Lee	c	sb, pl	Kamo 1994
Gladiolus x *grandiflorus*	st	se, pl	Stefaniak 1994
Green Bay, Wine and Roses, Top Brass, Mornlo	ab, tb	ms, r, pl	Sen and Sen 1995
Gladiolus x *grandiflorus* cv. Peter Pears	co, cs	se, pl	Remotti 1995
Gladiolus x *hybridus* cvs. Her Majesty, Aldebaram, Bright Eye	c, cos, inf	sb,r, pl	Kumar et al. 1999

Abbreviations: a=anther, ab=axillary bud, axs=axillary shoot bud, b=bud, c=callus, cb=corm bud, co=cormel, cos=corm section, cs=cell suspension, f=flower, inf=inflorescence, infs=inflorescence stem, ms= multiple shoots, mt=meristem, pl=plantlet formation, r=rooting, s=stolons, sb=shoot bud regeneration, se=somatic embryogenesis, st = shoot tip, tb=terminal bud.

and thicker leaves, and all formed a bulblet with a few strong contractile roots (Remotti 1995). Plants transferred to soil showed 100% recovery and developed normally. The germination rate of somatic embryos of gladiolus is substantially high, but could be further improved to 80-90% for commercial production of plants.

Orchids

Orchids belong to Orchidaceae, the largest family of flowering plants. Orchid flowers show an incredible range of diversity in flower size, shape, color, structure, number, and fragrance. Propagation of orchids is slow both by conventional as well as vegetative propagation methods. Normally vegetative propagation is done through division of clumps or rhizomes, cuttings and separation of offshoots and keikis produced from the stem or pseudobulbs. Knudson (1922) eves the first to develop *in vitro* propagation and cultivation of orchid and later Rotor (1949). Subsequently, Morel (1964) discovered that the protocorms when cut into small pieces and subcultured, could regenerate more protocorms and produced more than 4 million plants in a year. Murashige (1974) enumerated 22 genera and Arditti (1977) listed 35 genera including some intergeneric hybrids propagated *in vitro*. Many researchers established *in vitro* propagation protocol by using axillary bud as well as shoot tip explants (Table 13.6).

Inflorescence stalks have been useful explants for multiplication of *Phalaenopsis* (Rotor 1949; Tse et al. 1971; Intuwong and Sagawa, 1973; Intuwong et al. 1972). Young leaves from mature plants of *Ascocenda* and *Renantanda* (Fu 1979) and *Rhynchostylis retusa* (Vij et al. 1984) have been used successfully for mass propagation. In *Phalaenopsis*, leaves, leaf sections, stem tissues and roots formed numerous protocorm-like bodies, which further developed into plantlets (Zimmer and Pieper 1978). In *Epidendrum*, callus formation was followed by regeneration of single plantlets (Stewart and Button 1975). Kim and Kako (1984) reported that the floret parts of *Cymbidium* produced a large number of multiple shoots. Subsequently, Philip and Nainar (1986) observed root tips of *Vanilla planifolia* producing 40 plants within 9 months. The immature embryos of *Paphiopedilum philippinense* were successfully used for mass multiplication (Cho and Valmayor 1988). Lee et al. (1986) indicated that the *in vitro* rhizome growth and subsequent shoot bud regeneration of *Cymbidium* species was obtained on MS media containing 10 mg L^{-1} BA. Begum et al. (1994) detailed the protocol for shoot multiplication of *Cymbidium* x *Thanksgiving* cv. Nativity on MS medium supplemented with 0.1 mg L^{-1} NAA and 0.5 mg L^{-1} BA. Nodal and leaf explants of *Spathoglottis plicata* were used for rapid multiplication on charcoal amended MS medium containing 5.37 µM NAA and 0.44 µM BA (Teng et al. 1997). The rhizome explant produced multiple shoots of *Geodorum densiflorum*

Table 13.6 ***In vitro* studies of orchids**

Species/Cultivar	*Explant source*	*Response*	*Reference*
Phalaenopsis amabilis	inf	pl, r	Intuwong et al. 1972
Phalaenopsis	axb, ab	pl, r	Zimmer and Pieper, 1978
Vanda spp.	axb, ab	pl, r	Sagawa and Kunisaki, 1982
Cymbidium Sazanami, Harunoumi	fst	pl, r	Kim and Kako,1984 Kim et al. 1988
Rhyncostylis retusa	l	sb, r	Vij et al., 1984
Vanilla planifolia	r	pl, r	Philip and Nainar, 1986
Paphiopedilum philippinense	em	pl	Cho and Valmayor, 1988
Cymbidium	plb	se, pl	Begum et al. 1994
Cymbidium kanran	rh	sb	Lee et al. 1986
Cymbidium × *Thanksgiving* cv. Nativity	plb	sb, pl	Begum et al., 1994
Spathoglottis plicata	l, n	pl, r	Teng et al. 1997
Spathoglottis plicata	st	r	Bapat and Narayanaswamy, 1977
Cymbidium, Aloifolium, Dendrobium aphyllum, Dendrobium moschatum	axb	pl, pl	Nayak et al. 1997
Cymbidium ensifolium var. misericors	pse,rh, r	c, sb	Chang and Chang, 1998
Oncidium "Gower Ramsey"	ep, me	se, pl	Chen et al. 1999
Cattleya walkeriana	s	pl	Islam et al. 1999
Paphiopedilum hybrid (*Paphiopedilum callosum* 'Oakhi' × *P. lawrenceanum* 'Tradition')	plb	c, sb, pl	Lin et al. 2000
Oncidium "Sweet Sugar"	fst	se, pl	Chen and Chang 2000b
Phalaenopsis 'Nebula'	pr	c, sb, pl	Chen et al. 2000
Cymbidium sinense	rh	sb, r	Chang and Chang 2000a
Geodorum densiflorum	rh	pl, r	Sheelavantmath et al. 2000

Abbreviations: ab = apical bud, axb = axillary bud, c = callus, em = embryo, ep = epidermal cell, fst = flower stalk, inf = inflorescence, l = leaf segment, me = mesocotyl tissue, ml = multiple, n = nodal explants, pl = plantlet formation, plb = protocorm-like-bodies, pr = protocorm, pse = pseudobulbs, r = rooting, rh = rhizome, s = seed, sb = shoot bud regeneration, se = somatic embryogenesis, st = stem.

(Sheelavantmath et al. 2000), and *Cymbidium* species (Paek and Yeung 1991), which were rooted and transferred to the field.

Direct somatic embryogenesis from leaf explants of *Oncidium* and subsequent plant regeneration have been reported (Chen et al. 1999; Chen and Chang 2000b). Somatic embryos were obtained directly from a wounded surface or via nodular callus masses within 45 days of culture. The embryos were germinated on a hormone-free basal medium (Chen and Chang 2000b). They further reported efficient plant regeneration via somatic embryogenesis, induced from root-tip, stem and leaf-derived callus culture of *Oncidium,* which was compact and yellowish-white. The frequency of embryo formation of root-derived calli was higher than stem- and leaf-derived calli. The combinations of NAA and TDZ significantly promoted embryo formation. High frequency (93.8%) somatic embryogenesis and an average of 29% somatic embryos / callus (3 x 3 mm^2) were found in root-derived callus on a basal medium supplemented with 0.1 mg L^{-1} NAA and 3.0 mg L^{-1} TDZ. All the somatic embryos developed into plantlets with 100% survival rate. In orchids, somatic embryogenesis seems to be an ideal approach for fast plant regeneration compared to vegetative propagation. A temporary immersion system could be tested for somatic embryo production of orchids.

Carnation

Carnation (*Dianthus caryophyllus*) is one of the most important commercially grown cut flowers and ranks next to rose. It is suitable for cut flower, bedding, pots, edging and rock gardens. It has excellent keeping quality, wide range of forms, capacity to withstand long distance transportation, and ability to dehydrate after shipping. Carnation is preferred by growers in many flower-exporting countries. [Cultivation is very limited, in the open, and done in cool climate or during winter.] It requires cooler nights, temperature ranging from 8–10 °C, and warm day, temperature 18–24 °C. Carnation is grown under protection for producing good quality cut flowers for export and domestic markets. Conventionally, it is propagated through cuttings. *In vitro* culture technique is an alternative method of plant propagation on a large-scale.

Micropropagation of carnation was previously reviewed by Mii et al. (1989) and Table 13.7 presents *in vitro* studies on carnation. Multiple shoots have been produced from various explants including shoot tip (Hackett and Anderson 1967), petals (Kakehi 1978), anther culture (Villalobos 1981), and axillary bud culture (Ghosh and Mohan Ram 1986; Miller et al. 1991).

Well-developed shoots were rooted by reducing sucrose concentration in both semisolid and liquid medium. Even during multiple shoot formation, adventitious root formation occurred at the base of the shoot in

Table 13.7 ***In vitro*** **studies of Carnation (*Dianthus caryophyllus*)**

Explant source	*Response*	*References*
Meristem	sp	Shabde and Murashige 1977
Shoot tip	sp	Baker and Phillips 1962, Hollings 1965,Takeda 1974
Shoot tip	ms, pl	Hackett and Anderson 1967, Petru and Landa 1974, Earle and Langhans 1975, Davis et al. 1977, Hempel, 1979
Shoot tip	c	Petru and Landa 1974, Earle and Langhans,1975
Stem node	ms, pl	Roest 1977, Roest and Bokelman 1981
Stem internode	c, se	Kakehi 1970, Debergh 1972, Malczewska et al. 1979
Hypocotyl	sb, pl	Petru and Landa 1974
Leaf	sb, pl	Frey and Janick 1991
Petal	sb, Pl	Kakehi 1978
Anther	sb, Pl	Villalobos 1981, Murty and Kumar 1976
Mesophyll protoplast	c	Mii and Cheng 1982
Axillary bud	ms, pl	Ghosh and Mohan Ram 1986
Leaf, Petals, Floral segments	c, sb, pl	Messeguer et al. 1993
Leaf, Stem explants	c, sb, pl	Palai et al. 1996
Axillary bud	sb, pl	Miller et al. 1991
Stem	c, cs, se,	Chaudhary and Kokchin, 1995, Frey et al. 1992
Stem, Petal	sb, pl	Nugent et al. 1991

Abbreviations: c=callus, cs=cell suspension, ms=multiple shoot formation, pl=plantlet formation, sb=shoot bud regeneration, se=somatic embryogenesis, sp=single plantlet formation.

hormone-free medium within 2 weeks of culture (Earle and Langhans 1975). Choudhary (1991) found that the best rooting was obtained on liquid medium with 10 μM IBA. The rooted plants were established in the soil. For more effective rooting, shoot cultures could be rooted in the liquid medium, shaking at low speed. Alternatively, *in vivo* rooting could also be tried and may reduce costs of plant production.

Somatic embryogenesis and plantlet regeneration in *Dianthus caryophyllus* cv. White Sim was accomplished successfully by Choudhary and

Kokchin (1995). They noted that callus formation and subsequent formation of cell suspension cultures were obtained in MS medium supplemented with 20 µM 2,4-D. At a lower concentration of 2,4-D (5 µM), cell suspensions were composed of relatively large cell clumps. Somatic embryos were obtained in the absence of plant growth regulators. The embryos were germinated with multiple shoots and rooted and the embryo-derived plantlets established in soil.

Petunia

Petunia hybrida is a decorative annual or perennial plant grown for its beautiful flowers in beds, borders, pots, hanging baskets and containers. It is a self-pollinated plant and numerous cultivars exhibit a range of variation in flower shape, size, petal number and color. Petunia is primarily a seed-propagated crop but asexual propagation using terminal cuttings is used for maintenance of elite inbred double lines. *In vitro* culture techniques are an alternative for mass propagation and genetic manipulation. Sharma and Mitra (1976) cultured shoot apical meristems of *Petunia hybrida* for mass production of plants and succeeded in shoot multiplication on MS medium supplemented with 2.2 µM BA + 5.7 µM IAA within 4 weeks of culture. Supplementation of low concentrations of IAA or NAA to the basal medium resulted in rooting of microshoots. Plant regeneration via somatic embryogenesis in *Petunia* has been reported (Handro et al. 1973; Rao et al. 1973a, b; Sangwan and Harada 1976). Anther culture of *P. hybrida* produced haploid, diploid, and triploid plants (Jain 1978; Jain and Bhalla-Sarin 1995). Similar results were obtained in *P. violacea* (Gupta 1982) and plants grew well in the greenhouse and produced flowers.

CONCLUSION

There are some basic problems facing floriculture worldwide such as production technology, postharvest technology, patent laws, and marketing. Selection and procurement of the right type of planting material is important for successful floriculture trade. Multiplication of true-to-type and high-quality planting material or seed and making it available to the consumer is the primary responsibility of the nursery industry. High-quality ornamental plants, e.g. disease-free material, are in great demand for commercial growing, domestic gardens, and landscapes. Modern methods of plant propagation are important tools in bridging the gap between demand and supply. Plant tissue culture is an alternative method for large-scale propagation and production of ornamental plants, that are disease-free and carry improved traits such as flower color, longevity and form, plant shape, and architecture. Tissue culture of ornamental plants has made tremendous progress for large-scale plant multiplication and consequently saves considerable time.

Recent progress in genetic manipulation of plant cells has opened new vistas for improvement of ornamental crops. Transfer of foreign genes into plants is based upon the availability of an efficient *in vitro* regeneration system. Transformation can be achieved by several methods, including direct insertion of DNA into protoplasts by microinjection or electroporation and biolistic method. Useful genes have been introduced and expressed in, e.g. carnation and chrysanthemum, giving color modification and extended vase life (Table 13.8).

Table 13.8 Genetic transformation study of major floricultural crops

Species/Cultivar	*Foreign genes*	*Reference*
Dianthus caryophyllus cvs. Improved White Sim,	NPTII, GUS	Firoozabady et al. 1991a Firoozabady et al. 1991b
Crowley Sim, Red Sim, White Sim	NPTII, GUS	Lu et al. 1991
Dendranthema grandiflora cv.	NPTII, GUS	Lemieux et al. 1990
Moneymaker	Antisense CHS	Lemieux et al. 1990
Dendranthema indicum cv. Korean	NPTII	Ledger et al. 1991
Rosa hybrida cv. Royalty	NPTII,GUS	Firoozabady et al. 1991a
Hybrid tea rose	NPTII, GUS	Noriega and Sondahl,1991
Petunia hybrida var. Ultra Blue	delta-9, fatty acid desaturase	Choudhary et al. 1994
Cymbidium spp.	GUS-INT, NPTII	Yang et al. 1999
Phalaenopsis orchid	GUS	Belarmino and Mii 2000
Dendrobium spp.	NPT II, Papaya Ring spot, Coat protein gene	Kuehnle and Sugii 1992
Gerbera jamesonii cv. Frithy	NPTII, GUS	Elomaa et al. . 1993

Inclusion of molecular marker techniques and genetic modification of ornamental crops would add a new dimension to conventional breeding programs. The genetic fidelity of somatic seedlings can be determined with molecular markers such as RAPD, AFLP, and RFLP to help detect any genetic change at the early stage of development of a transgenic.

Furthermore, such an approach might be helpful in identifying genotypes of potential commercial value in genetic improvement programs as well as trait-specific molecular markers could be used for protecting varieties.

REFERENCES

Ahloowalia BS (1992) *In vitro* radiation induced mutants in *Chrysanthemum*. Mutation Breed. Newl , 39: 6.

Ahmed HA (1986) *In vitro* regeneration and propagation of meristem apices of chrysanthemum. Kerteszeti-egyetem-Kozlemenyei. 50: 199-214.

Ahmed KZ, Sagi F (1993) Use of somaclonal variation and *in vitro* selection for induction of plant disease resistance: prospects and limitations. Acta Phytopath. Entomol. Hung. 28: 143-159.

Aitken-Christie J (1991) Automation. *In*: PC Debergh, RH Zimmerman (eds.). Micropropagation Technology and Application. Kluwer Acad. Publ., the Netherlands, pp. 363-388.

Aitken-Christie J, Kozai T, Takayama S (1995) Automation *in* plant tissue culture: General Introduction and overview. *In*: JAitken-Christie, T Kozai, MAL Smith (eds.). Automation and Environmental Control in Plant Tissue Culture. Kluwer Acad. Publ., the Netherlands, pp. 1-18.

Ara KA, Hossa *in* MM, Quasem MA, Ali M, Ahmed JU (1997) Micropropagation of rose: *Rosa* sp. Cv. Peace. Plant Tiss. Cult. 7 (2): 135-142.

Arditti J (1977) Clonal propagation of orchids by means of tissue culture. *In*: J. Arditti (ed.). Orchid Biology: Review and Perspectives. Cornell Univ. Press, New York, NY, pp. 467-520.

Arello EF, Pasqual M, Pinto JEBP, Barbosa MHP (1991) *In vitro* establishment of explants and seedling regeneration *in Gerbera jamesonii* Bolus et Hook by tissue culture. Pesquisa-Agropecuaria-Brasileira 26 (2): 269-273.

Arene L., Pellegrino C, Gudin S (1993) A comparison of the somaclonal variation level of *Rosa hybrida* cv. Meirutral plants regenerated from callus or direct induction from different vegetative and embryonic tissues. Euphytica 71: 83-92.

Arnold NP., Binns MR, Cloutier CD., Barthakur NN, Pellerin R (1995) Auxins, salt concentrations and their interactions during *in vitro* rooting of winter-hardy and hybrid Tea roses. HortSci, 30 (7): 1436-1440.

Babu P, Chawla HS (2000) *In vitro* regeneration and *Agrobacterium* mediated transformation in gladiolus. J. Hort. Sci. Biotech. 75 (4): 400-404.

Bajaj YPS (1986) *In vitro* preservation of genetic resources. *In*: Nuclear Techniques and *In Vitro* Culture for Plant Improvement. Proc. Int. Symp., 19-23 rd August, Int. Atomic Energy Agency. Vienna. pp. 43-57.

Bajaj YPS, Sidhu MMs, Gill APS (1983) Some factors affecting the *in vitro* propagation of *Gladiolus*. Sci. Hort. 18: 269-275.

Baker R, Phillips DJ (1962) Obtaining pathogen-free stock by shoot tip culture. Phytopath., 52: 1242-1244.

Bapat VA, Narayanaswamy S (1977) Rhizogenesis in a tissue culture of the orchid *Spathoglottis*. Bull. Torry Bot. Club 104: 2-4.

Barbosa MHP, Pasqual M, Pinto JEBP, Arello EF, Barros I, de-Barros I (1993) Effects of benzylaminopurine and indole-3-acetic acid on *in vitro* propagation of *Gerbera jamesonii* Bolus et Hook cv. Appelbloesem. Pesquisa—Agropecuaria-Brasileira 28(1): 15-19.

Barve DM, Iyer RS, Kendurkar S, Mascarenhas AF (1984) An efficient method for rapid propagation of some budded rose varieties. Ind. J. Hort. 41: 1-7.

Begum AA, Takami M, Kako S (1994) Formation of protocorm-like bodies (PLB) and shoot development through *in vitro* culture of outer tissue of *Cymbidium* PLB. J. Jpn. Soc. Hort. Sci. 63 (3): 663-673.

Belarmino MM, Mii M (2000) *Agrobacterium*-mediated genetic transformation of a *Phalaenopsis* orchid. Plant Cell Rep., 19 (5): 435-442.

Benetka V (1985) Some experiences of methodology with the isolation of somatic mutations in the rose cultivar 'Sonia'. Acta Pruboniciana 50: 9-25.

Ben-Jaacov J, Langhans RN (1972) Rapid multiplication of chrysanthemum plants by stem tip proliferation. Hort. Sci. 7: 289-290.

Ben-Meir H, Scovel G, Ovadis M, Vainstein A (1997) Molecular markers in the Breed. of ornamentals. Acta Hort. 447: 599-601.

Bhattacharya P, Dey S, Das N, Bhattacharya BC, Bhattacharya P (1990) Rapid mass propagation of *Chrysanthemum morifolium* by callus derived from stem and leaf explants. Plant Cell Rep., 9: 439-442.

Bhojwani SS, Razdan MK (1996) Plant Tissue Culture: Theory and Practice. Elsevier, Amsterdam, Netherlands (rev. ed.).

Bian HW, Wang JH, Lin WQ, Han N, Zhu MY (2002) Accumulation of soluble sugars, heat-stable proteins and dehydrins in cryopreservation of protocorm-like bodies of *Dendrobium candidum* by the air-drying method. J. Plant Physiol. 159: 1139-1145.

Binsfeld PC, Wingender R, Schnab H (1999) An optimized procedure for sunflower protoplast (*Helianthus* ssp.) cultivation in liquid culture. Helia 22 (30): 61-70.

Blakesley D, Lenton JR, Horgan R (1991) Uptake and metabolism of 6-benzylaminopurine in shoot cultures. Physiol. Plant. 81: 343-348.

Bouman H, de Klerk GJ (1996) Somaclonal variation in Biotechnology of ornamental plants. In: R. Geneve, J. Preece, S. Merkle, Biotechnology of ornamental plants. Eds., CAB International, UK, pp.165-183.

Bouman H, Kuijpers AM, de Klerk GJ (1992) The influence of tissue culture methods on somaclonal variation in *Begonia*. Physiol. Plant. 85: A45.

Bressan PH, Kim YJ, Hyndman SE, Hasegawa PM, Bressan RA (1982) Factors affecting *in vitro* propagation of rose. J. Amer. Soc. Hort. Sci. 107 (6): 979-990.

Broertjes C, Van Harten AM (1978) Application of Mutation Breeding. Methods on the Improvement of Vegetatively Propagated Crops. Elsevier, Amsterdam, Netherlands,

Burger DW, Liu L, Zary KW, Lee CI (1990) Organogenesis and plant regeneration from immature embryos of *Rosa hybrida* L. Plant Cell, Tiss. Org. Cult. 21: 147-152.

Bush SR, Earle ED, Langhans RH (1976) Plantlets from petal segments, petals, epidermis and shoot tips of the periclinal chimera, *Chrysanthemum morofolium* 'Indianapolis'. Amer. J. Bot. 63: 729-737.

Cappadocia M, Chretien L, Laublin G (1988) Production of haploids in *Gerbera jamesonii* via ovule culture: Influence of fall versus spring sampling on callus formation and shoot regeneration. Can. J. Bot. 66: 1107-1110.
Chan AP (1966) *Chrysanthemum* and rose mutations induced by X-rays. Proc. Amer. Soc. Hort. Sci. 88: 613- 620.
Chang C, Chang WC (1998) Plant regeneration from callus culture of *Cymbidium ensifolium* var. Misericors. Plant Cell Rep. 17: 251-255.
Chang C, Chang WC (2000a) Effect of thidiazuron on bud development of *Cymbidium sinense* Willd *in vitro*. Plant Growth Regul. 30 (2): 171-175.
Chatterjee C, Corell MJ, Weathers PJ, Wyslouzil BE, Wacers DB (1997) Simplified acoustic window mist bioreactor. Biotech. Tech. 11: 155-158.
Chen JT, Chang C, Chang WC (1999) Direct somatic embryogenesis on leaf explants of Oncidium Gower Ramsey and subsequent plant regeneration. Plant Cell Rep. 19: 143-149.
Chen JT, Chang WC (2000b) Plant regeneration via embryo and shoot bud formation from flower-stalk explants of Oncidium Sweet Sugar. Plant Cell, Tiss. Org. Cult. 62: 65-100.
Chen YC, Chang C, Chang WC (2000) A reliable protocol for plant regeneration from callus culture of Phalaenopsis. *In vitro* Cell Dev. Biol.-Plant, 36: 420-423.
Chen YZ, He XD, Jiang PY, Wang CM (1985) *in vitro* propagation of Chrysanthemum leaves. J. Jiangsu Agrec. College Jiangsuu Nongxueyuan Xuebao. 6: 33-36.
Cho MS, Valmayor HL (1988) Effects of culture media on the growth of seedlings derived from embryo culture in *Paphiopedilum philippinense* (Tropical orchid). Korean J. Plant Tiss. Cult. 15: 103-110.
Choudhary ML (1991) Vegetative propagation of carnation *in vitro* through multiple shoot development. Indian J. Hort. 18: 177-181.
Choudhary ML, Chin CK, Polashock JJ, MartIn CE (1994) Agrobacterium-mediated transformation of *Petunia hybrids* with yeast Δ-9 fatty acid desaturase. Plant Growth Regul. 15: 113-116.
Choudhary ML, Kokchin C (1995) Somatic embryogenesis in cell suspension culture of carnation (*Dianthus caryophyllus* L.). Plant Growth Regul. 16: 1-4.
Chu CY, Knight SL and Smith MAL (1993) Effect of liquid culture on the growth and development of miniature rose (*Rosa chinensis* Jacq. 'Minima'). Plant Cell, Tiss. Org. Cult. 32: 329-334.
Clark E, Izhar S, Hanson MR (1986) Independent segregation of chloroplast DNA and cytoplasmic male sterility in *Petunia* somatic hybrids. Molec. Gen. Genet. 199: 440-445.
Condliffe PC, Davey MR, Power JB, Koehorst-van Putten H, Visser PB (2003) An optimised protocol for rose transformation applicable to different cultivars. Acta Hortic., 612: 115-120.
Corbeneau F, Engelmann F, Come D (1990) Ethylene production as an indicator of chilling injury in oil palm (*Elaeis guineensis*) somatic embryos. Plant Sci. 71: 29-34.
Corneanu GC, Corneanu M (1992) Preliminary studies about human bioenergy effect on *In vitro* vegetal cultures. Rev. Roum. Biol. 37: 113-117.
Cornish EC, Baudinette SC, Graham MW, Stevenson KR, et al. (1991) Expression of antisense ethylene forming enzyme in transgenic carnation (*Dianthus*

caryophyllus). 31st Ann. Gen. Mtg. Austr. Soc. Plant Physiol. Australian National University, Canberra, 2-4th October 1991, Abstract No. 65.

Corrie, S. Tandon P (1993) Propagation of *Cymbidium giganteum* Wall. through high frequency conversion of encapsulated protocorms under *in vivo* and *in vitro* conditions. Ind. J. Exp. Biol., 31: 61-64.

Dalsou V, Short KC (1987) Selection of sodium chloride tolerance in *Chrysanthemum*. Acta Hort. 212: 737-740.

Davies DR (1980) Rapid propagation of roses *in vitro*. Sci. Hort. 13: 385-389.

Davis MJ, Baker R, Hanan JJ (1977) Clonal multiplication of carnation by micropropagation. J. Amer. Soc. Hort. Sci. 102: 48-53.

de Donato M, Peruceo E (1984) Micropropagation of *Chrysanthemum* by means of lateral meristem stimulation. Ann. Fac. Sci. Agric. Univ. Stud. Torino (Italy) 13: 103-115.

de Jong J, Rademaker W, van Wordragen MP (1993) Restoring adventitious shoot formation of *Chrysanthemum* leaf explants following cocultivation with *Agrobacterium tumfaciens*. Plant Cell, Tiss. Org. Cult. 32: 263-270.

De Vries DP, Dubois LAM (1988) The effect of BAP and IBA on sprouting and adventitious root formation of 'Amanda' rose single-node soft wood cuttings. Sci. Hort. 34: 115-121.

de Wit JC, Esendam HF, Horkanen JJ, Tuominen U (1990) Somatic embryogenesis and regeneration of flowering plants in rose. Plant Cell Rep. 9: 456-458.

Debener T, Bartels C, Spethmann W (1997) Parentage analysis in interspecific crosses between rose species with RAPD markers. Gartenbauwissenschaft, 62: 180-184.

Debener T, Mattiesch L (1996) Genetic analysis of molecular markers crosses between diploid roses. Acta Hort. 424: 249-252.

Debergh P (1972) Root formation in *Dianthus caryophyllus in vitro*. Med. Fac. Kandwet., Rijk. Gent. 37: 41-46.

Dohm A, Ludwig C, Schilling D, Debener Th. (2002) Transformation of roses with genes for antifungal proteins to reduce their susceptibility to fungal diseases. Acta Hortic., 572: 105-111.

Douglas GC, Rutledge CB, Casey AD, Richardson DHS (1989) Micropropagation of floribunda ground cover and miniature roses. Plant Cell, Tiss. Org. Cult.., 19: 55-64.

Dubois LAM, deVries DP (1995) Preliminary report on the direct regeneration of adventitious buds on leaf explants of *in vitro* grown glasshouse rose cultivars. Gartenbauwissenschaft 60 (6): 249-253.

Earle ED, Langhans RW (1974a) Propagation of *Chrysanthemum in vitro*. I. Multiple plantlets from shoot tips and the establishment of tissue cultures. J. Amer. Soc. Hort. Sci. 99: 128-131.

Earle ED, Langhans RW (1974b) Propagation of *Chrysanthemum in vitro*. II. Production growth and flowering of plantlets from tissue cultures. J. Amer. Soc. Hort. Sci. 99: 352- 358.

Earle ED, Langhans RW (1975) Carnation propagation for shoot tips cultured in liquid medium. Hort. Sci. 10: 608-610.

Elomaa P, Honkanen J, Puska R, Seppanen P, et al. (1993) *Agrobacterium*–mediated transfer of antisense chalcone synthase cDNA to *Gerbera hybrida* inhibits flower pigmentation. Bio/Tech. 11: 508-511.

Etienne H, Berthouly M (2002) Temporary immersion systems in plant micropropagation. Plant Cell, Tiss. Org Cult. 69: 215-231.

Fiore S, Pasquale F de, Carimi F, Sajeva M.(2002) Effect of 2,4-D and 4-CPPU on somatic embryogenesis from stigma and style transverse thin cell layers of *Citrus.* Plant Cell Tiss. Org. Cult., 68:57-63.

Firoozabady E, Moy Y, Courtney-gutterson N, Robinson K (1994) Regeneration of transgenic rose (*Rosa hybrida*) plants from embryogenic tissue. Biotech. 12: 609-613.

Firoozabady E, Moy Y, Tucker W, Robinson K, Gutterson N (1995) Efficient transformation and regeneration of carnation cultivars using *Agrobacterium.* Molec. Breed. 1: 283-293.

Firoozabady E, Noriega C, Sondahl MR, Robinson KEP (1991a) Genetic transformation of rose (*Rosa hybrida* cv. Royalty) via *Agrobacterium tumefaciens.* in vitro 27: 154A.

Firoozabady E., Lemieux CS, Moy YS, Moll B, Nicholas JA, Robinson KEP (1991b) Genetic engineering of ornamental crops. In vitro, 27: 96A.

Frey L, Janick J (1991) Organogenesis in carnation. J. Amer. Soc. Hort. Sci. 116 (6): 1108-1112.

Frey L, Saranga Y, Janick Y (1992) Somatic embryogenesis in carnation. Hort. Sci. 32: 63-65.

Fu FML (1979) Studies on the tissue culture of orchids. II. Clonal propagation of *Aranda, Ascocenda, Cattleya* by leaf tissue culture. Orchid Rev. 87: 343-346.

Fujii Y, Shimzu K (1990) Regeneration of plants from stem and petals of *Chrysanthemum coccineum.* Plant Cell Rep. 8: 625-627.

Fukai S, Morii M, Oe M (1988) Storage of *Chrysanthemum* (*Dendrathema grandiflorum* Ramat.) plantlets *in vitro.* Plant Cell, Tiss. Org. Cult. 5: 20-25.

Fukai S, Oe M (1990) Morphological observations of *Chrysanthemum* shoot tips cultured after cryoprotection and freezing. J. Jpn. Soc. Hort. Sci. 59: 383-387.

Gamborg OL, Miller RA , Ujima K (1968) Nutrient requirement of suspension cultures of soyabean root cultures. Exp. Cell Res. 50: 151-158.

Ganapathi TR, Bapat TR, Rao PS (1994) *In vitro* development of encapsulated shoot-tips of cardamum. Biotech. Tech. 8: 234-239.

Gertsson UE, Andersson E (1985) Propagation of *Chrysanthemum* x *horotorum* and *Philodendron scandens* by tissue culture. Rapport, Institutionen for Tradgardsvetenskap. Sveriges Lantbruksuniver. Sitlet, 41: 17.

Ghashghaie J, Brenckmann F, Saugier B (1991) Effect of agar concentration on water status and growth of rose plants cultured *in vitro.* Physiol. Plant. 82 (1): 73-78.

Ghosh S, Mohan Ram HY (1986) Multiplication of spray-carnations by axillary bud culture. Curr. Sci. 55 (19): 966-971.

Gill R, Gerrath J, Saxena PK (1992) High-frequency direct embryogenesis in thin layer cultures of hybrid seed geranium *(Pelargonium).* Can J. Bot., 71:408-413.

Ginzburg C, Ziv M (1973) Hormonal regulation of cormel formation in *Gladiolus* stolons grown *in vitro.* Ann. Bot. 37: 219-224.

Graves ACF, Goldman SL (1987) *Agrobacterium tumefaciens*-mediated transformation of the monocot genus *Gladiolus*: Detection of expression of T-DNA-encoded genus. J. Bacteriol. 169: 1745-1746.

Griesbach RJ (1992) Incorporation of the GUS gene into orchids through embryo electrophoresis. HortSci. 27: 620.
Gudin S, Mouchotte J (1996) Integrated research in rose improvement. A breeder's experience. Acta Hort. 424: 285-292.
Gupta PP (1982) Genesis of microspore-derived triploid petunias. Theor. Appl. Genet. 61: 327-331.
Haberlandt G (1902) Kulturversuche mit isollierten pflanzenzellen. S.B. Weisen Wien Naturwissenschaften,111:69-92.
Haccius, B (1978) Question of unicellular origin of non-zygotic embryos in callus cultures. Phytomorph. 28: 74-81.
Hackett WP, Anderson JM (1967) Aseptic multiplication and maintenance of differentiated carnation shoot tissue derived from shoot apices. Proc. Amer. Soc. Hort. Sci. 90: 365-369.
Handro W, Rao PS, Harada H (1973) A histological study of the development of buds, roots and embryos in organ cultures of *Petunia Inflata* R.Fries. Ann. Bot. 37: 817-821.
Harti D, Kuzmicic I, Jug-Dujakovic M, Jelaska S (1993) The effect of genotype on *Gerbera* shoot multiplication *in vitro*. Acta Botanica Croatica 52: 25-32.
Hasegawa PM (1980) Factors affecting shoot and root initiation from cultured rose shoot tips. J. Amer. Soc. Hort. Sci. 105 (2): 216-220.
Hempel M (1979) Studies on *in vitro* multiplication of carnations. I. The influence of some cytokinins on the differentiation of shoot apices. Acta Hort. 91: 317-321.
Hempel M, Petos-witkowska B, Tymoszuk J (1985) The influence of cytokinins on multiplication and subsequent rooting of *Gerbera in vitro*. Acta Hort. 157: 301-304.
Hill GP (1967) Morphogenesis of shoot primordia in cultivated stem tissue of a Garden Rose. Nature, 216: 596-597.
Hill GP (1968) Shoot formation in tissue cultures of *Chrysanthemum* 'Bronze Pride'. Physiol. Plant. 21: 386-389.
Hitmi A, Barthomeuf C, Sallanon H (2000) Cryopreservation of *Chrysanthemum cinerariaefolium* shoot tips. J. Plant Physiol. 156: 408-412.
Hohe A, Winkelmann T, Schwenkel HG (1999) The effect of oxygen partial pressure in bioreactors on cell proliferation and subsequent differentiation of somatic embryos of *Cyclamen persicum*. Plant Cell, Tiss. Org. Cult. 59 (1): 39-45.
Hollings M (1965) Disease control through virus-free stock. Ann. Rev. Phytopath. 3: 367-396.
Hoshino Y, Nakano M, Mii M (1995) Plant regeneration from cell suspension-derived protoplasts of *Saintpaulia ionantha* Wendl. Plant Cell Rep. 114: 341-344.
Hsia CN, Korban SS (1996) Factors affecting *in vitro* establishment and shoot proliferation of *Rosa hybrida* L. and *Rosa chinensis minima. In Vitro* Cell, Dev. Biol.-Plant., 32:217-222.
Hsia CN, Korban SS (1996) Organogenesis and somatic embryogenesis in callus cultures of *Rosa hybrida* and *Rosa chinensis*-minima. Plant Cell Tiss. Org. Cult. 44(1): 1-6.
Huang JM, Ni YY, Lin MH (1987) The micropropagation of *Gerbera*. Acta-Hort. Sinica 14 (2): 125-128.

Huang MC, Chu CY (1985) A scheme for commercial multiplication of Gerbera (*Gerbera hybrida* Hort.) through shoot tip culture. J. Jpn. Soc. Hort. Sci. 54 (1): 94-100.
Huang SC., Tsai CC, Sheu CS (2000) Genetic analysis of *Chrysanthemum* hybrids based on RAPD molecular markers. Bot. Bull. Acad. Sinica 41 (4): 257-262.
Huitema JBM, Gussenhoven G, Dons JJM, Broertjes C (1986) Induction and selection of low-temperature tolerant mutants of *Chrysanthemum morifolium* Ramat. Nuc.Tech. *in vitro* Culture Plant Improv. 12: 321-327.
Hussey G (1975) Totipotency in tissue explants and callus of some members of the Liliaceae, Iridaceae and Amaryllidaceae. J. Exp. Bot. 26; 253-262.
Hussey G (1977) *In vitro* propagation of *Gladiolus* by precocious axillary shoot formation. Sci. Hort. 6: 287-296.
Hutchinson JF, Kaul V, Maheswaran G, Moran JR, Graham MW, Richards D (1992) Genetic improvement of floricultural crops using Biotechnology. Aust. J. Bot. 40: 765-787.
Hutchinson JF, Miller R, Kaul V, Stevenson T, Richards D (1989) Transformation of *C. morifolium* based on *Agrobacterium* gene transfer. J. Cell Biochem., p. 261 (abstract).
Hvoslef-Eide AK, Munster C (1998) Somatic embryogenesis of *Cyclamen persicum* Mill. in bioreactors. Comb. Proc. Int. Plant Prop. 47: 377-382.
Hyndman SE, Hasegawa PM, Bressan RA (1982) Stimulation of root initiation from cultured rose shoots through the use of reduced concentrations of mineral salts. HortSci. 17 (1): 82-83.
Intuwong O, Kunisaki JT, Sagawa Y (1972) Vegetative propagation of *Phalaenopsis* by flower stalk cuttings. Na Okika O Hawaii [Hawaii Orchid J.] 1: 13-18.
Intuwong O, Sagawa Y (1973) Clonal propagation of *Sarcanthus* orchids by aseptic culture of inflorescences. Amer. Orchid Soc. Bull. 42; 209-215.
Ishiooka N, Tanimoto S (1990) Plant regeneration from Bulgarian rose callus. Plant Cell, Tiss. Org. Cult. 22: 197-199.
Islam MO, Matsui S, Ichihashi S (1999) Effects of light quality on seed germination and seedling growth of *Cattleya* orchids *in vitro*. J. Jpn. Soc. Hort. Sci. 68 (6): 1132- 1138.
Izhar S, Schlichter M, Swartzberg D (1983) Sorting out of cytoplasmic elements in somatic hybrids of *Petunia* and the prevalence of the heteroplasmon through several meiotic cycles. Molec. Gen. Genet. 190: 468-474.
Izhar S, Tabib Y (1980) Somatic hybridization in *Petunia*. II. Heteroplasmic state in somatic hybrids followed by cytoplasmic segregation into male sterile and male fertile lines. Theor. Appl. Genet. 57: 241-246.
Jacobs G, Bornman CH, Allan P (1968) Tissue culture studies on rose: use of pith explants. S. Afric. J. Agric. Sci. 11: 673-678.
Jain SM (1978) *In vitro* production of haploids in higher plants. PhD thesis, Jawaharlal Nehru Uni., New Delhi, India.
Jain SM (1993a) Somaclonal variation in *Begonia* x *elatior* and *Saintpaulia ionantha* L. Sci. Hort. 54: 221-231.
Jain SM (1993b) Studies on somaclonal variation in ornamental plants. Acta Hort. 336: 365-372.

Jain SM (1993c) Growth hormonal influence on somaclonal variation in ornamental plants. Proc. XVII Eucarpia Symp. Creating Genomic Variation in Ornamentals. pp. 93-103.

Jain SM (1997a) Micropropagation of selected somaclones of *Begonia* and *Saintpaulia*. J. Biosci. 22: 585-592.

Jain SM (1997b) Somaclonal variation and mutagenesis for crop improvement. Maatalouden tutkimuskeskuksen julkaisuja MTTK, Jokioinen, Finland, vol.18, pp. 122-133.

Jain SM (2000) Mechanisms of spontaneous and induced mutations in plants. Radiation Res. Proceedings. vol 2, pp. 255-258.

Jain SM (2001) Tissue culture-derived variation in crop improvement. Euphytica 118: 153-166.

Jain SM (2002) Feeding the world—Biotech. induced mutations and breeding. *In*: Proc. INC'02, MINT, Malaysia. pp. 1-14.

Jain SM, Bhalla-Sarin N (1995) Haploidy in Petunia. *In*: *in vitro* Haploid Production in Higher Plants. SM Jain, SK Sopory, RE Veilleux (eds.). Kluwer Acad. Publ., Dordrecht, Netherlands, vol. 5, pp. 53-71.

Jain SM, Buiatti M, Gimelli F, Saccardo F (1997) Somaclonal variation in improving ornamental plants. *In*: SM Jain, DS Brar, BS Ahloowalia, (eds.). Somaclonal Variation and induced Mutation in Crop Improvement. Kluwer Acad. Publ., Dordrecht, Netherlands, pp. 81-105.

Jain SM, de Klerk GJ (1998) Somaclonal variation in Breed. and propagation of ornamental crops. Plant Call, Tiss. Cult. Biotech. 4 (2): 63-75.

Jain SM, Maluszynski M (2004). Induced mutations and biotechnology in improving crops. In: A. Mujib (Ed.) *In vitro* applications in crop improvement: Recent progress. Oxford & IBH Pub. Co. Pvt. Ltd., New Delhi, India (In Press).

Jain SM, Oker-Blom C, Pehu E, Newton RJ (1992) Genetic engineering: an additional tool for plant improvement. Agric. Sci. Finl. 1: 323-338.

Jain SM, Vitti D, Tucci M, Grassotti A, Rugini E, Saccardo F (1998) Biotechnology and agronomical aspects in gerbera improvements. Adv. Hort. Sci. 12: 47-53.

James J (1983) New roses by irradiation: An update. Amer. Rose Annuals pp. 99-101.

Janakiram, T (2000) Planting material of Ornamental crops: present status and future prospects. Floriculture Today, October, pp. 26-31.

Janeiro, L.V., Ballester, A., Vietiez, A.M (1995) Effects of cold storage on somatic embryogenesis of *Camellia*. J. Hort. Sci. 70: 665-672.

Jang J, Raemaker W, Huitema JBM (1987) *In vitro* characterization of low-temperature tolerant mutants of *Chrysanthemum morifolium* Ramat. Acta Hort. 197: 97-102.

Jayasree N, Devi BP (1997) Production of synthetic seeds and plant regeneration in *Rosa hybrida* L. Cv. King's ransom. Ind. J. Exp. Biol. 35: 310-312.

Jenes B, Morre H, Cao J, Zhang W, Wu R (1993) Techniques for gene transfer. *In*: S Kung, R Wu, (eds.). Transgenic Plants. Acad. Press, San Diego, A, USA, vol.1, pp. 125-146.

Jerzy M, Lubomski M (1991) Adventitious shoot formation on *ex vitro* derived leaf explants of *Gerbera jamesonii*. Sci. Hort. 47: 115-124.

Kaeppler SM, Kaeppler HF, Rhee Y (2000) Epigenetic aspects of somaclonal variation in plants. Plant Molec. Biol. 43: 179-188.

Kaicker US (1982) Mutation Breeding in roses. Ind. Rose Annu. 2: 35-42.
Kakehi M (1970) Studies on tissue culture of carnation. I. Relationships between the growth and histodifferentiation of callus and each tissue on aseptic culture. Bull. Hiroshima Agric. Coll. 4: 40-49.
Kakehi M (1978) Studies on tissue culture of carnation. V. induction of redifferentiated plants from petal tissue. Bull. Hiroshima Agric. Coll. 6: 159-166.
Kamo K (1994) Effect of phytohormones on plant regeneration from callus of *Gladiolus* cultivar 'Jenny Lee'. In vitro Cell Dev. Biol. 30P: 265-331.
Kamo K (1997) Factors affecting *Agrobacterium tumefaciens* mediated gusA expression and opine synthesis in *Gladiolus*. Plant Cell Rep. 16: 389-392.
Kamo K, Chen J, Lawson R (1990) The establishment of cell suspension cultures of *Gladiolus* that regenerate plants. In vitro Cell Dev. Biol.–Plant 26: 425-430.
Kaul V, Miller RM, Hutchinson JF, Richards D (1990) Shoot regeneration from stem and leaf explants of *Dendrathema grandiflora* Tzvelev (syn. *Chrysanthemum morifolium* Ramat.). Plant Cell, Tiss. Org. Cult. 21: 21-30.
Kawata M., Oono K (1997) Protoclonal variation in crop improvement. In: SM Jain, DS Brar, BS Ahloowalia, (eds.). Somaclonal variation and induced mutation for crop improvement. Kluwer Acad. Publ. Dordrecht, Netherlands.
Kevers C, Boyer N, Courduroux J, Gaspar T (1992) The influence of ethylene on proliferation and growth of rose shoot culture. Plant Cell, Tiss. Org. Cult. 28: 175-181.
Khosh-khui M, Sink KC (1982a) Micropropagation of new and old world rose species. J. Hort. Sci. 57 (3): 315-319.
Khosh-khui M, Sink KC (1982b) Callus induction and culture of *Rosa*. Sci. Hort. 17: 361-370.
Khosh-khui M, Sink KC (1982c) Rooting enhancement of *Rosa hybrida* for tissue culture propagation. Sci. Hort. 17: 371-376.
Kim KW, Kako S (1984) Studies on clonal propagation in the *Cymbidium* floral organ culture *in vitro*. J. Kor. Soc. Hort. Sci. 25; 65-71.
Kim YJ, Hong YP, Park YK, Cheung SK, Kim EY (1988) *In vitro* propagation of *Cymbidium* Kanran. Research Reports Rural Development Administration, Horticultural. (Korea) 30 (3): 77-82.
Kintzios S, Manos C, Makri O (1999) Somatic embryogenesis from mature leaves of rose (*Rosa* sp.). Plant Cell Rep. 18 (6): 467-472.
Kirby EG (1988) Recent advances in protoplast culture of Hort. crops: Conifers. Sci. Hort. 37: 267-276.
Knudson L (1922) Flower production by orchid grown non-symbiotically. Bot. Gaz. 89: 192.
Korbin M, Podwyszynska M, Komorowska B, Wawrzynczak D (2002) Transformation of gerbera plants with tomato spotted wilt virus (TSWV) nucleoprotein gene. Acta Hort. 572: 149-157.
Kozai T (1988) High technology in protected cultivation. Horticulture in new era. Tokyo: Organizing Committee of Int. Sym. High Tech. in Protected Cultivation. pp. 1-49.
Kozai T (1990a) Autotropic (sugar-free) tissue culture for promoting the growth of plantlets *in vitro* and for reducing biological contamination. Proc. Bangkok. Thailand: International Sym. Appl. Biotech. Small industries. pp. 39-51.

Kozai T (1990b) Micropropagation under photoautotropic conditions. *In*: P Debergh, RH Zimmerman, (eds.). Micropropagation: Tech. and Application. Kluwer Acad. Publ. Dordecht, Netherlands, pp. 449-471.

Kozai T (1991a) Autotropic micropropagation. *In*: YPS Bajaj, (ed.). Biotechnology in Agriculture and Forestry, vol.17. High-tech and Micropropagation. Springer -Verlag, New York, NY, pp. 313-343.

Kozai T (1991b) Controlled environments in conventional and automated micropropagation. *In*: R Levin, IK Vasil, (eds.). Cell Culture and Somatic Cell Genetics of Plants. vol.8. Scale-up and Automation in plant tissue culture. Acad. Press Inc., London, pp. 213-230.

Kozai T, Kubota C, Watanabe I. (1988) Effect of basal medium composition on the growth of carnation plantlets in auto- and mixotropic tissue culture. Acta Hort. 230: 159-166.

Kozai T, Kubota C., Watanabe I (1990a) The growth of carnation plantlets *in vitro* cultured photoautotropically and photomixotropically on different media. Environ. Control Biol. 28: 21-27.

Kozai T, Lee H, Hayashi M (1990b) Photoautotropic micropropagation of *Rosa* plantlets under CO_2 enriched and high photosynthetic photon flux conditions. Italy: Abstracts 23rd Int. Hort. Con. Oral. Int. Soc. Hort. Sci. Wageningen, Netherlands, 173.

Kudo S, Shibata N, Kanno Y, Suzuki M (2002) Transfromation of chrysanthemum [*Dendranthema grandiflorum* (Ramat.) Kitamura] via *Agrobacterium tumefaciens*. Acta Hort. 572: 139-147.

Kuehnle AR, Sugii N (1992) Transformation of *Dendrobium* orchid using particle bombardment of protocorms. Plant Cell Rep. 11: 484-488.

Kumar A, Kumar VA (1995) High-frequency *in vitro* propagation in *Chrysanthemum maseimum*. Ind. Hort. Jan-March: 37-38.

Kumar A, Sood A, Palni LMS, Gupta AK (1999) *In vitro* propagation of *Gladiolus hybridus* Hort.: Synergistic effect of heat shock and sucrose on morphogenesis—Micropropagation of *Gladiolus*. Plant Cell, Tiss. Org. Cult. 57: 105-112.

Kumar A, Sood A, Palni UT, Gupta AK, Palni LMS (2001) Micropropagation of *Rosa damascena* Mill. from mature bushes using thidiazuron. J. Hort. Sci. Biotech. 76 (1): 30-34.

Kunitake H, Imamizo H, Mii H (1993) Somatic embryogenesis and plant regeneration from immature seed-derived calli of rugosa rose (*Rosa rugosa* Thurb.). Plant Sci. 90: 187-194.

Laliberte S, Chretien L., Vieth J (1985) *In vitro* plantlet production from young capitulum explants of *Gerbera jamesonii*. Hort Sci. 20: 137-139.

Langford PJ, Wainwright H (1987) Effects of sucrose concentration on the photosynthetic ability of Rose shoots *in vitro*. Ann. Bot. 60: 633.

Larkin PJ, Scowcroft WR (1981) Somaclonal variation—A novel source of variability from cell cultures for plant improvement. Theor. Appl. Genet. 60: 197-214.

Lazar M, Cachita C (1983) Micropropagation of *Chrysanthemum*. III. *Chrysanthemum* multiplication *in vitro* from capitulum explants. Prod. Veg. Hort. 32: 44-47.

Ledger SE, Deroles SC, Given NK (1991) Regeneration and *Agrobacterium*-mediated transformation of *Chrysanthemum*. Plant Cell Rep. 10: 195-199.

Lee JK, Pack KY, Chun CK (1979) *in vitro* propagation of chrysanthemum through shoot apical meristem culture. J. Korean Soc. Hort. Sci. 20: 192-199.

Lee JS, Shim KK, Yoo MS, Lee JS, Kim YJ (1986) Studies on rhizome growth and organogenesis of *Cymbidium kanran* cultured *in vitro*. J. Korean Soc. Hort. Sci. 27: 174-180.

Lemieux CS, Firoozabady E, Robinson KEP (1990) *Agrobacterium*-mediated transformation of chrysanthemum. *In*: J DeJong, (ed.). Proc. Eucarpia Symp. Integration *in vitro* Tech. Ornamental Plant Breed. Pudoc, Wageningen, Netherlands, pp. 150-155.

Levin R, Gaba V, Tal B, Hirsch S, Denola D, Vasil IK (1988) Automated plant tissue culture for mass propagation . Biotech. 6: 1035-1040.

Li X, Krasnyanski SF, Korban SS (2002) Somatic embryogenesis, secondary somatic embryogenesis, and shoot organogenesis in *Rosa*. J. Plant Physiol. 159: 313-319.

Li XQ, Gasic K, Cammue B, Broekaert W, Korban SS (2003) Transgenic rose lines harbouring an antimicrobial protein gene, *Ace-AMP1*, demonstrate enhanced resistance to powdery mildew *(Sphaerotheca pannosa)*. Planta, 218:226-232.

Lilien-Kipnis H, Kochba M (1987) Mass propagation of new *Gladiolus* hybrids. Acta Hort. 212: 631-638.

Lim S, Seon JH, Paek KY, Son SH, Han BH, Drew RA (1998) Development of pilot scale process for mass production of *Lilium* bulblets *in vitro*. Acta Hort. 461: 237-241.

Lin YH, Chang C, Chang WC (2000) Plant regeneration from callus culture of *Paphiopedilum* hybrid. Plant Cell, Tiss. Org. Cult. 62: 21-25.

Lloyd D, Roberts AV, Short KC (1988) The induction *in vitro* of adventitious shoots in *Rosa*. Euphytica, 37: 31-36.

Logan AE, Zettler FW (1985) Rapid *in vitro* propagation of virus indexed *Gladioli*. Acta Hort. 164: 169-180.

Lowe JM, Davey MR, Power JB, Blundy KS (1993) A study of some factors affecting *Agrobacterium* transformation and plant regeneration of *Dendrathema grandiflora* Tzvelev (syn. *Chrysanthemum morifolium* Ramat.). Plant Cell, Tiss. Org. Cult. 33: 171-180.

Lu CY, Nugent G, Wardley T (1990) Efficient, direct plant regeneration from stem segments of chrysanthemum (*Chrysanthemum morifolium* Ramat cv. Royal Purple). Plant Cell Rep. 8: 733-736.

Lu CY, Nugent G, Wardley-Richardson T, Chandler SF, Young R, Dalling MJ (1991) *Agrobacterium*-mediated transformation of carnation (*Dianthus caryophyllus* L.). Bio/Tech. 9: 864-868.

Malaure RS, Barclay G, Power JB, Davey MR (1991) The production of novel plants from florets of *Chrysanthemum morifolium* using tissue culture. I. Shoot regeneration for ray florets and somaclonal variation exhibited by the regenerated plants. J. Plant Physiol. 139: 8-13.

Malczewska E, Molas R, Skrzyczack CZ, Hempel M (1979) Studies on *in vitro* multiplication of carnations. V. Callus, growth and differentiation. Acta Hort. 91: 345-351.

Maluszynski M, Ahloowalia BS, Sigurbjornsson B (1995) Application of *in vivo* and *in vitro* mutation techniques for crop improvement. Euphytica 85: 303-315.
Mandal AKA, Chakrabarty D, Datta SK (2000) *In vitro* isolation of solid novel flower colour mutants from induced chimeric ray florets of chrysanthemum. Euphytica 114: 9-12.
Marchant R, Davey MR, Lucas JA, Lamb CJ, Dixon RA, Power JB (1998b) Expression of a chitinase transgene in rose (*Rosa hybrida* L.) reduces development of blackspot disease (*Diplocarpon rosae* Wolf.). Molec. Breed. 4: 187-194.
Marchant R, Power JB, Lucas JA, Davey MR (1998a) Biolistic transformation of rose (*Rosa hybrida* L.). Ann. Bot. 81: 109-114.
Martin C, Carre M, Vernoy R (1981) La Multiplication Vegetative *in vitro* des vegetaux ligneux cultivees: Cas des Rosiers. C. R. Acad. Sci. Paris III, 293: 175-177.
Mathis NL, Hinchee MAW (1994) Agrobacterium inoculation techniques for plant tissues. *In*: SB Gelvin, RA Schilperoort (eds.). Plant Molec. Biol. Manual, B6: 1-9, Kluwer Acad. Publ. Dordrecht, Netherlands.
Matsumoto H, Onozawa Y (1989) Development of non-chimaeric mutation lines through *in vitro* culture of florets in *Chrysanthemum*. Sci. Rep. Fac. Agric., Ibaraki Univ. 37: 55-61.
Matthews D, Mottley J, Horan I, Roberts AV (1991) A protoplast to plant system in roses. Plant Cell, Tiss. Org. Cult. 24: 173-180.
Matthews D, Mottley J, Yokoya K, Roberts AV (1994) Regeneration of plants from protoplasts of *Rosa* species (Roses). *In*: YPS Bajaj, (ed.). Biotechnology in Agriculture and Forestry, vol. 29. Plant Protoplasts and Genetic Engineering. Springer Verlag, Heidelberg.
May RA, Trigiano RN (1991) Somatic embryogenesis and plant regeneration from leaves of *Dendrathema grandiflora*. J. Amer. Soc. Hort. Sci. 116: 366-371.
Messeguer J, Arconada MC, Mele E (1993) Adventitious shoot regeneration in carnation (*Dianthus caryophyllus* L.) Sci. Hort. 54: 153-163.
Meyer HJ, Van-Staden J (1988) The *in vitro* culture of *Gerbera aurantiaca*. Plant Cell, Tiss. Org. Cult. 14 (1): 25-30.
Meynet J, Sibi M (1984) Haploid plants from *in vitro* culture of unfertilized ovules in *Gerbera jamesonii* . Z. Pflanzenzuccht. 93: 78-85.
Mii M, Buiatti M, Gimelli F (1989) Carnation. *In*: D. Evans, WR Sharp, PV Ammirato, Y Yamada, (eds.). Handbook of Plant Cell Culture. vol.5, Ornamentals. McGraw Hill Company, New York, NY, pp. 284-318.
Mii M, Cheng SM (1982) Callus and root formation from mesophyll protoplasts of carnation. *In*: A Fujiwara, (ed.). Plant Tissue Culture, Maruzen, Tokyo, pp. 585-586.
Miller RM, Kaul V, Hutchinson JF, Richards D (1991) Adventitious shoot regeneration in carnation (*Dianthus caryophyllus*) from axillary bud explants. Ann. Bot. 67: 35-42.
Miyazaki S, Tashiro Y, Shimada T (1976) Tissue culture of *Chrysanthemum morifolium* Ramat. I. Cultivar differences in organ formation. Agric. Bull. Saga Univ. 40: 31-44.
Morel G (1964) Tissue culture: A new means of clonal propagation of orchids. Amer. Orchid Soc. Bull. 33: 473-478.

Mulin M, Tran Thanh Van K. (1989) Obtention of *in vitro* flowers from thin epidermal cell layers of *Petunia hybrida* (Hort.). Plant Sci., 62: 113-121.

Murashige T (1974) Plant propagation through tissue culture. Ann. Rev. Plant Physiol. 25: 135-166.

Murashige T, Serpa M, Jones JB (1974) Clonal multiplication of *Gerbera* through tissue culture. Hort Sci. 9: 175-180.

Murashige T, Skoog F (1962) A revised medium for rapid growth and bioassays with tobacco tissue cultures. Plant Physiol. 15:473-497.

Murty YS, Kumar V (1976) *In vitro* production of plantlets from the anthers of *Dianthus caryophyllus* L. Acta Bot. 4: 172-173.

Nayak NR, Rath SP, Patnaik S (1997) *In vitro* propagation of three ephityic *Cymbidium, Aloifolium* (L.)SW, *Dendrobium aphyllum* (Roxb.) Fisch and *Dendrobium moschatum* (Buchham)SW through thidiazuron-induced high frequency shoot proliferation. Sci. Hort. 71: 416-426.

Negi, J.P (2000) Current status and development in floriculture. Floriculture Today, December, pp. 12-14.

Nhut DT, Teixeira da Silva JA, Bui VL, Thorpe T, Tran Thanh Van K. (2003a) Woody plant micropropagation and morphogenesis by thin cell layers. In: Nhut DT, Van Le B, Tran Thanh Van K, Thorpe T. (eds.) Thin cell layer culture system: regeneration and transformation applications. Kluwer Acad. Publ., Dordrecht, The Netherlands, pp.473-493.

Nhut DT, Teixeira da Silva JA, Bui VL, Tran Thanh Van K. (2003b) Thin cell layer (TCL) morphogenesis as a powerful tool in woody plant and fruit crop micropropagation and biotechnology, floral genetics and genetic transformation. In: Jain SM, Ishii K. (eds.) Micropropagation of woody trees and fruits. Kluwer Acad. Publ., Dordrecht, The Netherlands, pp.783-814.

Nikaido T, Onogawa Y (1989) Establishment of a non-chimeric flower colour mutation through *in vitro* cultures of florets from a sport on *Chrysanthemum* with special references to the genetic background of the mutation line obtained. Sci. Rep. Fac. Agric. Ibaraki Univ. 37: 63-69.

Noriega C, Sondahl MR (1991) Somatic embryogenesis in hybrid tea roses. Bio/ Tech. 9: 991-993.

Novak FJC (1991) *In vitro* mutation system for crop improvement. *In*: FLC novac (ed.). Plant Mutation Breed. for Crop Improvement. IAEA, Vienna, Austria, Vol. 2, pp. 327-342.

Nugent G, Wardley T, Richardson, Lu CY (1991) Plant regeneration from stem and petal of carnation (*Dianthus caryophyllus* L.). Plant Cell Rep. 10: 477-480.

Oh MH, Kim SG (1994) Plant regeneration from petal protoplast culture of *Petunia hybrida*. Plant Cell, Tiss. Org. Cult. 36 (3): 275-283.

Ohishi K, Sakurai Y (1988) Morphological changes in *Chrysanthemum* derived from petal tissue. Res. Bull. Aichiken Agric. Res. Cent. 20: 278-284.

Ohki S (1994) Scanning electron microscopy of shoot differentiation *in vitro* from leaf explants of the African violet. Plant Cell Tiss. Org. Cult., 36:157-162.

Orlikowska T, Nowak E, Marasek A, Kucharska D (1999) Effects of growth regulators and incubation period on *in vitro* regeneration of adventitious shoots. Plant Cell, Tiss. Org. Cult. 59 (2): 95-102.

Osternack N, Saare-Surminski K, Preil W, Lieberei R. (1999) Induction of somatic embryos, adventitious shoots and roots in hypocotyls tissue of *Euphorbia pulcherrima* Willd. Ex Klotzsch : comparative studies on embryogenic and organogenic competence. J. Appl. Bot., 73:197-201.

Paek KY, Hahn, EJ Son SH (2001) Application of bioreactors for large scale micropropagation systems of plants. in vitro Cell, Dev. Biol.- Plant 37 (2): 149-157.

Paek KY, Yeung EC (1991) The effect of 1-naphthaleneacetic acid and N^6-benzyladenine on the growth of *Cymbidium forrestii* rhizomes *in vitro*. Plant Cell, Tiss. Org. Cult.24: 65-71.

Palai SK, Rout GR, Das P (1996) *In vitro* plant regeneration of *Dianthus caryophyllus* cv. Alas Red through callus culture. Ind. J. Hort. 53 (1): 1-7.

Pavingerova D, Dostal J, Biskova R, Benetka V (1994) Somatic embryogenesis and *Agrobacterium*-mediated transformation of *Chrysanthemum*. Plant Sci. 97: 95-101.

Petru E, Landa Z (1974) Organogenesis in isolated carnation plant callus tissue cultivated *in vitro*. Biol. Plant. 16: 450-453.

Philip VJ, Nainar SAZ (1986) Clonal propagation of *Vanilla planifolia* (Salisb.) Ames using tissue culture. J. Plant Physiol. 122: 211-215.

Pierik LM, Jansen JLM, Maasdam A, Binnendijk CM (1975) Optimization of *Gerbera* plantlets production from excised capitulum explants. Sci. Hort. 3: 351-357.

Pierik RLM, Segers Th A (1973) *In vitro* culture of mid-rib explants of *Gerbera* : adventitious root formation and callus induction. Z. Pflanzenphysiol. 69 (3): 204-212.

Pierik RLM, Steegmans HHM, Marelis JJ (1973) *Gerbera* plantlets from *in vitro* cultivated capitulum explants. Sci. Hort. 1: 117-119.

Podwyszynska M, Olszewski T (1995) Influence of gelling agents on shoot multiplication and the uptake of macroelements by *in vitro* culture of rose, cordyline and *Homalomena*. Sci. Hort. 64: 77- 84.

Power JB, Berry SF, Chapman JV, Cocking EC, Sink KC (1979) Somatic hybrids of unilateral cross-incompatible *Petunia* species. Theor. Appl. Genet. 55: 97-99.

Power JB., Frearson EM, Hayward C, George D, et al. (1976) Somatic hybridization of *Petunia hybrida* and *P. parodii*. Nature 263: 500-502.

Prasad RN, Chaturvedi HC (1988) Effect of explants on micropropagation of *Chrysanthemum morifolium* . Biol. Plant. 30: 20-24.

Prasad RN, Sharma AK, Chaturvedi HC (1983) Clonal multiplication of *Chrysanthemum morifolium* 'Otome zakura' in long-term culture. Bangladesh J. Bot. 12: 96-102.

Preil W (1991) Application of bioreactors in plant propagation. *In*: PC Debergh, RH Zimmerman, (eds.). Micropropagation technology and application., Kluwer Acad. Publ., Dordrecht, Netherlands pp. 425-445.

Preil W.(2003) Micropropagation of ornamental plants. In: Laimer M, Rucker W.(eds.) Plant tissue culture 100 years since Gottlieb Haberlandt. Springer-Verlag, New York, pp.115-133.

Preil W, Hoffmann M (1985) *In vitro* storage in chrysanthemum Breed. and propagation . *In*: A Schafer-Menuhr (ed.). *In vitro* Techniques. Martinus Nijhoff, Dordrecht, Netherlands, pp. 161-165.

Puite KJ (1992) Progress in plant protoplast research. Physiol. Plant. 85: 403-410.

Quatrano RS (1968) Free preservation of cultured flax cells utilising DMSO. Plant Physiol. 43: 2057-2061.

Rademaker W, Jong J, DeJong J (1990) Genetic variation in adventitious shoot formation in *Dendranthema grandiflora* (*Chrysanthemum morifolium*) explants. *In*: J Jong (ed.). integration of *in vitro* Techniques in Ornamental Plant Breed. Proceedings Symposium. Wageningen, Netherlands, pp. 34-38.

Rajagopalan (2000) Export potential of Indian Floriculture and need of policy environment. Floriculture Today, September, pp. 29-33.

Rajapakse S, Hubbard M, Kelly JW, Abbott AG, Ballard RE (1992) Indentification of rose cultivars by restriction fragment length polymorphism. Sci. Hort. 52: 237-245.

Rani V., Raina SN (1996) PCR technology. Botanica 46: 78-81.

Rani V., Raina SN (2000) Genetic fidelity of organized meristem-derived micro-propagated plants: A critical reappraisal. In vitro Cell. Dev. Biol.-Plant 36: 319-330.

Rao PS, Handro W, Harada H (1973a) Hormonal control of differentiation of shoots, roots and embryos in leaf and stem cultures of *Petunia inflata* and *Petunia hybrida*. Physiol. Plant. 28: 458-463.

Rao PS, Handro W, Harada H (1973b) Bud formation and embryo differentiation in *in vitro* cultures of *Petunia*. Z. Plfanzenphysiol. 69: 87-90.

Remotti PC (1995) Primary and secondary embryogenesis from cell suspension cultures of *Gladiolus*. Plant Sci. 107: 205-214.

Remotti PC, Loffler HJM (1995) Callus induction and plant regeneration from *Gladiolus*. Plant Cell, Tiss. Org. Cult. 42:171-178.

Reynoird JP, Chriqui D, Noin M, Brown S, Marie D (1993) Plant regeneration from *in vitro* leaf culture of several *Gerbera* species. Plant Cell, Tiss. Org. Cult. 33: 203-210.

Roberts AV, Horan I, Mathews D ,Mottley J (1990) Protoplast technology and somatic embryogenesis in *Rosa*. Euphytica. Proc. Symp. 10-14 Nov. J deJong (ed.) Integration of *in vitro* techniques in Ornamental Plant Breeding. CPO Centre for Plant Breed. Research, Wageningen, Netherlands.

Roberts AV, Smith EF (1990) The preparation *in vitro* of chrysanthemum for transplantation to soil. I. Protection of roots by cellulose plugs. Plant Cell, Tiss. Org. Cult. 21: 129-132.

Roberts AV, Walker S, Horan I, Smith EF, Mottley J (1992) The effects of growth retardants, humidity and lighting at stage III on stage IV of micropropagation in chrysanthemum and Rose. Acta Hort. 319:135-138.

Roberts AV, Yokoya K, Walker S, Mottley J (1995) Somatic embryogenesis in *Rosa* spp. In: SM Jain, PK Gupta, RJ newton (eds.). Somatic Embryogenesis in Woody Plants. Kluwer Acad. Publ. Dordrecht, Netherlands vol.2, pp. 277-289.

Robinson KEP, Firoozabady E (1993) Transformation of floriculture crops. Sci. Hort. 55: 83-99.

Roest S (1977) Vegetative propagation *in vitro* and its significance for mutation Breeding. Acta Hort. 78: 349-359.

Roest S, Bokelmann GS (1973) Vegetative propagation of *Chrysanthemum cinerariaefolium in vitro*. Sci. Hort. 1: 129-132.

Roest S, Bokelmann GS (1975) Vegetative propagation of *Chrysanthemum morifolium* Ram. *in vitro*. Sci. Hort. 3; 317-330.

Roest S, Bokelmann GS (1976) The storage of *Chrysanthemum morifolium* in test tubes. Vakblad-voor-de-Bloemisterij 31: 55.
Roest S., Bokelmann GS (1981) Vegetative propagation of carnation *in vitro* through multiple shoot development. Sci. Hort. 14: 357-366.
Rogers RM, Smith MAL (1992) Consequences of *in vitro* and *ex vitro* root initiation for miniature rose production. J. Hort. Sci. 67: 535-540.
Rosu A, Skirvin RM, Bein A, Norton MA, Kushad M, Otterbacher AG (1995) The development of putative adventitious shoots from a chimeral thornless rose (*Rosa multiflora* Thurb. ex J. Murr.) *In vitro*. J. Hort. Sci. 70 (6): 901-907.
Rotor G, Jr (1949) A method for vegetative propagation of *Phalaenopsis* species and hybrids. Amer. Orchid Soc. Bull. 18: 738-739.
Rout GR, Das P (1997) Recent trends in the Biotechnology of *Chrysanthemum* : a critical review. Sci. Hort. 69: 239-257.
Rout GR, Debata BK, Das P (1989) *In vitro* mass-scale propagation of *Rosa hybrida* cv. Landora. Curr. Sci. 58(15): 876-878.
Rout GR, Debata BK, Das P (1990) *In vitro* clonal multiplication of roses. Proc. Natl. Acad. Sci. (India), 60 (3): 311-318.
Rout GR, Debata BK, Das P (1991) Somatic embryogenesis in callus cultures of *Rosa hybrida* L. cv. Landora. Plant Cell, Tiss. Org. Cult. 27: 65-69.
Rout GR, Debata BK, Das P (1992) *In vitro* regeneration of shoots from callus cultures of *Rosa hybrida* cv. Landora. Indian J. Exp. Biol. 30:15-18.
Rout GR, Palai SK, Panday P, Das P (1996) Direct plant regeneration of *Chrysanthemum morifolium* Ramat cv. Deep Pink: Influence of explant source, age of explants, culture environment, carbohydrates, nutritional factors and hormone regime. Natl. Acad. Sci. India. 67: 57-66.
Rout GR, Samantaray S, Mottley J, Das P (1999) Biotechnology of the rose: a review of recent progress. Sci. Hort. 81: 201-228.
Sagawa Y, Kunisaki JT (1982) Clonal propagation of orchids by tissue culture. W *In*: Furusawa (ed.). Proc. 5th Plant Tissue Culture. Tokyo, pp. 683-684,
Sakai A (1960) Survival of the twigs of woody plants at –196°C. Nature, 185: 392-394.
Sakamoto Y, Onishi N, Hirosawa T (1995) Delivery systems for tissue culture by encapsulation. *In*: JA Christie, T Kozai, ML Smith (eds.). Automation and Environmental control in Plant Tissue Culture, Kluwer Acad. Publ., Dordrecht, Netherlands, pp. 215-243.
Salehi H, Khosh-Khui M (1997) Effects of explant length and diameter on *in vitro* shoot growth and proliferation rate of miniature roses. J. Hort. Sci. 72: 673-676.
Sallanon H, Maziere Y (1992) Influence of growth room and vessel humidity on the *in vitro* development of rose plants. Plant Cell, Tiss. Org. Cult. 30: 121-125.
Sangwan RS, Harada H (1976) Chemical factors controlling morphogenesis of *Petunia* cells cultured *in vitro*. Biochem. Physiol. Pflanzen. 170: 77-84.
Sauvadet MA, Brochard P, Boccon-Gibod J (1990) A protoplast-to-plant system in *Chrysanthemum*: differential responses among several commercial clones. Plant Cell Rep. 8: 692-695.
Schum A, Hoffman K, Felten R. (2002) Fundamentals for integration of somatic hybridization in rose breeding. Acta Hort. 572: 29-36.
Schum A, Preil W (1981) Regeneration of callus from *Chrysanthemum morifolium* mesophyll protoplast. Gartenbauwissenochaft 46 (2): 91-93.

Scorza R, Cordts JM, Mante S (1990) Long-term somatic embryo production and plant regeneration from embryo-derived peach callus. Acta Hort. 280: 183-190.
Scotti Campos P, Pais MS (1990) Mass propagation of the dwarf rose cultivar 'Rosa mini'. Sci. Hort. 43: 321-330.
Sen J, Sen S (1995) Two step bud culture technique for a high frequency regeneration of *Gladiolus* corms. Sci. Hort. 64; 133-138.
Seon JH, Kim YS, Son SH, Paek KY, van Plas LHW, deKlerk GJ (2000) The fed-batch culture system using bioreactor for the bulblets production of oriental lilies. Acta Hort. 520: 53-59.
Shabde M, Murashige T (1977) Hormonal requirements of excised *Dianthus caryophyllus* L. shoot apical meristem *in vitro*. Amer. J. Bot. 64: 433-448.
Sharma AK, Mitra GC (1976) *In vitro* culture of shoot apical meristem of *Petunia hybrida* for mass production of plants. Indian J. Exp. Biol. 14: 348-350.
Sharma AK, Tandon P, and Kumar A (1992) Regeneration of *Dendrobium wardianum* Warner (Orchidaceae) from synthetic se(eds.). Ind. J. Exp. Biol. 30: 747-748.
Sheelavantmath SS, Murthy HN, Pyati AN, Kumar HGA, Ravishankar BV (2000) *In vitro* propagation of the endangered orchid, *Geodorum densiflorum* (Lam.) Schltr. through rhizome section culture. Plant Cell, Tiss. Org. Cult. 60: 151-154.
Simmonds J, Werry T. (1987) Liquid shake cultures for improved micropropagation of *Begonia* x *hiemalis*. Hort. Sci., 22: 122-124.
Simonsen J, Hildebrandt AC (1971) *In vitro* growth and differentiation of *Gladiolus* plants from callus cultures. Can J. Bot. 49: 1817-1819.
Singh, F (2001) United structure will benefit rapid growth in India. Flower TECH 4 (1): 16-18.
Sink KC., Power JB., Natarella NJ (1978) The interspecific hybrid *Petunia parodii* x *P. inflata* and its relevance to somatic hybridization in the genus *Petunia*. Theor. Appl. Genet. 53: 205-208.
Skirvin RM (1978) Natural and induced variation in tissue culture. Euphytica 27: 241-266.
Skirvin RM, Chu MC (1979) *In vitro* propagation of 'Forever Yours' rose. Hort Sci. 14(5): 608-610.
Skirvin RM, Chu MC (1984) The effect of light quality on root development on *in vitro* grown miniature roses. Hort Sci. 19: 575 (abstract).
Skirvin RM, Janick J (1976) Tissue culture induced variation in scented *Pelargonium spp*. J. Amer. Soc. Hort. Sci. 101: 281-290.
Skirvin RM, Norton M, McPheeters KD (1993) Somaclonal variation: has it proved useful for plant improvement. Acta Hort. 336: 333-340.
Skirvin RM., Chu MC, Walter JC (1984) Tissue culture of the rose. Amer. Rose Ann. 69: 91-97.
Smart NJ., Fowler MW (1984) An airlift column bioreactor suitable for large-scale cultivation of plant cell suspensions. J. Exp. Bot. 35: 531-537.
Smilansky Z, Uniel N, Zieslin N (1986) Mutagenesis in roses (cv. Mercedes). Environ. Exp. Bot. 26: 279-283.
Soczek U, Hempel M (1988) The influence of some organic medium compounds on multiplication of *Gerbera in vitro*. Acta Hort. 226: 643-646.
Stefaniak B (1994) Somatic embryogenesis and plant regeneration of *Gladiolus* (*Gladiolus* hort.). Plant Cell Rep. 13: 386-389.

Steward FC, Mapes M.O Mears K (1958) Growth and organized development of cultured cells. II. Organization in cultures grown from freely suspended cells. Amer.J. Bot. 45: 705-707.

Stewart J, Button J (1975) Tissue culture studies in *Paphiopedilum*. Amer. Orchid Soc. Bull. 44: 591-599.

Takayama S (1991) Mass propagation of plants through shake and bioreactor culture techniques. *in*: YPS Bajaj, (ed.). Biotechnology in Agriculture and Forestry. Springer-Verlag, Berlin, pp. 1-46.

Takayama S, Akita M (1998) Bioreactor techniques for large-scale culture of plant propagules. Adv. Hort. Sci. 12: 93-100.

Takeda Y (1974) Studies on the establishment of pathogen free stocks by use of shoot tip culture and its application to the commercial carnation production. Spe. Rep. Shiga Pref. Agric. Exp. Sta. 11: 1-124.

Tanaka K, Kanno Y, Kudo S, Suzuki M (2000) Somatic embryogenesis and plant regeneration in chrysanthemum (*Dendranthema grandiflorum*) (Ramat) Kitamura. Plant Cell Rep. 19: 946-953.

Tautorus TE., Dunstan DI (1995) Scale-up of embryogenic plant suspension cultures in bioreactors. *In*: SM Jain, PK Gupta, RJ Newton, (eds.). Somatic Embryogenesis in Woody Plants. Kluwer Acad. Publ., Dordrecht, Netherland, vol.1, pp. 265-269.

Teixeira da silva JA. (2003) Thin Cell Layer technology in ornamental plant micropropagation and biotechnology. African J. Biotech., 2:683-691.

Teixeira de Silva JA, Nhut DT. (2003) Cells:functional units of TCLs. In: Nhut DT, Van Le B, Tran Thanh Van K, Thorpe T. (eds.) Thin cell layer culture system : regeneration and transformation applications. Kluwer Acad. Publ., Dordrecht, The Netherlands, pp.65-134.

Teng WL, Nicholson L, Teng MC (1997) Micropropagation of *Spathoglottis plicata*. Plant Cell Rep. 16: 831-835.

Topoonyanont N, Dillen W (1988) Capitulum explants as a start for micro-propagation of *Gerbera*: culture technique and applicability. Mededelingen-van-de-Faculteit-Landbouwwetenschappen,- Rijksuniversiteif-Gent. 53 (1): 169-173.

Tran Thanh Van K, Bui VL. (2000) Current status of thin cell layer method for the induction of organogenesis or somatic embryogenesis. In: Jain SM, Gupta PK, Newton RJ. (eds.) Somatic embryogenesis in woody plants. Vol.6, Kluwer Acad. Publ., Dordrecht, The Netherlands, pp.51-92.

Tse ATY, Smith RJ, Hackett WP (1971) Adventitious shoot formation of *Phalaenopsis* nodes. Amer. Orchid Soc. Bull. 40: 807-810.

Tweddle D, Roberts AV, Short KC (1984) *In vitro* culture of roses. In: FJ Novak, L. Havel, J Dolegel, (eds.). Plant Tissue and Cell Culture Application to Crop Improvement. Czechoslovak Acad. Sciences, Prague.

Urban LA, Sherman JM, Moyer JW, Daub ME (1992) Regeneration and *Agrobacterium*-mediated transformation of *Chrysanthemum. in vitro* Cult. Hort. Breed., 28th June- 2nd July 1992, Baltimore, MD, p. 49 (abstract).

Urban LA, Sherman JM, Moyer JW, Daub ME (1994) High frequency shoot regeneration and *Agrobacterium*-mediated transformation of *Chrysanthemum*. Plant Sci. 98: 69-79.

Vainstein A, Ben-Meir H, Zuker A, Watad AA, et al. (1995) Molecular markers and genetic transformation in the Breed. of ornamentals. Acta Hort. 420: 65-67.

Valles M, Boxus PH (1987) Regeneration from *Rosa* callus. Acta Hort. 212: 691-696.

van der Salm TPM, Van der Toorn CJG, Hanischten-cate CH, Dons HJM (1996) Somatic embryogenesis and shoot regeneration from excised adventitious roots of the root stock *Rosa hybrida* cv. Money Way. Plant Cell Rep. 15: 522-526.

Van Wordragen MF, Ouwerkerk PBF, Dons HJM (1992) *Agrobacterium rhizogenesis* mediated induction of apparently untransformed roots and callus in *Chrysanthemum*. Plant Cell, Tiss. Org. Cult. 30: 149-157.

Vasil IK (1994) Automation of plant propagation. Plant Cell, Tiss. Org. Cult. 39: 105-108.

Vij SP, Sood A, Plaha KK (1984) Propagation of *Rhynchostylis retusa* Bl. (Orchidaceae) by direct organogenesis from leaf segment cultures. Bot. Gaz. 145: 210-214.

Villalobos V (1981) Floral differentiation in carnation (*Dianthus caryophyllus* L.) from anthers cultivated *in vitro*. Phyton (Argentina) 41: 71-75.

von Arnold S, Sabala I, Bozhkov P, Dyachok J, Filonova L (2002) Developmental pathways of somatic embryogenesis. Plant Cell, Tiss. Org. Cult. 69: 233-249.

Votruba R, Kodyteck K (1988) investigation of genetic stability in *Chrysanthemum morifolium* " Blanche Poitevine Supreme" after meristem culture. Acta Hort. 226: 311-319.

Voyiatzi C, Voyiatzi DG, Tsiakmaki V (1995) *In vitro* shoot proliferation rates of the rose cv. (Hybrid tea) 'Dr. Verhage', as affected by apical dominance regulating substances. Sci. Hort. 61: 241-249.

Walther F, Sauer A (1986) *In vitro* mutagenesis in roses. Acta Hort. 189: 37-46.

Wang SO, Ma SS (1978) Clonal multiplication of *Chrysanthemum in vitro*. J. Agric. Assoc. China, 32: 64-73.

Wated AA, Raghothama KG, Kochba M, Nissim A, Gaba V. (1997) Micropropagation of *Spathiphyllum* and *Syngonium* is facilitated by use of Interfacial membrane rafts. Hort Sci., 32:307-308.

Wilfret GJ (1971) Shoot tip culture of *Gladiolus*: An evaluation of nutrient media for callus tissue development. Proc. Fla. State Hort. Soc. 84: 389-393.

Williams JGK, Kubelik AR, Livak KJ, Rafalski IA, Tingey SV (1990) DNA polymorphisms amplified by arbitrary primers are useful as genetic markers. Nucl. Acids Res. 18: 6531-6535.

Winkelmann T, Grunewaldt J (1995a) Analysis of protoplast-derived plants of *Saintpaulia ionantha* (H. Wendl.). Plant Breed. 114: 346-350.

Winkelmann T, Grunewaldt J (1995b) Genotypic variability for protoplast regeneration in *Saintpaulia ionantha* (H. Wendl.). Plant Cell Rep. 14: 704-707.

Wising K, Schell J, Kahl G (1988) Foreign genes in plants: transfer, structure, expression and applications. Ann. Rev. Genet. 22: 421-497.

Woltering EJ (1990) Beneficial effects of carbon dioxide on development of gerbera and rose plantlets grown *in vitro*. Sci. Hort. 44: 341-345.

Woodson WR, Goldsbrough PB (1989) Genetic transformation of carnation using *Agrobacterium tumefaciens*. Hort Sci. 24: 80.

Yang J, Lee HJ, Shin DH, Oh SK, et al. (1999) Genetic transformation of *Cymbidium* orchid by particle bombardment. Plant Cell Rep. 18: 978-984.

Young PS, Murthy HN, Yoeup PK (2000) Mass multiplication of protocorm-like bodies using bioreactor system and subsequent plant regeneration in *Phalaenopsis*. Plant Cell, Tiss. Org. Cult. 63: 67-72.

Zieslin N, Gavish H, Ziv M (1987) Growth interactions between calli and explants of rose plants *in vitro*. Plant Sci. 49: 57-62.

Zimmer K, Pieper W (1978) Clonal propagation of *Phalaenopsis* by excised buds. Orchid Rev. 86: 223-227.

Zito SW, Teo CD (1990) Constituents of *Chrysanthemum cinerariaefolium* in leaves of regenerated plantlets and callus. Phytochemistry 29: 2533-2534.

Ziv M (1970) Transplanting *Gladiolus* plants propagated *in vitro*. Sci. Hort. 11: 257-260.

Ziv M (1979) Enhanced shoot and cormlet proliferation in liquid cultured *Gladiolus* buds by growth retardants. Plant Cell, Tiss. Org. Cult. 17: 101-110.

Ziv M (1990a) The effect of growth retardants on shoot proliferation and morphogenesis in liquid cultured gladiolus plants. Acta Hort. 280: 207-214.

Ziv M (1990b) Morphogenesis of gladiolus buds in bioreactors-implication for scaled-up propagation of genotypes. *In*: HJJ Nijkamp, LHW van der Plas, J van Atrijk, (eds.). Progress in Plant Cellular and Molecular Biology. Kluwer Acad. Publ., Dordrecht, Netherlands, pp. 119-124.

Ziv M (1991) Morphogenic patterns of plants micropropagated in liquid medium in shaken flasks or large-scale bioreactor cultures. Israel J. Bot. 40: 145-153.

Ziv M (1992) Morphogenetic control of plants micropropagated in bioreactor cultures and its possible impact on acclimatization. Acta Hort. 319: 119-124.

Ziv M (1995) The control of bioreactor environment for plant propagation in liquid culture. Acta Hort. 393: 25-38.

Ziv M (1999) Organogenic plant regeneration in bioreactors. *In*: A Altman, M Ziv S Izhar (eds.). Plant Biotechnology and *in vitro* Biology in the 21st Century. Kluwer Acad. Publ., Dordrecht, Netherlands, pp. 673-676.

Ziv M (2000) Bioreactor technology for plant micropropagation. *In*: J Janick, (ed.). Horticultural. Reviews. John Wiley & Sons Inc., New York, NY, vol. 24, pp. 1-30.

Ziv M, Halevy AH, Shilo R (1970) Organs and plantlet regeneration of *Gladiolus* through tissue culture. Ann. Bot. 34: 671-676.

Ziv M, Kipnis L (1990) *Gladiolus*. *In*: PV Ammirato, DA Evans, WR Sharp, YPS Bajaj, (eds.). Handbook of Plant Cell Culture. MacMillan, New York, NY, vol.5, pp. 461-678.

Ziv M, Lilien-Kipnis H, Borochov, A, Halevy, AH (1997) Bud cluster proliferation in bioreactor cultures of *Ornithogalum dubium*. Acta Hort. 430: 307-310.

Ziv M, Ronen G, Raviv M (1998) Proliferation of meristematic clusters in disposable presterilized plastic bioreactors for large-scale micropropagation of plants. In vitro Cell. Dev. Biol.- Plant 34: 152-158.

Index